The Merchant Ship in the British Atlantic, 1600–1800

Technology and Change in History

Series Editors

Adam Lucas
Steven A. Walton

VOLUME 18

The titles published in this series are listed at *brill.com/tch*

The Merchant Ship in the British Atlantic, 1600–1800

Continuity and Innovation in a Key Technology

By

Phillip Reid

BRILL

LEIDEN | BOSTON

Cover illustrations: (front) The main mast of a Jamestown replica; (back) One of the Jamestown replicas, Jamestown Festival Park, Jamestown, Virginia. Photographs by the author.

Library of Congress Cataloging-in-Publication Data

Names: Reid, Phillip, 1967– author.
Title: The merchant ship in the British Atlantic, 1600–1800 : continuity and innovation in a key technology / Phillip Reid.
Description: Leiden ; Boston : Brill, [2020] | Series: Technology and change in history, 1385–920X ; vol. 18 | Includes bibliographical references and index.
Identifiers: LCCN 2020011388 (print) | LCCN 2020011389 (ebook) | ISBN 9789004424081 (hardback) | ISBN 9789004426344 (ebook)
Subjects: LCSH: Merchant ships—Great Britain—History—17th century. | Merchant ships—Great Britain—History—18th century. | Sailing ships—Design and construction—History—17th century. | Sailing ships—Design and construction—History—18th century.
Classification: LCC VK57 .R45 2020 (print) | LCC VK57 (ebook) | DDC 623.822/4094109032–dc23
LC record available at https://lccn.loc.gov/2020011388
LC ebook record available at https://lccn.loc.gov/2020011389

Typeface for the Latin, Greek, and Cyrillic scripts: "Brill". See and download: brill.com/brill-typeface.

ISSN 1385-920X
ISBN 978-90-04-42408-1 (hardback)
ISBN 978-90-04-42634-4 (e-book)

This book is printed on acid-free paper and produced in a sustainable manner.

for Carl Swanson

Contents

Contents

Acknowledgements

It will come as no surprise to the reader that a study whose scope is as broad as the title suggests was possible only with access to expertise and assistance from those both within and beyond its author's own areas of professional competence. I have thus incurred debts, and I am happy to acknowledge them. First, to my advisor and friend, Neil Kennedy, who has been supporting this for years now, and mentoring faculty at Memorial University of Newfoundland whose consistent support and encouragement were, literally, remarkable. Next, to the late Lewis "Skip" Fischer, doyen of maritime history. I wish he could have seen this book published, as that would have pleased him. Jerry Pocius, Scott Jamieson, Jeff Webb, Dan Walker, and Olaf Janzen were consistently accommodating and taught me much in my time with them. In addition to Olaf, I must thank Warren Riess and Barry Gaulton for their more abbreviated but important contribution to the project in their capacity as examiners of my doctoral thesis. Warren was also one of a trio of veteran ship archaeologists, along with Nick Burningham and Fred Hocker, who answered lots of e-mails, answered the phone or called me back, gave me reading assignments, and taught me much of what I can now contribute to the present volume, all purely out of generosity and devotion to the advancement of knowledge.

Also generous were the distinguished historians whose correspondence made a material contribution to furthering this effort: John McCusker, Christopher French, Thomas Truxes, Joseph Goldenberg, Ab Hoving, Adrian Leonard, and Larrie Ferreiro. Steve Klomps of the Peabody-Essex Museum shared his years of research on Merrimac River shipbuilding with me over a two-year period. Eleanor Breen, City Archaeologist of Alexandria, Virginia, shared all the information she had about the vessel they excavated in 2017, and invited me to give a talk there to help raise money for the conservation of the remains, which I am happy to say is underway.

Archivists and librarians did more than retrieve materials for me. They pointed me to collections I needed to examine. Deserving of special mention are Anna Shumilak of the Archives of Manitoba; Kathy Flynn, Catherine Robertson, Anne Deschaine, and Hannah Swan of the Peabody-Essex Museum's Phillips Library; Malia Ebel of the New Hampshire Historical Society; Kim Nusco of the John Carter Brown Library; Nancy Egloff of the Jamestown-Yorktown Foundation; and Marsha Hayes of the New Hanover County Public Library.

The masters, hands, and shipwrights who sat for interviews and exchanged correspondence with me so that I could incorporate their insights and experiences into the study were: Jim Graczyk, William T. "Chip" Reynolds, Sharon

Dounce, Drew McMullen, Walter Rybka, Eric Speth, and Joakim Severinson. The information they can provide in a sentence or two can be had no other way, and is indispensable.

My friend Steve Kinsey, marine biologist at UNC-Wilmington, engaged with me repeatedly in discussions of evolutionary theory as I sorted out the intellectual methodology of the study that eventually gave rise to the present volume. The perspective and tutelage of an accomplished "natural philosopher," as someone from my period would have called him, were enriching.

I thank the North American Society for Oceanic History for allowing me to present papers related to this study at their annual conferences in 2015 and 2017, and for comments and encouragement. The Society for the History of Technology accepted a session on doing pre-industrial history of technology at their annual meeting in 2018, during which I presented a paper on this topic. I appreciate that and I appreciate the participation and contributions of Anne McCants, Adam Lucas, and Pam Long. Validation goes a long way toward sustaining the will to pursue lonely work with no certain outcome, and I am grateful to the prize committees at NASOH and T2M (the International Association for the History of Transport, Traffic & Mobility) for awarding me their Clark G. Reynolds Student Paper Prize and John Scholes Transport History Research Essay Prize, respectively.

I must also thank the editors and anonymous reviewers at *The Northern Mariner/Le Marin du Nord*, the *Journal of Transport History*, the *International Journal of Maritime History*, and *The Mariner's Mirror* for comments and suggestions and for publishing articles based on earlier work on key aspects of this study; Victoria Barnett-Woods for including my chapter on the British Atlantic merchant ship in the edited collection *Cultural Economies of the Atlantic World*; and Christian Rollinger of the University of Trier for including a chapter on risk and sailors' culture for an edited collection currently under review.

For permission to use images, I thank Chris Dostal at Texas A&M University, Pic Lee Song at the *Duyfken* 1606 Replica Foundation, the *Sjöhiskoriska museet*, Stockholm, for making the priceless Chapman collection of plans and drawings available online copyright-free; photographer Jon Caspar and Jan Ross of the Kalmar Nyckel Foundation, Drew McMullen of the Sultana Education Foundation, and photographer Michael Wootton; Capt. Reynolds, formerly of the *Half Moon*, photographer John W. Mangrum, and Eduard Van Breen of the New Netherland Museum; and Scarlet Faro, Royal Museums Greenwich.

Finally, I thank Stefan Einarson at Brill for initially accepting this book for review, the anonymous reviewers who made helpful suggestions for revising it, Rosanna Woensdregt for seeing it through the publishing process, the Brill production team, and the series editors, Adam Lucas and Steven Walton, for their support and advice. I am delighted to see this book included in their Technology and Change in History series.

These contributions made the interpretations in this book possible, but I am solely responsible for those interpretations as well as any errors they may contain. For any aspiring scholars who may be reading this, I will point out that I have not mentioned grants and awards I did *not* get, nor those who questioned the merits of this project. The privilege of writing acknowledgements for a published work comes after persevering through challenges and obstacles as well as taking advantage of encouragement and open doors. The wise writer will turn constructive criticism into better work, and rejection and naysaying into redoubled determination.

Early research travel was funded by the Department of History and the School of Graduate Studies, Memorial University of Newfoundland; and by a short-term fellowship from the Phillips Library of the Peabody-Essex Museum. More recently, a small grant from The Society for Nautical Research funded research assistance in the UK, for which I heartily thank both them and the capable and knowledgeable Joanna Thomas of the University of Exeter, as well as Professor Maria Fusaro, who connected us. I would have liked to examine more UK archival materials for this project, and in person, but funding for that was not forthcoming. I welcome any suggestions for specific UK collections holding promise for future research on questions raised by the study.

Food, clothing, and shelter were provided by my wife, Andie. That is important, as only live humans with free time can write history books. At my age, I am lucky to have healthy parents, and I thank them for a lifetime of support and genuine interest in what I do, this project certainly not excepted.

Finally, if the person to whom this book is dedicated finds it worthwhile, its author will be satisfied.

P.F.R.
Wilmington, North Carolina
December 2019

Figures and Tables

Figures

Tables

Introduction

The merchant ship carried the people, products, and plagues that created the New Worlds of the European-American empires. The British Atlantic developed as one of those New Worlds between 1600 and 1800, as part of an empire that, by the end of that period, was global. Even after a significant portion of the early empire won political independence from Great Britain in the early 1780s, the empire pulled its former colonies back into an orbit of mutual benefit, although at the same time the new republic of the United States entered into far-reaching maritime competition with the empire to which it had formerly, and largely beneficially, belonged.

The British Empire of this period was a maritime empire. It was established and sustained by sailors on ships, and the merchants who owned those ships. The merchants and their property were, in the mid- to late eighteenth century at least, increasingly well-protected during the chronic warfare of the period by the ships and sailors of the Royal Navy, which could project the power of the imperial state effectively so long as British possessions remained maritime possessions.[1] It was when the British American colonists, determined to acquire more and more land for agriculture, pushed far enough west to clash with more indigenous nations and with Britain's arch-rival, France, that the British ability to exert control began to weaken. As British Americans looked westward—landward—and turned their backcountry into their homeland, they acquired both the motivation and the means to resist, and ultimately overthrow, an imperial power whose interests were increasingly challenged by their American subjects' relentless push away from the coast. When open conflict came, the British were well-equipped to attack and secure seaboard port towns. They could not subdue what lay behind those towns.

The new United States would increasingly turn its focus toward westward expansion, but it retained the original basis of its economy—maritime trade—well beyond the end of this period. British Americans on the eastern seaboard, in the West Indies, and on the Great Lakes had been building and sailing ships, and growing and making products for those ships to carry, since the earliest

1 N.A.M. Rodger and Richard Harding trace the development of British naval infrastructure and logistical support in the eighteenth century and credit them for much of the later fighting and commerce-policing effectiveness of the Royal Navy. See Rodger, *The Wooden World: An Anatomy of the Georgian Navy* (Glasgow: Collins, 1986); and *The Command of the Ocean* (London: Penguin Books, 2004); Harding, *Seapower and Naval Warfare, 1650–1830* (London: UCL Press, 1999); and *The Emergence of Britain's Global Naval Supremacy: The War of 1739–1748* (Woodbridge, UK: The Boydell Press, 2010).

 | DOI:10.1163/9789004426344_002

permanent English settlements. British Atlantic trade travelled in a fleet built and manned on both sides of the Atlantic and throughout the Caribbean basin. After the War of American Independence, ships flying both the Union Jack and the Stars and Stripes carried on the old trades that had generated so much wealth—and so much suffering. Ships under those flags crossed wakes with those flying the ensigns of France, Spain, and the Netherlands, as they always had, for the British Atlantic never existed in isolation from the empires of other European powers, but in a web of inter-imperial connections rife with clashes and enriched by constructive collusions.

The merchant ships employed by those empires, throughout the period, both reflected and facilitated those clashes and collusions. Historians have written much on the development of the British Atlantic in the seventeenth and eighteenth centuries, including its specifically maritime aspects.[2] Economic and business historians have made great progress over the past half-century or so in understanding how the British Atlantic maritime economy generated great wealth and power on both sides of the ocean.[3] Social historians have produced important studies on shipboard labor, the work culture of British Atlantic mariners, and the all-important Atlantic slave trade.[4] Historians

2 Of special significance is Ian K. Steele, *The English Atlantic, 1675–1740: An Exploration of Communication and Community* (Oxford: Oxford University Press, 1986). A 2007 *festschrift* to Steele provides a recent sampling of work inspired by him and a summation of his influence. Nancy Rhoden, ed., *English Atlantics Revisited* (Kingston, ON: McGill-Queen's University Press, 2007).

3 Douglass C. North, "Sources of Productivity Change in Ocean Shipping, 1600–1850," *Journal of Political Economy* 76:5 (September–October 1968): 953–70; James F. Shepherd and Gary M. Walton, *Shipping, Maritime Trade, and the Economic Development of Colonial North America* (Cambridge: Cambridge University Press, 1972); John J. McCusker and Russell R. Menard, *The Economy of British America, 1607–1789* (Chapel Hill: University of North Carolina Press, 1991). On merchant shipping, one volume towers over the subject even today: Ralph Davis, *The Rise of the English Shipping Industry in the Seventeenth and Eighteenth Centuries* (London: Macmillan, 1962). The book was reprinted in 1972 with a new introduction by the author. The International Maritime Economic History Association (now the International Maritime History Association) reprinted it in 2012 as No. 48 in its *Research in Maritime History* series, with an introduction by Lewis R. Fischer and David M. Williams, in which they refer to the book as "one of the most significant books in maritime history published in the twentieth century" (iii). That edition is currently published by Liverpool University Press. Unless otherwise noted, citations from Davis here will be from the 1972 edition.

4 The literature on the social and labor history of early modern seafarers is extensive. For an introduction, see Davis, 105–26 (Ch. 6). Important recent work includes W. Jeffrey Bolster, *Black Jacks: African American Seamen in the Age of Sail* (Cambridge, MA: Harvard University Press, 1997); Peter Earle, *Sailors: English Merchant Seamen 1650–1775* (London: Methuen, 1998); Paul Gilje, *Liberty on the Waterfront: American Maritime Culture in the Age of Revolution*

focused on the effects of extra-human forces and organisms on human history have brought needed attention to transplanted diseases, their vectors, and their effects.[5] They have examined the historical role of the major ocean currents, Caribbean hurricanes, and—especially recently—the fisheries, from an ecological perspective.[6] Today, historians are paying new attention to interactions between British Americans and indigenous powers. Those studies are exploring, among other things, the interactions—especially trade—that took place on the water.[7]

Indeed, historians are paying more attention to water in general as a theater for human activity, and a shaper of human perspectives and decisions.[8] A historical work on the British Atlantic is more likely now to pay closer attention to the Atlantic itself than was once the case, and this one is no exception. Such

---

(Philadelphia: University of Pennsylvania Press, 2004); Marcus Rediker, *Between the Devil and the Deep Blue Sea: Merchant Seamen, Pirates and the Anglo-American Maritime World, 1700–1750* (Cambridge: Cambridge University Press, 1987); Daniel Vickers with Vince Walsh, *Young Men and the Sea: Yankee Seafarers in the Age of Sail* (New Haven, CT: Yale University Press, 2005). Crews of slavers faced special labor and risk conditions. See Emma Christopher, *Slave Ship Sailors and their Captive Cargoes* (Cambridge: Cambridge University Press, 2006); Robert Harms, *The* Diligent*: A Voyage Through the World of the Slave Trade* (New York: Basic Books, 2002); and Marcus Rediker, *The Slave Ship: A Human History* (New York: Viking, 2007). The classic study of life in the Royal Navy in the eighteenth century is Rodger's *Wooden World*. For the Atlantic merchant service in this period, we are even fortunate enough to have some published primary sources. Some important examples are Robert Barker, *The Unfortunate Shipwright, Or Cruel Captain* (London, 1795), Google Books digital facsimile; Edward Barlow, *Barlow's journal of his life at sea in king's ships, East & West Indiamen & other merchantmen from 1659 to 1703*, two volumes, transcribed by Basil Lubbock (London: Hurst & Blackett, 1934); and Samuel Kelly, *Samuel Kelly: An Eighteenth Century Seaman*, ed. Crosbie Garstin (New York: Frederick A. Stokes Company, 1925).

5 See Alfred W. Crosby, *Ecological Imperialism: The Biological Expansion of Europe, 900–1900* (Cambridge: Cambridge University Press, 2004); and J.R. McNeill, *Mosquito Empires: Ecology and War in the Greater Caribbean, 1620–1914* (Cambridge: Cambridge University Press, 2010).

6 See Matthew Mulcahy, *Hurricanes and Society in the British Greater Caribbean, 1624–1783* (Baltimore, MD: Johns Hopkins University Press, 2006).

7 See for example Andrew Lipman, *The Saltwater Frontier: Indians and the Contest for the American Coast* (New Haven, CT: Yale University Press, 2017); Ann M. Carlos and Frank D. Lewis, *Commerce by a Frozen Sea: Native Americans and the European Fur Trade* (Philadelphia: University of Pennsylvania Press, 2010); and Jessica Yirush Stern, *The Lives in Objects: Native Americans, British Colonists, and Cultures of Labor and Exchange in the Southeast* (Chapel Hill: University of North Carolina Press, 2017).

8 See Lincoln Paine, *The Sea and Civilization: A Maritime History of the World* (New York: Alfred A. Knopf, 2013); David Armitage, Alison Bashford, and Sujit Sivasundaram, eds., *Oceanic Histories* (Cambridge: Cambridge University Press, 2000), the first book in a series of the same title; and Helen M. Rozwadowski, *Vast Expanses: A History of the Oceans* (London: Reaktion Books, 2018).

a perspective must in large part be a technological one, as humans cannot live on, in, or under water without technology.

Be that as it may, the key technology without which the British Atlantic could not have been established or maintained—the merchant ship—has remained in peripheral view in British Atlantic history. Archaeologists have worked on it, as have a few ship historians concerned with technical matters, but a scholarly appreciation of this central technology has not yet taken its place on the shelf.[9]

A technological history of the British Atlantic merchant ship in this period is properly situated amongst literature on the Atlantic World, maritime economic history, and maritime archaeology, as well as the history of technology. Such a history has the potential to link all four fields more closely than they have been so far. That requires incorporating insights from the latter two fields into the first two. It requires reminding maritime archaeologists that the first two fields are asking questions that maritime archaeology can help to answer, and demonstrating to technological historians that early modern technological history, including that of the ship, has something significant to offer a field dominated by more modern interests.

The best Atlantic World literature cannot be accused of ignoring what merchant ships did. The *weltanschauung* of the field is one of connections, networks, and trade patterns over water, linking coastal settlements whose denizens look seaward at least as much as they look inland. Atlantic historians have considered the use of merchant ships in British expansion, the settlement and growth of American colonies and plantations, the relationships between the Atlantic worlds of the principal European maritime rivals, slavery and the slave trade, the social history of sailors, smuggling, piracy, and naval operations.

When it comes to linking their arguments to ship technology, however, Atlantic historians have tended to rely on earlier maritime economic history—most usually Ralph Davis's *Rise of the English Shipping Industry* and Shepherd's and Walton's *Shipping, Maritime Trade, and the Economic Development of Colonial North America*—and, in a few cases, on technical ship histories,

---

9 For expert treatment of the merchant ship in this period, see Alan McGowan, *Tiller and Whipstaff: The Development of the Sailing Ship, 1400–1700* (London: Her Majesty's Stationery Office, 1981); and *The Century Before Steam: The Development of the Sailing Ship 1700–1820* (London: Her Majesty's Stationery Office, 1980); Robert Gardiner, ed., *The Heyday of Sail: The Merchant Sailing Ship, 1650–1830* (Edison, NJ: Chartwell Books, 2002), a volume in the Conway's History of the Ship series; David R. MacGregor, *Merchant Sailing Ships 1775–1815: Their Design and Construction* (Watford, UK: Argus Books, 1980) and *Fast Sailing Ships: Their Design and Construction, 1775–1875* (Lymington, UK: Nautical Pub. Co. 1973); and Howard I. Chapelle, *The Search for Speed Under Sail, 1700–1855* (New York: W.W. Norton, 1967).

an important example of which is Ian Steele's citing of Alan McGowan and John Harland in his *English Atlantic*, as support for remarks on steering improvements.[10] McGowan and Harland are important historians of the ship. Davis's and Shepherd's and Walton's work are as important to their field as any. It will be shown, however, that the technological issues they addressed in their work were not, and have not been, resolved (as Davis, especially, acknowledged). Without the technical expertise needed to grapple directly with those technological issues, Atlantic historians have relied on the interpretations of others—only some of whom have such technical expertise themselves—to form their understanding of ship technology and how it contributes to their arguments.

As for maritime economic history, Ralph Davis unquestionably stands out here, and he explicitly called for attention to ship technology by maritime economic historians. He also expressed the hope that archaeology would contribute much to the technological understanding necessary to further test his conclusions, rightly suspecting that the state of technological understanding impeded our understanding of the subject of his magnum opus.

Davis was actually making two claims that were easy to confuse as one, and to dismiss as one. The first claim was that we could learn more about the English-British merchant shipping industry in this period by studying the technology of the ordinary merchant ship. The second was that technological change in the merchant ship was significant, although we did not as yet know how, and played a significant role in the growth of productivity in the industry. Most maritime economic historians, then and since, have muddled those together, and in rejecting the second claim, felt comfortable ignoring the first.

Even ship historian McGowan, however, with his technical understanding far surpassing that of almost any maritime economic historian, dismisses any suggestion that some still-unknown change in basic ship design played a role in increased merchant ship productivity in the period, writing that it was "inconceivable that any such development could have occurred without its having left a single trace. Hull design per se is not a factor in this evidence of greater economic efficiency, although hull size is contributory." He adds that the only technical innovations that could have contributed to an increase in ton-per-man ratios in this period were the adoption of wheel steering and an increasing emphasis on using fore-and-aft sails on larger ships.[11] These last are the two technical innovations Steele accepted in *The English Atlantic*.

---

10 See Steele, *The English Atlantic*, 50, n. 39.

11 McGowan, *The Century Before Steam*, 31.

McGowan was almost certainly correct that no single dramatic alteration in hull design, as yet undiscovered, occurred during this period. As for his denial that adaptations in hull design over time were not a contributing factor to "economic efficiency," however, the evidence now at our disposal strongly suggests otherwise. The evidence also discourages us from accepting his claim that changes in steering systems and sailing rig were the "only technical innovations that could have contributed to an increase in ton-per-man ratios." Those particular innovations were important enough to warrant the attention they are currently receiving—and more—though recent evidence calls into question McGowan's assumption about their relationship—the assumption cited by Steele.

A bulky literature wrestles with how to measure productivity in the economy of the early British Atlantic, what the trends in productivity were, and what factors made what contributions to it. The consensus is that productivity increased through the period—that in spite of the wars and their disruptions, the overall trend was for growth in the British Atlantic economy.[12] The debate is more about what weight to assign each of the contributing factors. Another consensus within that debate is that technological change in merchant ships—changes in their design and construction—was not a primary factor. That leaves the question: Was it a factor at all? On that, there is no firm consensus.

The seminal economic history on the productivity of early modern British merchant shipping started with Nobel laureate Douglass North and continued with his students Gary Walton and James Shepherd, in the heyday of econometric history.[13] While North seems more willing to consider technology as a

12 The latter part of the period considered in this book straddles a significant economic milestone: 1750, before which economic growth has been attributed to "Smithian growth," spurred by the very processes of contact, connection, and commerce facilitated by the ordinary merchant ship; and the beginnings of sustained economic growth of the modern sort, spurred by the conditions, agendas, processes, and products we traditionally refer to collectively as the Industrial Revolution. According to Joel Mokyr, technology in the earlier period was "the auxiliary source of power" for economic change, with commerce and "institutional improvement" providing the "main engine," whereas after 1750 technology became "the main engine." This shift was, however, far more obvious after 1830, and it is important to note that in neither scenario is the ordinary merchant ship a less-important technology. See Joel Mokyr, *The Enlightened Economy: An Economic History of Britain, 1700–1850* (New Haven, CT: Yale University Press, 2009), 5.

13 North, "Sources of Productivity Change in Ocean Shipping, 1600–1850;" Gary M. Walton, "Obstacles to technical diffusion in ocean shipping, 1675–1775," *Explorations in Economic History* 8:2 (Winter 1970): 123–40; Shepherd and Walton, *Shipping, Maritime Trade, and the Economic Development of Colonial North America.*

productivity factor than Shepherd and Walton, the thrust of all their work is that the security of the seas and more efficient business organization predominate as the factors bringing about productivity growth in shipping, and that technical change in merchant ship design and construction was somewhere between insignificant and irrelevant. Some important scholars, however, took issue with this assumption, including John McCusker, Russell Menard, and Frederic Lane, who criticized the methodology of data analysis from more than one angle, and while none of them claimed innovation in ship technology as a factor of paramount importance in productivity growth, none was willing to assume it was insignificant.[14] Neither was Richard Unger, whose work on the subject is more recent and who has paid serious attention to ship technology.[15]

Economists measure productivity in terms of input and output. The more output per unit of input, the greater the productivity. This is a straightforward concept with a steam engine. If one engine produces 20 horsepower per pound of fuel burned, and the other only 15, the first engine is more efficient—more productive—by an exact amount. It is of course vastly more complicated when scholars are attempting to measure overall productivity in a complex economy over a period of decades or centuries. As a recent collection of conference papers edited by Richard Unger makes clear, while economic historians are still working on shipping productivity, they largely agree on how to measure that productivity, and they agree that there was slow but steady growth in overall shipping productivity before, during, and after the period of this study. They also agree that technical change was much more incremental in this period than in the ones immediately preceding and succeeding it. They do not, however, write off those incremental changes as irrelevant to productivity growth in the period.[16]

The productivity measure most central to my purpose here is the ton-per-man ratio. Economic historians have scrutinized the idea closely, and debated how best to employ and interpret it, but it remains their favorite index. The basic idea is that the fewer humans it took to move a ship of a given capacity,

14 Kenneth Morgan juxtaposes North, Walton, and Shepherd on the one hand with McCusker, Menard, and Nathan Rosenberg on the other. Rosenberg is much more concerned with the history of technology than the other two, so he will be discussed in that context. See Kenneth Morgan, *Bristol & The Atlantic Trade in the Eighteenth Century* (Cambridge: Cambridge University Press, 1993), 46; and Frederic Lane, « Progrès technologique et productivité dans les transports maritimes de la fin du Moyen Age au début des Temps Modernes », *Revue Historique* 251 (1974): 277–302.

15 Richard W. Unger, "Ship Design and Energy Use, 1350–1875," in Richard W. Unger, ed., *Shipping and Economic Growth 1350–1850* (Leiden: Brill, 2011), 267.

16 Unger in Unger, *Shipping and Economic Growth*, 252.

the more efficient a cargo conveyance it was, since the costs of the crew, water, and victuals—and the space they occupied—militated against the profit of the shippers, and those shippers identified these costs as ones over which they had more control than they did over others. Davis was the first to argue convincingly that ton-per-man ratios in this period increased over time, a trend interrupted, but never reversed, by war. Subsequent scholarship has reinforced his conclusion. What remains to be accomplished is a more comprehensive explanation for why it happened.

One important consideration when examining ton-per-man ratios in this period, and one which has received scholarly attention, is the defensive armament carried aboard, which influenced the size of the crew. The more guns the ship carried, the more sailors she needed to man those guns in action while operating the vessel too. The installation of guns penalized the ship as a profit-making commercial tool in several ways. The ship was more expensive to build if she were stoutly-constructed enough to carry the weight of guns and absorb the violence of their recoil. She was also heavier as a result of that stouter construction, and thus either slower or in need of a more powerful, and thus more expensive and more labor-intensive, sailing rig to move her as fast as a lighter ship of the same size, especially in light air (little wind). The guns took up what could have been cargo space, as did their heavy and bulky ammunition and gunpowder.

Owners, however, had good reason not to trust cargoes to unarmed vessels in the Atlantic World during much of this period. In the seventeenth century, European naval power in the western Atlantic and Caribbean was weak. Navies had not developed the bureaucratic structures or the physical infrastructure to maintain warships and their crews so far from home in good enough condition to police American waters, so merchant ships were vulnerable to predation by privateers and pirates.[17] The marine insurance industry, too, would only fully develop late in the period. So, for most of the period—though this would eventually change in response to the changing security environment—the merchant ship was a compromise between capacity for cargo and the ability to defend itself and that cargo.

North, Shepherd, and Walton focus on the increase in security at sea as the most important factor allowing for an increase in shipping productivity. Shepherd's and Walton's practice of treating privateering and piracy as synonymous confuses the issue, but Christopher French attempts to clarify it. Piracy ceased to be the threat that it had been to Atlantic Basin commerce after the

17 Rodger, *The Wooden World* and *The Command of the Ocean*; Harding, *Seapower and Naval Warfare* and *The Emergence of Britain's Global Naval Supremacy*.

1720s, as the bulk of recent work has shown, he argues, but privateering was a major component of naval warfare throughout the period. The difference, to French, is the increased effectiveness of wartime convoys after the mid-eighteenth century.[18]

Russell Menard weighs safer seas versus better ships quite differently than North, Shepherd, and Walton—at least for the period 1725–65 in the Atlantic tobacco trade, for which, he argues, we have better data than for other trades in much of the period. He is confident enough in his methodology to attribute 40% of productivity growth to safer seas and 60% to better ships.[19] Drawing specifically on the tobacco trade, Menard claims that Walton did not restrict his data set exclusively to ships making the London-Chesapeake run, so he included coastal and inter-island vessels, which were quite different in terms of ton-per-man ratios, operating costs, and passage times. Menard also claims that Walton overestimated the decline in port times, which thus exaggerated the contribution of that factor at the expense of others, and that he "missed a decline in passage time by relying on evidence from American coastal voyages."[20] French, too, is skeptical of a general decline in port turn-around times, claiming that there was no general decline in the time ships spent in colonial ports between voyages in the first three quarters of the eighteenth century, at least for ships trading from London to New York, Jamaica, and Virginia, and that Walton's demonstration of a decline in port times in the Chesapeake hinges on data from the Scottish trade which grew to be more efficient than that based in London.[21] Menard continues: "It was not that ships improved their speed under constant conditions but rather that they shifted their routes and thus encountered different conditions."[22] This was made possible, he believes, by the introduction of more weatherly and maneuverable ships, largely thanks to the adoption of the triangular jib headsail, and the steering wheel. Here, he relies on Steele's *English Atlantic*, which in turn, as we have seen, relies on McGowan—thus highlighting the fact that to some extent, this

18 Christopher J. French, "Productivity in the Atlantic shipping industry: a quantitative study," *Journal of Interdisciplinary History* 17:3 (Winter 1987): 613–38.

19 Russell R. Menard, "Transport Costs and Long-Range Trade, 1300–1800: Was there a European 'transport Revolution' in the Early Modern Era?" in James D. Tracy, ed., *The Political Economy of Merchant Empires* (Cambridge: Cambridge University Press, 1991), 262.

20 Menard in Tracy, 258. Menard is discussing two specific works here: Shepherd's and Walton's *Shipping, Maritime Trade, and the Economic Development of Colonial North America*, and Walton's "A Measure of Productivity Change in American Colonial Shipping," *Economic History Review*, 2nd series, 21 (1968): 268–82.

21 French, "Productivity," 623–24.

22 Menard in Tracy, 259.

literature is caught in a self-referential loop. For Menard, then, ships were able to make shorter passages by heading more directly to their destinations and by exercising greater navigational control. He faults Walton's evidence for being "particularly misleading on the characteristics of ships, the key to his argument that technical improvements were not important in the tobacco trade."[23] Evaluating Menard's argument, then, will require examining the technological claims of McGowan in light of all available evidence. This is something highly unlikely to happen within maritime economic history.

John McCusker, Menard's co-author of the landmark *The Economy of British North America, 1607–1789*, the best compendium of work on that subject to 1985, entered this debate in his 1968 doctoral dissertation, by objecting to Walton's claim that ships did not grow in the century prior to the American Revolution.[24] McCusker identified the importance of understanding tonnage. He showed that the term meant different thing at the same time, that those meanings changed over time, and that they are not interchangeable.[25] He faults Walton for assuming that two different types of tonnage measurement were the same, and for basing his conclusions on that assumption, when in fact, understanding the difference between the two demonstrates that ships did indeed grow during this period.[26]

Maritime economic history, then, has indeed grappled with the role of merchant ship technology as it relates to important questions about the development of the British Atlantic economy. It has raised questions about ship technology that all studies of this technology should attempt to answer. It cannot, however, answer those questions on its own. That will require work in the history of technology itself.

Frederic Lane and Nathan Rosenberg provide a transition from maritime economic to technological history, by fully-engaging both. They focus on technological innovation—what it means, to whom, and when. Lane faults what he considers the traditional definition of technological innovation, which he attributes to Shepherd and Walton, specifically. For Lane, their definition, "an

---

23 Menard in Tracy, 260–61.

24 McCusker and Menard, *The Economy of British America, 1607–1789*; John J. McCusker, *Rum and the American Revolution: The Rum Trade and the Balance of Payments of the Thirteen Continental Colonies* (London: Routledge, 1997), a published version, with minor revisions, of McCusker's University of Pittsburgh dissertation.

25 McCusker, "The Tonnage of Ships Engaged in British Colonial Trade During the Eighteenth Century," originally published in *Research in Economic History* 6 (1981): 73–105, and republished in the collection *Essays in the Economic History of the Atlantic World* (London: Routledge, 1997), 26–46.

26 McCusker, *Rum and the American Revolution*, 836–38, note 27.

advance in knowledge that allows fewer inputs to produce a given output," does not take into account a technique already known but not previously utilized, then employed as conditions change in such a way that its employment becomes practical or desirable. So, the adoption of a previously-developed technology in a new temporal or spatial context would not be a technological innovation to Shepherd and Walton, but would be to Lane.[27]

Lane criticizes North and Walton for failing to give adequate credit to innovation in their assessment of shipping productivity. After presenting examples of technological discoveries that could not, at the time of their discovery, overcome political or economic barriers to their employment, he suggests that historians should reverse what North and Walton did, by looking first for the technological innovation, and then examining the factors hostile to it—insecure seas, high costs—to explain its absence in a given subsequent time and place—a methodology that goes hand-in-hand with his call for including the adoption of previously-known technology in the definition of innovation.[28]

Another way to express Lane's objection to the North-Walton approach is as a failure to distinguish between the discovery of a technology and its diffusion.[29] Technologies are mere footnotes of curiosities if they are not adopted, so the study of diffusion is just as central to the history of technology as that of invention. Nathan Rosenberg examines diffusion as much more than the adoption of a technology, though. He argues that what we tend to consider diffusion—the dissemination of an innovation—is actually a process of "secondary inventions" in which the original technology is modified, improved, and adapted to better-suit those employing it—especially those employing it outside the time and place of its original setting.[30] To Rosenberg, understanding this process is just as important to understanding major inventions—whose importance, he says, is by no means immediately apparent but requires time to realize, as costs come down, technical obstacles to commercial feasibility are overcome, and enough people adopt the invention. So, even "Eureka!" moments are more processes than moments, at least in terms of real social benefit.[31]

---

27 Lane, 284.

28 Lane, 286–87.

29 Lane, 284–85. To return to the consensus, though: Lane may have objected to North's and Walton's methods, and to Shepherd's and Walton's definition, but in the end he, like them, concludes that the technological potential of the merchant ship in this time and place went unrealized due to the conjuncture of political and economic barriers (Lane, 302).

30 Nathan Rosenberg, "Factors Affecting the Diffusion of Technology," *Explorations in Economic History* 10:1 (Fall 1972): 7.

31 Rosenberg, 8.

Rosenberg, too, objects to North's assessment of the role of technological change in shipping productivity. While lauding him for, and agreeing with him on, his argument for the overwhelming importance of organizational and marketing improvements, Rosenberg finds North's "attempt to downgrade the contribution of technological change to the growth in productivity of ocean shipping ... more questionable."[32] His specific objection is very close to Lane's. To Rosenberg, the existence of a successful pure cargo ship type by 1600 (the Dutch *fluit*) means that, security-related obstacles to its employment in the Atlantic aside, the fact that it was eventually adopted (in modified form) for Atlantic service constitutes technological change. He gives North the benefit of the doubt by saying that North's impression of technological change as "scarcely of any significance whatever" was "doubtless unintentional," but that it ultimately serves to obscure the process of diffusion of that technology in the Atlantic.[33] Rosenberg, then, joins the chorus of economic historians who insist that we take technology seriously in attempting to assess the place of the merchant ship in the economic history of the period.

Understanding a people's technology in the broadest possible context has much to suggest about how those people used the materials and skills they had, to accomplish a purpose they had in mind, and later, to cope with unanticipated developments. How a people develop and use a technology both reflects and shapes their experiences, their prejudices, their talents, and their limitations. Technology, it should be noted, encompasses more than a manufactured material contrivance. It also includes the skills necessary to make and utilize that contrivance in its operating environment—in this case, the Atlantic and Caribbean basins—and the motivation and will to do so. Knowledge enables practice, which in turn can produce an artefact. Technology, then, is both tangible and intangible, and it is only comprehensible in its specific operational context.

Technology is also technical; it requires the acquisition of a specialized body of knowledge to understand it. To some extent, understanding a merchant ship from 1740 requires knowing much—though certainly not all—of what a shipwright or master would have known about it. For the historian, it also requires at least some mastery of the skills and tools of analysis at our disposal today, and an apprehension of how scholars in multiple disciplines have applied

---

32 Rosenberg, 30.

33 Rosenberg, 29–33. The specific work of North's Rosenberg refers to is "Sources of Productivity Change in Ocean Shipping, 1600–1850." Rosenberg separates North from Walton by citing the latter's "Obstacles to Technical Diffusion in Ocean Shipping," which, given its title alone, is immune from the charge of ignoring diffusion (Rosenberg, 31).

those tools and skills in the past. None of that is included in the training most of us receive as historians. I believe that is largely why this technology has not yet been thoroughly examined or appreciated by historians—even those preoccupied with period maritime affairs.[34]

Another reason, however, is likely related to the history of the history of technology. Nowhere has our modern western tendency to view history as linear progress been more evident than in traditional western histories of technology, which in fact have tended more toward the hagiography of successful inventors and both the celebration and bemoaning of the rapid and accelerating "progress" of modern technology—a phenomenon generally exaggerated—than toward an accurate historically-informed appreciation of the variety and vagary of human technological development.[35] Good historians are aware of this, and in the past rightly looked askance at the specialty, ignoring or dismissing it as deterministic.[36] However, much good technological history has been written since then, and is being written now. It is largely preoccupied, however, with explaining industrialization, mass marketing, and the social structures and cultural ramifications of those post-1800 phenomena. That points to the fact that the problem actually goes beyond the historiographical trends, to modern history itself.

We have been profoundly affected by the history of the history of technology since 1800. In fact, as I discuss in this volume, the movement to denigrate pre-industrial technology and celebrate the promise of the dawning age of rational progress and technological mastery of nature began before 1800, while the technology being so denigrated was still dominant. By 1850, that view was so well-ensconced as to be unchallenged. To put that in perspective: in 1850, commercial sail was still viable in most trades. To a much lesser extent, it was still viable in 1900, but those huge ships were built of steel, not wood, and no one in 1800 could have built one.

Nor would they have needed to, and understanding that is crucially important to this endeavor. Perhaps we in today's West are faced with a greater challenge

34 Lack of technical knowledge is a common impediment in academic history. Willem F.J. Mörzer Bruyns makes this point about the history of navigation in "Research in the History of Navigation: Its Role in Maritime History," *International Journal of Maritime History* 21:2 (December 2009): 285. We need more academically qualified maritime historians who know as much about ships, navigation, and oceanography as they do about economic theory and statistics.

35 This is a primary theme of one of the most important recent monographs in the history of technology, David Edgerton's *The Shock of the Old: Technology and Global History since 1900* (Oxford: Oxford University Press, 2011).

36 See Don Leggett and Richard Dunn, eds., *Re-inventing the Ship: Science, Technology and the Maritime World, 1800–1918* (Farnham, UK: Ashgate, 2012), 6.

in understanding past technologies than at any other time in our history, even with all the tools and cumulative knowledge at our disposal with which to do so. That is because our technological worldview has been so distorted by the rapid and ever-accelerating pace of technological change brought about by the digital computer that it is difficult for us to conceive of a contraption like a wooden sailing ship from 1740 as technology at all.[37] We are heavily predisposed to view such technology as archaic, quaint, limiting, hidebound. When we read period treatises whose ambitious authors denounce their fellow shipwrights as ignorant, benighted traditionalists, incapable of seeing the light of reason and science, our predisposition is reinforced. We do not understand complex artisanal craft. We understand R&D (research and development), factories, engineering degrees, computer-aided drafting systems, wind tunnels, 3D simulations, blueprints.

And yet, none of those things made possible the creation of the Atlantic world we live in. Artisanal craft did. Pre-industrial technology did.[38] We need to understand that if we want to fully understand the people who created that world and the world they created. If Michael Adas is correct, this is perhaps even more important for Americans than for Western Europeans, who much more thoroughly internalized the technological disillusionment of the First World War, while their counterparts across the Atlantic became more, rather than less, enamored of the promise of technological progress and "modernization" as key to overall human improvement. Adas is surely correct in identifying such a bias as a product, and perpetuation, of an ideology of Western hegemony, and any such ideology necessarily distorts historical understanding.[39]

It will be worth the effort for historians—not just archaeologists and material-culture anthropologists—to learn about how past peoples developed and used their own technologies, which were different from ours, for their own purposes, which were also different from ours. That is what this volume endeavors to do. I hope to show that, by bringing together the insights of several disciplines, where they intersect, we can add materially to our understanding of the development of the British Atlantic. I intend to present that argument in language accessible to those conversant in history, but not necessarily conversant in nautical matters. I also hope to show that there is still much we have

37 Edgerton, *The Shock of the Old.*

38 See Brooke Hindle, ed., *America's Wooden Age: Aspects of its Early Technology* (Tarrytown, NY: Sleepy Hollow Press, 1985).

39 Michael Adas, *'Machines as the Measure of Men': Science, Technology, and Ideologies of Western Dominance* (Ithaca, NY: Cornell University Press, 1989), especially 402–18.

to learn, and that only by continuing to utilize all the tools and perspectives at our disposal will we learn any of it—because this is a challenging subject.

Much of that challenge can be attributed to the historical and archaeological record—or lack thereof. On the historical side, we must contend with the fact that ordinary merchant ships were built by people who did not need anything on paper to do it, and owned by people whose interest in them was commercial. For the most part, merchants were interested in the money a ship could make or lose, not in the ship itself.[40] Their correspondence and record-keeping reflect that. The contracts between shipwrights and their merchant clients specified basic dimensions and, to varying degree, other aspects of construction, but never enough that we could reconstruct a vessel from a contract. These records do, however, help us to understand the overall approach to risk management taken by merchant shipowners, and that approach, as we shall see, has much to suggest to us about the technological choices they made. Merchants' papers are, therefore, worthy of serious attention here.

Official shipping records provide much less information on the vessels themselves. They are useful for compiling the statistics that economic historians use, but they will not, on their own, help us learn anything new about the merchant ship as a technological contrivance. When combined with other source material, however, they will continue to prove useful.

One priceless exception to the general paucity of paper records is a small assortment of plans, drawn up by the Royal Navy, of ordinary merchant vessels taken into naval service during the period. It was standard for the naval dockyards to record plans and specifications of these vessels, and some have survived. They contain information about which we would otherwise have no access. We also have a few sets of plans of mid- to late eighteenth-century British vessels, and we should make use of them in the marine engineers' laboratory.[41]

The archaeology side is hampered by the poor survivability of wood in warm water. Those examining wrecks in cold water such as the Baltic have a distinct advantage.[42] In general, what relevant wrecks we do find in the temperate

40 I qualify this statement because the evidence does indicate that owners took an aesthetic interest in their ships and conceived of them as extensions of themselves and as objects of personal and professional pride.

41 Worthy of special mention are plans for eighteenth-century merchantmen found by David MacGregor in the archives of shipyards in the UK, plates of which he included in *Merchant Sailing Ships*.

42 There are period wreck sites in the waters around Newfoundland that have not yet been fully investigated. These might add much to what we eventually learn. I thank Prof. Barry Gaulton at Memorial University of Newfoundland for bringing this to my attention.

FIGURE 1 Archaeological remains of Browns Ferry vessel, from a riverbank in South Carolina
PHOTOGRAPH BY THE AUTHOR

Atlantic and Caribbean are almost completely disarticulated; or, if more intact, only the lowermost structural timbers and planking remain—those timbers which sank below the sand or mud and were thus protected from marine life, weather, and human activity.

Our best archaeological remains are the hulls found in terrestrial coastal sites—vessels that were abandoned on riverbanks, or deliberately buried as infill on the waterfront, and spent the intervening centuries completely underground rather than underwater. These promise much, and we are still finding them. They are, however, rare; and, as the above photograph of the Browns Ferry remains shows, they are far from complete hulls (Figure 1). Every one we can find, preserve, and analyze, however, will add to what we know about design and construction practices.

To overcome, at least in part, the incomplete historical and archaeological records requires, first, combining the two. That may seem obvious, but it has not been done to a sufficient extent. This volume is intended as a step toward rectifying that. It also requires bringing to bear the work of historians, antiquarians, and curators from outside the academy on questions of interest to academic historians, translated, as it were, so as to be accessible.

Finally, there are two approaches to research and interpretation that have much to offer that we will not acquire any other way. The first draws on the experiences of shipwrights and mariners who build and operate period replica vessels, taking an experimental-archaeology approach.[43] Used with caution, these experiences can help us understand how—if not why—the people we are interested in built and used their ships the way they did. They can strongly suggest the "why" component, but without explicit confirmation from the historical records left by the original users, such suggestions must remain in the class of theory.

Experimental archaeology, on both land and water, has a well-established history and methodology, and has answered its skeptics, whose reservations highlight the potential pitfalls of the enterprise.[44] The idea is to reconstruct an artefact and learn its use through hands-on trial and error. Because replica ships are so complex and expensive to build, they are never built solely for experimental archaeology; utilizing them for that purpose requires recognizing the potential of knowledge gained through the experience of normal vessel operation, gleaning that knowledge from the operators, and factoring out any differences between the replica and the original, and between the operation of the replica and the original, as best we can.

---

43 For a conceptual introduction to the experimental archaeology of ships, see Ole Crumlin-Pedersen, "Experimental archaeology and ships—principles, problems and examples," in Lucy Blue et al., eds., *Connected by the Sea: Proceedings of the Tenth International Symposium on Boat and Ship Archaeology, Roskilde, 2003* (Oxford: Oxbow, 2006), 1–7; Seán McGrail, "Experimental Archaeology: Replicas and Reconstructions," in Jenny Bennett, ed., *Sailing Into the Past: Learning from Replica Ships* (Barnsley, UK: Seaforth, 2009), 16–23; and Nicholas]Nick] Burningham, "Experimental Maritime Archaeology," in Claire Smith, ed., *Encyclopedia of Global Archaeology* (New York: Springer, 2014), 2716–18. My earliest argument for bringing an experimental-archaeology approach to this subject using existing operational replicas and reconstructions may be found in Phillip Reid, "The Time Machine? Using Replica Analysis to Understand Merchant Ships and the Development of the British Atlantic, 1600–1800," *The Northern Mariner/Le Marin du Nord* 26:3 (July 2016): 299–316.

44 For a skeptic's view particularly relevant to this topic, see Timm Weski, "The value of experimental archaeology for reconstructing ancient seafaring," in Blue et al., *Connected by the Sea*, 63–67.

The other modern tool at our disposal is the computer, powerful iterations of which run marine engineering applications capable of closely—though not exactly—simulating the behavior of three-dimensional objects in water. Naval architects and marine engineers can construct digital models of vessels based on plans and experiment with hypothetical scenarios impossible to reproduce in the real world, and can run calculations on capacities and structural forces at work. This is specialized and expensive work, and it is beyond the skill set of historians and even conventionally-trained archaeologists. That, and the fact that these tools have become only recently available, explains why only a start has been made in bringing them to bear on old ships. The little insight such work has already revealed, however, is tantalizing.

This study, then, pulls together evidence from disparate enterprises run by groups of people who do not typically talk to each other or write for each other. The subject demands such an approach. It is to be hoped that it will encourage more cross-disciplinary comparing of notes, and more communication between those inside and outside the academy, when different sets of questions come together around the same subject. When that happens, we are not only assured of a more comprehensive understanding of that subject; we are better-protected against a distorted understanding of it.

For me, the important questions on this subject started with this one: How can a technology so frequently depicted in the historical literature as static, if not stagnant, have adequately served a world of growth and change for two centuries?[45] In the course of the study, it became clear that the technology was not static, let alone stagnant, but neither did it undergo any revolutionary changes in this period; it was adapted through "microinvention" rather than

45 Narratives of "revolutionary" technological change emerged from a modern *zeitgeist*, beginning in the early modern period and continuing into the mid-twentieth century. They have historically depended upon the positing of "bookend" periods of relative stasis; such constructs require adherence to a macro view that, typically, oversimplifies historical reality and lends itself to the formulation and maintenance of monocausal theories of change. Monocausal theories do not tend to survive rigorous evidentiary testing, as Lucas shows, and as I will show in this book, but they have taken hold in both the scholarly and public imaginations, and have yet to be displaced by more recent, and more sophisticated, assessments. See Adam Lucas, "Narratives of Technological Revolution in the Middle Ages," in Albrecht Classen, ed., *Handbook of Medieval Studies: Terms, Methods, Trends* (Berlin: De Gruyter, 2010), 967–90. Lucas does not argue that there have never been significant changes in relatively brief spans of time, but that a) the most prominent earlier arguments for "revolutionary" change in the Middle Ages have been discredited by more recent scholarship; and b) apprehending and explaining "revolutionary" cultural—as opposed to political—change pose a serious challenge to our skills as scholars (see p. 989).

"macroinvention," to use Joel Mokyr's terms.[46] This raised other questions having to do with the nature of technological stasis and change vis-à-vis stasis and change in the societies utilizing that technology. Could we locate connections between continuities and changes in the technology and what was going on economically, politically, socially, and culturally?

As it turns out, that is more difficult than it is in the later period, where the history of technology is more comfortable. The people who "invented" changes to the merchant ship in this period are anonymous to us. The process by which those adaptations diffused over time is opaque. We are forced to work with inference, and to acknowledge inference as such, remaining vigilant not to claim we have discovered causality when the evidence does not allow that. We must be content with "This is how this worked, and this is perhaps why it was done this way, and this is why we think that." As always, the more evidence we bring to bear that corroborates our conclusions over time, the more confident we can be in those conclusions. That is why it is important to keep looking for wrecks, analyzing the ones we find, experimenting with replicas and virtual models, and returning to the archives for those elusive, passing references to the technology of the merchant ship by those whose livelihoods depended on them.

Drawing on all the sources available thus far will allow us to take a walk-around tour of the ordinary British Atlantic merchant ship. First, however, we will familiarize ourselves with the actual Atlantic part of the British Atlantic world—its physical forces and their interactions, its size and power, its utility and its dangers. We will consider the main variants of the merchant ship in this time and place, and gain a basic understanding of their design principles, construction, and operation. We will explore the work of shipwrights and how they practiced their skilled craft. From the logs and ledgers and letters they left, we will come to know the risks their owners, masters and crews took to buy, fit out, and operate these things, and how those risks characterized business and labor in this world. To most of those people, these vessels meant taking a chance on a better life. To others, especially those loaded aboard in chains on the African coasts, they were prisons and death traps, and for most of those who survived them, conveyances to a worse life and early death. We will discover how these ships were worked, and what it was like for different people in different circumstances to experience them, either up close and in person, or across the sea, by anxiously-awaited correspondence, depending on whether the ship held the person's life, fortune, or both. We will investigate how

46 Joel Mokyr, *The Lever of Riches: Technological Creativity and Economic Progress* (Oxford: Oxford University Press, 1990), Kindle edition, location 299.

shipwrights tried to meet both the desires of their clients and the demands of the sea, using an inherited skill set that encouraged prudent improvisation but mitigated strongly against drastic departure from proven principles. We will consider how the political environment of imperial rivalry, manifested in trade laws and armed conflict, influenced the fitting-out and operation of the merchant ship, and how the nature of that technology affected the decision-making parameters of those in power. By paying particular attention to some core continuities and some important changes in the merchant ship over the course of the period, we will attempt to better-understand an approach to expensive and complex technology both related to and different from our own. That can only advance our understanding of the British Atlantic, human life on the ocean, and the history of technology.

•••

As England began to pursue overseas adventures beyond Ireland around 1600, she was decidedly late to the game compared to the Iberian kingdoms, and would soon trail the other major Protestant maritime power, the Netherlands, as well. What both the English and the Dutch had in common, however, was that their American maritime ventures began as harassment operations against Spain. The first ships sent from England to the Caribbean and then the eastern seaboard were small, stout *galleons*—multi-purpose, well-armed vessels manned by both mariners and soldiers. The small English galleon was not the best vessel at anything—fighting, sailing fast, carrying the most cargo—but it was a proven compromise of those competing imperatives, and it was what the English adventuring companies needed as they made their first tentative ventures into unknown and hostile waters.[47] Spain still claimed dominion over the Americas (except for Portuguese Brazil), and any foreign interloper could expect to be attacked if noticed.

English shipyards developed a reputation for building stout, well-armed, and thus defensible merchant vessels in the seventeenth century. The prime motivation was the Mediterranean trade, already well-established, lucrative—and dangerous, given the ubiquitous presence and power of the Barbary corsairs—maritime predators sponsored and harbored by the Muslim principalities of North Africa, much of whose income they provided.[48] The suitability of such vessels for trading in hostile waters offset their principal drawback—their cost.

47 Jonathan Adams, "*Sea Venture*: A second interim report—part 1," *International Journal of Nautical Archaeology and Underwater Exploration* 14:4 (1985); 297.

48 Davis, 247.

The Dutch, on the other hand, enjoyed fishing and trading in the Baltic and North Seas, where the risks of predation were normally low, and they learned to build relatively cheap, capacious, unarmed or lightly armed cargo vessels, using wind-powered sawmills. Those ships, along with the Amsterdam stock exchange and the Dutch mercantile emphasis, allowed the Netherlands to undercut the shipping rates of its European rivals. Try as they might, neither the English nor the French could beat the Dutch at this game. The Dutch emerged from the Thirty Years War in 1648, and, beginning in the 1650s, the English decided to compete with them by force, using both protectionist legislation and war.

While the Navigation Acts and the Anglo-Dutch Wars hurt the Netherlands, and especially their prospects for an American empire comparable to their rivals', neither could eliminate the Dutch from Atlantic trade nor negate the fact that the Dutch were building merchant ships that could make more money on secure routes than English ships. During the Anglo-Dutch Wars, the English captured so many Dutch ships that there was little to no incentive for English yards to copy Dutch designs. As those ships began to wear out, however, evidence does suggest a Dutch influence on new English construction.

This highlights a central paradox of the imperial, "mercantilist" world of which the British Atlantic was part.[49] On the one hand, the "mercantilist" theory of empire emphasized separate imperial spheres, in constant competition with each other in a zero-sum game. The mother country was to keep its colonies in an exclusive and dependent trade relationship with itself, defending that closed loop from any interference by rival powers. On the other hand, those theoretical boundaries were always porous, to say the least. Goods crossed them, beginning with the Spanish neglect of their West Indies possessions in favor of conquest and extraction on the Central and South American mainland. People crossed them, bringing those goods, along with news and ideas and diseases. Technology crossed them; Dutch-English hybridization in ship design seems to have had its counterpart in the French merchant fleet, and French, English, and Dutch shipwrights visited the others' yards and took detailed notes on the others' methods.[50] Sometimes this was done openly, and

49 On the history and current status of the venerable idea of mercantilism, see Jonathan Barth, "Reconstructing Mercantilism: Consensus and Conflict in British Imperial Economy in the Seventeenth and Eighteenth Centuries," *William and Mary Quarterly* 73:2 (April 2016): 257–90.

50 An especially noteworthy example resulted in a treatise by Blaise Ollivier, *18th Century Shipbuilding: remarks on the navies of the English & the Dutch from observations made at their dockyards in 1737*, trans. and ed. David H. Roberts (Rotherfield, UK: Jean Boudriot Publications, 1992).

that is even more paradoxical—as is the trade that carried on between warring powers in the Atlantic—the norm, not the exception. We would call much of that trade "smuggling," but they did not.[51]

Those warring powers, however, made war on each other so often in this period that war was the rule, not the exception. With war came maritime predation, by naval cruisers and privateers, and that necessitated defensible merchant ships. A thorough understanding of the true costs of defense is necessary if we are to understand the relationship between merchant ship technology and the maritime economy as it developed in the British Atlantic. Closely related to a merchant ship's ability to defend itself against attack are two other forms of risk mitigation: insurance and naval protection. Considering all three, and any identifiable trends in their use and success, will underscore the necessity of treating material technology in the context of external factors shaping it. The ship, in a proper technological history, is no more instructive in and of itself than a cluster of joined pieces of a jigsaw puzzle.

The costs of armament may also include extra labor costs, as it took more hands to effectively fight a ship than to sail it. A sailor could not haul on a line and work a gun at the same time. Caution is required when interpreting manning data, however, as so many factors—the availability of sailors, the anticipated risks of casualties to violence or sickness, the preferences and means of owners and masters—as well as minimal requirements for operation influenced how many men we might find aboard any given ship on any given voyage.

The changing nature of fighting at sea may well have influenced a significant change in hull design over the course of the period, as the high fighting platforms on deck, already lower than they had been in the sixteenth century, got lower and lower. Other structural changes may be directly related to that evolution. As we will see, however, it is unwise to think we have ever isolated an observable change—or continuity—to one factor in isolation. Even if, hypothetically speaking, one factor—in this case, the increasing emphasis on broadside cannonade in ship-to-ship combat, vis-à-vis grappling and boarding—provided the initial impetus for the development, unintended consequences are inevitable, and those will bear heavily on how a development proceeds from its inception—or if it proceeds at all. Before we begin

---

51 Smuggling is an important subject in the maritime history of this time and this world. The place to start is Wim Klooster's *Illicit Riches: Dutch Trade in the Caribbean, 1648–1795* (Leiden: KITLV Press, 1998). On American smuggling in the West Indies during the Seven Years War, see Thomas Truxes, *Defying Empire: Trading with the Enemy in Colonial New York* (New Haven, CT: Yale University Press, 2008).

to examine specific continuities and innovations, we will need to be familiar with as many forces acting upon the merchant ship as possible—both human and natural.

That means we cannot consider a change in typical sail plan as solely a matter of aerodynamics, or of reducing the cost of rigging, or of labor saving. It will not do to paper over what is likely to be a combination of influences with the convenient term "efficiency." We will see that a ship is, more than anything else, the result of a set of compromises, in which some imperatives have been given precedence over others. "Efficiency" in one respect will reduce "efficiency" in another. An important—and relatively accessible—example is speed versus capacity. We will be spending some time with that one, as it is central to ship design and ship profitability.

In fact, we have to think of influences beyond the rational and empirical. Good technological history teaches us that human ingenuity offers us any number of ways to pursue the same goal with a contrivance, and that tradition, cultural preference or prejudice, and aesthetics have a guiding influence on what we choose and what we do not.[52] That truth still meets resistance among those who find the notion of strictly rational, empirical technological decision-making comforting. This is most germane to the topic at hand, as the most vehement criticism leveled at the merchant shipbuilding industry, both during the period and after, was that it was hidebound by tradition and cultural prejudice, to the detriment of its "progress," and that if it were to embrace rational, empirical criteria ("science"), it would free itself from such regressive limitations and make better ships. The problems with that argument are manifold, and worth exploring in depth. That does not mean we dismiss the argument; it too is a historical phenomenon demanding investigation and interpretation. To begin with, it assumes that a technological art cannot be both traditional and guided by cultural prejudice on the one hand, and on the other hand, serve its users as effectively as one resulting from the application of "science." This is a prejudice itself, of course, one that belongs to the sensibilities of the Enlightenment. Generally speaking, that term refers to a set of intellectual predilections that urged inquiry, that valued the questioning of old assumptions and ways of doing things, and that inspired Europeans to connect intellectual enterprise to the sensible, tangible world in new ways. Indeed, as Joel Mokyr put it, the Enlightenment's contribution to economic growth "was the drive to expand the accumulation of useful knowledge and

52 For a compelling case—and one that had a powerful influence on this study—see Eric Schatzberg, *Wings of Wood, Wings of Metal: Culture and Technical Choice in American Airplane Materials, 1914–1945* (Princeton: Princeton University Press, 1998).

direct it toward practical use."[53] However, Mokyr also points out that "[t]he impact of the Enlightenment on the economy was slow and subtle. It would be impossible to discern its influence in the fields and shops of Britain in the eighteenth century."[54]

Mokyr's observation applies just as well to the ordinary shipyard. Larrie Ferreiro has shown that, while the foundational principles of modern, "scientific" naval architecture were laid down in the period we are considering, they did not begin to have a material influence on everyday ship design until after this period.[55] As a modern naval architect as well as a historian, Ferreiro's perspective on the designs and performance of the ships we are considering will be useful, as we keep in mind this guiding question: Would the principles of ship design to which we adhere—nineteenth-century principles—have improved any aspects of the performance of period vessels *without costing them in other areas*? That last bit is crucial, as there is strong evidence for the argument that natural factors rather than the lines of the hull wielded primary control over the real-world performance of a merchant ship at sea.

The preceding may seem somewhat arcane, but it is part of an overall attempt to push past the deep-seated prejudices with which we tend to approach technological history—especially the technological history of a period in which technology was the product of artisans rather than "engineers" and mathematicians. That the most eminent mathematicians of the day wrestled seriously with problems of hydrodynamics in ship design should make clear to us how seriously the subject was taken by "natural philosophers," and how complex and difficult it was—so complex and difficult, in fact, that none of the mathematicians' work found its way into mainstream period ship design. Was that because the stubborn, tradition-bound, insular fraternity of shipbuilders shut out innovations as a matter of course? The evidence suggests not. The evidence suggests that such innovations were not needed. Considering why that might be so will highlight defining characteristics of this world as well as what changed afterward. Contrast is essential to understanding something. That is one of the most important reasons we do history.

The chapters that follow will first introduce the natural and human forces creating, challenging, and sustaining the British Atlantic. Next, we will situate the ordinary merchant ship where those forces intersected, and discuss the basics of the contrivance itself, including essential—and only

53 Mokyr, *Enlightened Economy*, 10.

54 Mokyr, *Enlightened Economy*, 30.

55 Larrie D. Ferreiro, *Ships and Science: The Birth of Naval Architecture in the Scientific Revolution, 1600–1800* (Cambridge, MA: MIT Press, 2006), 23.

essential—terminology, all of which will be italicized, and may be found in the Glossary. A discussion of how ships were designed and built, and by whom, will lead to one about who owned them and paid for them, profited and lost by them. Considering the work environment and skills of the people who formed an integral part of the ship will add a fuller appreciation of what went into operating one of these things, especially when we consider the relationship between the work routines on board and alternative arrangements of rigging, steering, and armament, as well as specific routes and trades. We will consider the human threats to the merchant ship, and how owners, crews, and the state reacted to those threats.

This book is by no means an attempt to put this subject to bed. In fact, if anything, it intends to do the opposite. The chapters that follow will raise specific problems and questions still under investigation, what we know or think we know about them, and what we hope to learn by ongoing research. How can we explain stasis and change in hull design and construction, rigs, steering systems, and armament, given the general silence of the documentary record and the imperative to avoid simplistic, teleological, deterministic interpretations? While there is much work still to be done on these and similar questions, extant evidence does suggest where that work is likely to lead. As we consider multiple possible answers to the important lingering questions—all of which have some support from the evidence—it will become apparent that these are not multiple-choice questions with one correct answer; that questions about a complex technology have complex answers. We will also notice, however, that a common thread runs through those possible answers, indicating that the industry pursued what we would call efficiency in the design, construction, and operation of its ships—constrained by externally-imposed limitations, but able to adapt in useful ways nonetheless. Ongoing and future research, both traditional and experimental, will test whether that common thread holds together or frays.

CHAPTER 1

A Ship's Atlantic

For four days in July, 1609, the aptly-named small English galleon *Sea Venture* had survived a battering from an Atlantic hurricane—an experience most of us cannot imagine. The ship was on her way to the new Virginia colony, with settlers and desperately-needed provisions on board. Leaking badly, she ran up on the ring of reefs around the island of Bermuda, 500 miles (804.67 km) off the eastern seaboard. Her people salvaged their possessions and the ship's cargo, and, using her wreckage and timber from the Bermuda cedars growing in abundance on the uninhabited island, they built two new vessels, which they named *Deliverance* and *Patience*, and in those, finally reached Virginia in May of the following year.[1]

The first context in which to consider the British Atlantic merchant sailing ship is the Atlantic itself. It was a challenging, dangerous place for humans and their contrivances (and still is). Its weather is fickle. It is subject both to frequent tropical hurricanes in the warmer months and strong, cold winter gales moving east from the North American continent. Its middle section, traditionally called The Doldrums or the Torrid Zone, is an expanse of calm, hot, humid air. To pick up the prevailing westerly winds, the only reliable way to get back to Europe or Africa from the Americas, requires sailing up the east coast of North America, taking advantage of the Gulf Stream, a river of warm water within the colder ocean that originates in the Gulf of Mexico and begins to curve toward the British Isles off the coast of Newfoundland, where it meets the cold, southward-flowing Labrador Current. That mixture of opposites creates dense fog and cold gales on the Grand Banks, where the great cod fishery developed as the first extractive industry by Europeans in the Americas, long before 1600.[2] By the time a vessel reached the Banks, however, chances were good that those aboard had already endured similar foul weather, induced

1 Critics believe that a contemporary account of the wreck, William Strachey's *A True Reportory of the Wracke and Redemption of Sir Thomas Gates, Knight* was one of the principal sources of inspiration for Shakespeare's *The Tempest*, believed to have been written in 1610–1611.

2 For an introduction to the Newfoundland cod fishery and its wider context—the trade network of which it was an integral component—see Peter E. Pope, *Fish Into Wine: The Newfoundland Plantation in the Seventeenth Century* (Chapel Hill: University of North Carolina Press, 2004). A well-respected popular history of cod as an important food commodity is Mark Kurlansky's *Cod: A Biography of the Fish That Changed the World* (New York: Penguin, 1997).

 DOI:10.1163/9789004426344_003

by clashes of winds and currents off the capes of what would become known as North Carolina and Virginia. The sea floor around each of those capes is a graveyard for hundreds of ships and thousands of people.

If the vessel had come from the Caribbean, the master had successfully negotiated one of the two narrow passages through the Bahamas, swept constantly by sudden squalls, easily strong enough to blow a ship onto one of the sharp reefs lying just below the surface. The Gulf Stream itself, whose power carried a ship northward like a conveyor belt, could quickly become a ship-killer too, if a northerly gale blew southward against the strong north-flowing current, setting up huge square standing waves into which the ship would slam as into a wall. The danger of a powerful gale, especially in winter, dogged the ship throughout her return voyage to Britain, the European continent, or Africa.

The Atlantic's best feature by far, from a sailor's perspective, is the southwest-flowing wind-and-current highway from the Old World to the Caribbean that English sailors called the Trade Winds, or simply the Trades. This component of the Atlantic gyre, the clockwise ellipse that generally characterizes Atlantic circulation, made it easy for vessels that made it as far south as the West African coast to turn seaward and ride downwind to the islands they knew as the West Indies. During winter and spring in the Northern Hemisphere, before excessive heat and evaporation set up the convection systems that generate unsettled weather and tropical cyclones, the Trades provided an efficient, reliable, and even reasonably comfortable passage to the American tropics. Because the Trades continued to predominate within the Caribbean basin, they permitted ships to sail between the islands—so long as the destination lay downwind from the starting point. To beat against the Trades—to try to sail against them—was an arduous, if not futile, effort.

It is important to understand the extent to which a sail-powered vessel can sail into the wind, as that, as much as anything else, dictated where these ships could go and where they could not. There is a certain zone on either side of the direction *from* which the wind is blowing—usually between 30 degrees and 60 degrees—into which the ship cannot sail. For most ocean-going ships in our period, that number was usually between 45 and 60.[3] Sailing upwind at all, though—any closer to the wind's direction than 90 degrees, or perpendicular to it—tends to be uncomfortable, slow, and stressful for the ship, her rig, and her people. That is because, as the crew pulls in the lines that bring the sails closer in toward the ship's centerline, the pressure on the sails builds, and tries

3 See John Harland, *Seamanship in the Age of Sail* (Annapolis, MD: Naval Institute Press, 1984), 11.

to *heel*—lean—the ship over. That lean changes the shape of the part of the ship that is underwater, to an asymmetrical shape that resists going straight, but tries to turn the ship either toward or away from the wind. The person at the *helm*, steering the ship, has to work harder to compensate for that. There is more strain on the *rudder* (the blade—or, in this case, more like a barn door—hanging on the stern that steers the ship), the steering gear, the rigging, and the sails. Since most waves are wind-driven, the ship is now bashing into them rather than riding with them. That increases strain on the hull—which is already being strained by rig forces transferred to the hull, to which the rig is attached—and makes for a rough, wet ride. It also further slows down the ship. The increased pressure from wind and waves also increases side-slippage, called *leeway*, pushing the ship downwind of her intended heading.

Sailing vessels, then, are better off sailing no closer to the wind than perpendicular to it. The fastest and smoothest sailing is generally to be had between that direction and dead downwind (with the wind directly behind the ship). Sailing dead downwind can be tricky, however, as a small wind shift can move the wind from one side of the ship to the other quickly and unpredictably, and with no forward water pressure on one side of the hull, the *stern* (rear) of the ship can get into a nausea-inducing corkscrew motion—a *yaw*—especially in a large *following sea*.[4] Splitting the difference between dead downwind and perpendicular to the wind, a point of sail modern sailors call *broad reaching*, is where most vessels, and certainly these vessels, were at their best. Trade Wind voyages were mostly a combination of dead downwind sailing (called *running*) and broad-reaching, and the *square-rigged* ocean-going merchant ships of the time were ideally suited for it, as they carried their driving sails over the centerline of the ship where they contributed to balance, rather than to one side or the other, as *fore-and-aft-rigged* vessels did.

That basic explanation should make it easier to understand why the major sailing routes were downwind, but not necessarily directly downwind. We should add, too, that ships did indeed *beat* (struggle) *to weather* ("to weather" is the older phrase for "to windward," or toward the direction from which the wind is blowing)—because at times they had to, especially if they found themselves with land close by and downwind—a *lee shore* ("lee" means "downwind"). With a heavy surf running or a rocky coastline, a ship driven ashore by wind and wave was a ship destroyed, a cargo lost, and most if not all of her people killed. One of the most common loss scenarios for ships in this period was *embayment*, in which the ship was either blown into an indentation in the

4 "Nausea" shares a root with "nautical," for good reason. The Spanish referred to what in English was called a *carrack*, a sixteenth-century large merchant ship, as a *nau*.

coast by bad weather, or inadvertently brought there by poor visibility or faulty navigation (the two generally went hand-in-hand, actually, as both coastwise piloting by landmark and open-ocean navigation by celestial sights depended on visibility)—and unable to escape because the wind and seas were driving into the bay, and the protruding land at either end blocked the way out. The ship's only hope was to drop her anchors, let out cable, and hope that conditions did not deteriorate so much that those anchors dragged or those cables snapped. Often, that hope was in vain, and those on board had time to contemplate their imminent and inevitable destruction.

The routes these ships had to use in the Atlantic were full of such coastal indentations and lee shores. Considering those, and the capes, and the reefs, and the storms, and the fog (and we left out ice), it can seem rather remarkable that any regular, reliable maritime trade could take place. Yet it not only took place; it grew and thrived. The obstacles and hazards of the Atlantic were real, but the opportunities it presented made the risks worth it for those who had the will and the skill, the means and the nerves, to venture forth—and who decided that venturing forth was the best of the limited options they saw before them.[5]

Aside from the North Atlantic fisheries, Europeans did not have much interest in seeing how far west or south they could sail into the Atlantic, until they hit upon the idea of using it as an alternative route to the Far East. Columbus' Caribbean landfall was an accident, and from then on, Europeans either stumbled upon or created their own opportunities in the Atlantic. Trial-and-error was the rule. The Spanish took over established precious-metals extraction from the indigenous American empires they conquered, and determined efforts by their European counterparts to find their own sources of such wealth proved fruitless time and again, as did the search for the Northwest Passage, a hypothetical route to the Pacific through, or above, the North American continent. What the early voyages by the Northern European maritime powers, including England, really accomplished was reconnaissance. Small, stoutly-built ships, whose designs were hybrids of Northern European and Mediterranean technological traditions blended in the fifteenth century, permitted long, self-sufficient voyages into unknown waters. Their capacity was sufficient to carry stores for the ocean crossing. They could be repaired by their own crews underway or on any foreign shore with a stand of timber—or even replaced, if it came to that, as with *Sea Venture* in Bermuda. They were, in fact, better able to

5 On the merchants' side of sea-venturing, see Thomas M. Doerflinger, *A Vigorous Spirit of Enterprise: Merchants and Economic Development in Revolutionary Philadelphia* (Chapel Hill: University of North Carolina Press, 1986). For an introduction to the merchant seaman of the early modern British Atlantic, see Earle, *Sailors*.

survive the hardships of such work than their crews, who were debilitated or killed by scurvy (Vitamin-C deficiency), typhus (called "gaol fever" and transmitted by lice), dysentery, and tropical fevers.

Scholarship over the past few decades has made it clear how profoundly disease shaped what we know of as the New World.[6] The deadliest mosquito-borne pathogens from tropical Africa, malaria and yellow fever, traveled to the American tropics on the Trade Wind run. Smallpox had almost surely done much of its evil work even before European settlement in the Americas took hold, first transmitted by escaped shipborne livestock. Indigenous Americans' lack of immunity to smallpox depopulated some areas of the Americas, and left surviving populations gravely weakened and much less able to resist European encroachment than they would have been otherwise. Europeans themselves, however, were just as vulnerable to tropical fevers as the indigenous Americans were to smallpox, and found it difficult to survive in the tropical and sub-tropical Atlantic.

The deadliness of the warmer climes, combined with the presence of the exclusive Spanish, strongly encouraged any Northern Europeans interested in settling the Americas to look northward. To their great cost, early English colonists were wrong to assume that equivalent latitude between England and New England meant equivalent winters. Even Virginia was a challenge to survive in the colder months. It would take time, and experience, to learn that the Gulf Stream warmed the British Isles by carrying warm water northeastward, just as it carried English ships, and that the same polar continental air that spun off the great North Atlantic gales made winters on the northern eastern seaboard much harsher than they were in England.[7] It was also generally colder during this period than it is now; harbors and river mouths that remain ice-free in our own time frequently froze over then, putting pressure on ship owners and masters to plan sailings accordingly.[8]

The northern settlers did survive—barely at first—because ships were able to resupply them and bring new colonists. They worked small farms, traded with the natives, fished, and built ships—mostly small craft suitable for fishing, personal transport, and local trading. As is to be expected, those craft were based on both European traditions and indigenous American technology. Such was the case throughout the settled Americas.

6 See Introduction, n. 5.

7 Benjamin Franklin sketched a roughly-accurate chart of the Gulf Stream, based on his own shipboard experiences. He completed his study of the phenomenon in 1775.

8 That statement was true before the current climate changes began notably manifesting themselves; it is of course only truer now.

It did not take long, however, for the New England shipbuilding industry to expand into ocean-going commercial vessels, building for both local owners and merchants based in England.[9] With no prospects for high-profit plantation agriculture, New Englanders took up maritime commercial pursuits—most of which provided services to the plantation economy developing to the south. They also enslaved indigenous people they captured in war, and sold them into slavery in the Caribbean. Those indigenous Americans boarded merchant ships as prisoners, to be sold into miserable servitude in places they could not imagine—places they would in most cases not long survive.

Tobacco did not grow well everywhere Europeans tried to grow it, but it did grow well in Virginia, and quickly became that colony's staple export. While settlers pushed inland for new land to plant, as the older plots quickly wore out, more and more ships needed to get into the Chesapeake Bay, up the James and York Rivers, and back out to sea on a regular basis. They encountered an unseen, quiet, serious threat—as did all ships operating in the warmer Atlantic: the *teredo navalis* "shipworm," a mollusk that bores into wood and can render a ship's timber little more than a honeycomb of twisting channels. It was best for tobacco ships waiting to load cargo in Virginia to wait upriver rather than in the Bay, as the creature only lives in salt and brackish water.

Fouling with grasses, algae, and barnacles posed another serious problem to vessels operating in warmer water. Today, we use expensive bottom coatings that leach biocide to prevent this growth, which could slow a ship down to half its normal speed with only a season's exposure. Given that a ship's bottom could only be cleaned by getting it out of the water, removing this growth required either a drydock, which was not available to most merchant ships for most of this period, or *careening*—gently beaching the ship on a suitable gently-sloping beach at high tide, and using stout lines from the tops of her masts to secure points ashore to lean her hard over, one side at a time. This operation required a suitable beach, and exposed a vessel and her crew to attack by enemies ashore and afloat. (Later in the period, crews had more opportunities to accomplish this task at a wharf.) The twin threats of shipworms and fouling limited operations for both merchant and naval vessels in the temperate and tropical Atlantic Basin, and put pressure on maritime powers to acquire and defend bases for servicing and harboring their ships. A lack of adequate repair

9 On the early New England shipbuilding and shipping industry, see Bernard Bailyn and Lotte Bailyn, *Massachusetts Shipping, 1697–1714: A Statistical Study* (Cambridge, MA: Belknap, 1959).

facilities limited what a contender could accomplish even with a formidable fleet, as the French would discover to their chagrin in the eighteenth century.[10]

While no overnight sensation, the Americas in time proved to be worth fighting over and investing in. Trial and error gave way to success, especially after Europeans learned to grow sugarcane in the tropics, and, later, rice in the Carolina low country. Both crops are skill- and labor-intensive to grow and process for market. Beginning in Brazil under the Portuguese, sugar cultivation spread throughout the Caribbean, with the Guianas, Barbados, Jamaica, and Saint-Domingue becoming the centers of production and the generators of great wealth for the empires.[11] That was a mid- to late-seventeenth century development, while the Carolina rice plantations were a phenomenon of the eighteenth century.

The success of both of these crops depended on two assets: technology and slave labor. In fact, as with the crew of a ship, skilled labor was integral to the technology. The cultivation and processing of sugar and rice required know-how, experience, and the operation of relatively complex and dangerous equipment. West Africans, not English planters, knew how to grow rice. Without their knowledge and experience, let alone labor, there would have been no South Carolina rice fortunes in this period. In the tropics, as in Virginia, European migrants and enslaved indigenous Americans proved an unsuitable labor force, vulnerable as they were to tropical disease, and unwilling as they were to submit to the brutal work regimes of the plantations. Whether in Brazil, Jamaica, Virginia, or South Carolina, the ever-increasing importation of African slaves fueled these engines of wealth.

Although more resistant to tropical disease and physical labor in tropical conditions than other Atlantic peoples, these Africans still died in droves from the conditions they were subjected to, from the slave pens on the African coast, to the stinking hot holds of ships at sea, and to the brutality of the plantation. They were overworked and underfed, especially on the sugar plantations, for whose owners the enslaved workers generated so much wealth that they were cheaper to replace with new slaves than to keep healthy with adequate food and shelter and a more survivable work regime. As the large-scale alterations to tropical land increased populations of disease-vector mosquitoes, exhausted and malnourished African laborers suffered their share of the consequences.

---

10 On the early French West Indies, see Philip P. Boucher, *France and the American Tropics to 1700: Tropics of Discontent?* (Baltimore, MD: Johns Hopkins University Press, 2008).

11 Sidney W. Mintz, *Sweetness and Power: The Place of Sugar in Modern History* (New York: Penguin Books, 1985); Richard S. Dunn, *Sugar and Slaves: The Rise of the Planter Class in the English West Indies, 1624–1713* (Chapel Hill: University of North Carolina Press, 1972).

To transport that slave labor force, to bring in building and food supplies, and to carry off the plantations' produce to market, required a growing fleet of merchant ships from Spain, Portugal, France, England, the Netherlands, Denmark, Sweden, and, increasingly, the Americas. Since the various empires were constantly jostling for access to lands and markets, and trying to keep the others out of the ones they had already claimed, they fought each other and preyed on each other. The need for naval power in the Americas was pressing. With the Iberians fully engaged on and south of the Equator, the North Atlantic and much of the West Indies were contested ground for the northern European powers. By the closing decades of the two-century period considered here, the dominant power in that contest would prove to be Great Britain, but it took that long for the British to clinch that position. Much of the credit for that achievement must go to British America, but in an important sense, British America was *too* successful; just as the British achieved near-complete dominance in the North Atlantic, the bulk of the eastern seaboard colonies, emboldened by a now-mature economy many felt was under dire threat from an overweening, domineering, and distant master, banded together and revolted. In London, the administration and the King—though by no means all of Parliament—agreed that British America had been the primary beneficiary of the costly Seven Years War, and with its wealth established and now protected from the French, it should contribute to paying off that debt.[12] What began as arguments and import sanctions deteriorated into polemic, mob violence, then local armed conflict, a civil war, invasion, and ultimately a world war, as Britain's rivals took advantage of her misfortune to attack her preeminent position. The American War of Independence perpetuated the naval conflict and maritime predation long endemic in the Atlantic; there was no Pax Britannica yet. Shortly after that conflict wound down, the wars of the French Revolutionary era ramped up, with a new maritime power—the United States—engaging in a Quasi-War with France in 1798 and a new maritime war with its former master in 1812.[13]

Thus, from raids on Spanish settlements and treasure fleets, to conflicts over who would take and keep a certain island or patch of coastline or natural resource, to all-out imperial warfare, maritime conflict remained a fact of life

12 See Edmund S. and Helen M. Morgan, *The Stamp Act Crisis: Prologue to Revolution* (Chapel Hill: University of North Carolina Press, 1953), and Patrick Griffin, *The Townshend Moment: The Making of Empire and Revolution in the Eighteenth Century* (New Haven, CT: Yale University Press, 2017).

13 The Quasi-War with France, as it is called, is not generally well-known. See Michael A. Palmer, *Stoddert's War: Naval Operations During the Quasi-War with France, 1798–1801* (Annapolis, MD: Naval Institute Press, 1987).

in the Atlantic, and merchant ships had much more to contend with than the already-formidable natural hazards and challenges of this volatile ocean. To understand the relationship between the merchant ship and war in this period requires understanding the important ways in which war itself was different from what it is in our world. That has much to do with why the merchant ship was different, too.

It would be an indefensible oversimplification to claim that the imperial Atlantic wars in the early modern period were not ideologically-motivated, as if there were some sharp demarcation between the European religious wars of the sixteenth and seventeenth centuries and their successors. Catholic-Protestant enmity certainly played a central role both in inciting and fueling the long wars between the Iberian powers and their principal Protestant adversaries, England and the Netherlands, also giving rise to the "Black Legend" of Spanish cruelty, intolerance, and greed that has persisted so tenaciously in English culture.[14] Indeed, there is no understanding British history under the Stuarts and Hanoverians without recognizing how that ideological conflict played out both at home and abroad, and how inextricable it was from struggles over the ideas of constitutional versus absolute monarchy and from the ambitions and exertions of European states on land and at sea. At the end of the period, the French Revolution, to which is traditionally attributed the birth of modern nationalism, offered a secular ideological banner to which adherents could offer the sort of fanatical devotion previously reserved for religious causes, and citizens of the new United States were sorting out what citizenship in a young republican nation-state meant, and among other things, that certainly included patriotism.[15]

Nevertheless, the Atlantic maritime violence that concerns us here was in large part a manifestation of a different sort of conflict, one that grew out of a mercantilist imperial competition for wealth and prestige, and one that, for the most part, can be distinguished from what we call "total war" whose goal is the complete subjection of the enemy at any cost. The Anglo-Dutch Wars were fierce and costly, but they had nothing to do with religion or the attempted conquest of one power's homeland by the other. Although tied to European power struggles, they were also about the control of trade, and the wealth it produced. There was an ideological component to the series of conflicts

---

14 See the magisterial comparative study of empires by John H. Elliott, *Empires of the Atlantic World: Britain and Spain in America, 1492–1830* (New Haven, CT: Yale University Press, 2007). Elliott discusses the Black Legend in its original context, and the ways in which it distorted the truth.

15 On merchant ships, the development of that nascent national identity is explored by Paul Gilje in *Liberty on the Waterfront.*

between England-Britain and France, given the Bourbon pursuit of a Catholic absolutism and the resolute pride the British took in identifying themselves as the chief antagonist and alternative to that, but in the Americas especially, the ambitions of both sides were focused on the acquisition of wealth from New World enterprises.

The nature of conflict in the Atlantic in this period, then, had much to do with material gain. It had a profiteering character—and by that, I mean something less pejorative than what we mean by "war profiteering" in our own time, as a denouncement of financial gain by major defense industries producing war materiel. Seventeenth- and eighteenth-century maritime warfare offered participants direct personal financial gain; there was incentive to capture, rather than destroy, the enemy's maritime assets—his ships and their cargoes. Naval personnel, from the ordinary seaman to the Admiralty, received shares of prize money from captured vessels. Merchant owners could apply for letters of marque or privateers' commissions from their governments, re-fit faster vessels for maritime predation, and share in the spoils of any captures they might accomplish.[16]

This sort of maritime warfare was possible because merchant vessels and naval vessels were much more closely related, technologically, than they would later become. While is it true that the divergence between a merchantman and a *ship of the line* was clear in the early seventeenth century, ships of the line were larger, specialized battleships heavily-built to fight in large formation against other large groups of similar ships. The ships we are concerned with were smaller warships, assigned to cruise against the enemy's commerce, singly or in squadrons of a few vessels, or to escort convoys; or privateers—privately-owned merchant vessels whose construction, speed, armament, and extra crew equipped them to attack other merchant vessels with some advantage. All of these vessels, however, operated the same way. They exploited the same motive power. They were equipped with the same types of weapons, used the same way. They were of roughly comparable sizes and capacities. Their similarities meant that the crew of one could operate another, should they find

16 A letter of marque was issued to a merchant owner whose vessel was intended primarily for commerce, not predation. It granted permission to attack and capture an enemy merchant vessel should the opportunity to do so present itself in the normal course of mercantile operations. A privateer's commission was issued to an owner who intended to use the vessel primarily for hunting enemy merchant ships and taking prizes. A privateer, then, was a private warship, licensed by the state. On privateering, see Carl E. Swanson, *Predators and Prizes: American Privateering and Imperial Warfare, 1739–1748* (Columbia: University of South Carolina, 1991); and David J. Starkey, *British Privateering Enterprise in the Eighteenth Century* (Liverpool: Liverpool University Press, 1990).

themselves put aboard one, to take her as a prize. There was no analogy to, say, a submarine versus an oil tanker in the twentieth century.

The profit motive in period maritime warfare went hand-in-hand with the willingness of belligerents to carry on trade with one another during a conflict. In the Americas, mercantile interests did their best to see to it that products in demand got where that demand was, regardless of their or their customers' ostensible political loyalties. To accomplish trade with an enemy in wartime required a sophisticated set of precautions and subterfuges, but those were already well-established as the means of circumventing the Navigation Acts when merchants decided that the Acts put them in an unfavorable trading position. A ship might pretend to be English or French or Spanish or Dutch, depending on which flag would best facilitate the successful landing or loading of cargo at a specific destination; and with the polyglot nature of the Atlantic—especially of the Atlantic's pool of sailors—she might even sail with a foreign master for the same purpose, and even a mostly-foreign crew. Owners had to weigh the potential payoff of success versus the possible consequences of getting caught.

Avoiding repercussions might well involve providing remuneration to certain officials in return for looking the other way as cargoes were loaded or unloaded and the ship's papers examined. Given the modest salaries paid and the considerable expenses incurred by officials assigned to the colonies, from customs inspectors to governors, what we would condemn as bribery was usually considered a normal component of the compensation due to those who might otherwise find their situations financially untenable. Colonial officials were caught between a rock and a hard place. They faced pressure from the seat of government to enforce state policy, but they lived in the colonies, and faced equal or greater pressure to conform to the interests of those in whose communities they lived. The wharf, or the deck of the ship, was frequently where those conflicting interests collided.[17]

So, in a porous, interconnected American maritime mercantile economy, far away from the capitals of Europe, political loyalties came with a strong element of expediency. There was a self-interest component, a notion of the rightness of reciprocity, in a subject's concept of his loyalty to a sovereign. There was no contradiction in serving the King and seeking one's fortune on the same voyage. In fact, it was understood that serving the King should carry a reward, and that reward should be legal tender or something convertible to it.

17 See Wim Klooster, "Inter-Imperial Smuggling in the Americas, 1600–1800," in Bernard Bailyn and Patricia L. Denault, eds., *Soundings in Atlantic History: Latent Structures and Intellectual Currents, 1500–1830* (Cambridge, MA: Harvard University Press, 2009), 141–80.

Sometimes, political loyalty dissipated completely. In the seventeenth century, into the opening years of the eighteenth, the Americas were much farther away from the Old World—and from each other, than they would be later, in every respect but geographically. Populations were smaller and thus their trades were smaller. Fewer ships called at any given port. The direct projection of state power was close to impossible. Survival required resourcefulness. Sometimes, that resourcefulness led mariners to what the state considered a capital crime, but for many in the small towns and hamlets along the American coasts, it was a normal part of commercial intercourse.[18] We call it piracy, as did those at the time who spoke for the established power structures or against their enemies' privateers in wartime—but that term, loaded with opprobrium, seems to have been less commonly used by those who trucked and traded with the "freebooters" who stretched the limits of what was tolerated in maritime predation. "Pirates" were usually, but not always, privateers who preyed more indiscriminately than others, or who either had not heard or chose to ignore the news that a war was over and thus privateering commissions had expired. Sometimes they were just criminals, in what is perhaps the most widely-agreed-upon sense of that term—those willfully and even maliciously engaged in predatory behavior against their own society, with no regard for its norms, for their own gain and, just as likely, for revenge against a society from which the criminal felt alienated. These maritime robbers became the stuff of legend; even in their own time, romantic fiction and quasi-fiction included them in the outlaw genre that glorifies the quixotic rebel's private war against a hierarchical and unjust society. In reality, these people perpetrated sadistic and indiscriminate violence, and as regular commerce and other networks of communication and community grew with populations, the imperial state was better-able to suppress them.[19]

All of these ship-borne behaviors that took place outside the pale of official imperial sanction took place in and around Europe, too, but of course on that side of the Atlantic, the resources available to the imperial states to enforce their wills and suppress extra-legal activity were far more potent. The British Atlantic, and its imperial counterparts, grew up in a two-century period during

18 This is a central argument in Guy Chet's *The Ocean is a Wilderness: Atlantic Piracy and the Limits of State Authority, 1688–1856* (Amherst: University of Massachusetts Press, 2014). For criticism of Chet, see Adrian Leonard's review in the *International Journal of Maritime History* 27:3 (August 2015): 583–84; Christopher Kingston's in the *Journal of Economic History* 75:3 (2015): 933–36; and Christopher P. Magra's in the *American Historical Review* 120:3 (June 2015): 969.

19 "Communication and community" is a phrase borrowed from Ian Steele, from the title of *The English Atlantic*.

which the Old-World centers of power attempted, when they could afford it, to consolidate a tentative hold on overseas sea routes, territory, and people. They had achieved much toward that end by the late eighteenth century, but in the meantime, the new American societies had grown up under strictly limited European control. There were times and circumstances in which some Americans wished for more of such control, and other times and circumstances in which they wished for less. Power was constantly negotiated by all the means to which people resort.

Those means, it should be clear by now, routinely included violence, and the merchant ship, which is something we rarely associate with violence today, was no exception, with her complement of cannon staring out from a row of gunports in her sides.[20] Even vessels too small for some version of the broadside arrangement found on warships typically carried a few pieces on deck, some aimed over the side rails and others—called *chasers*—pointing forward from the bow and rearward from the stern. For sailors signing on to these ships for wages or shares, the presence of guns and the possibility that they might have to use them in the normal course of their work was something they accepted along with the discomforts and dangers of the sea itself and the physically-demanding nature of the work. The viability of this technology depended on an adequate supply of skilled laborers willing to bear the risks and tribulations of working sail on the open sea. Normally—and not surprisingly—it was a young man's work. Most of those who survived it either left the sea for other work after a few years, or got promoted to mate and, perhaps, master. Masters typically had at least some stake in the vessel, either as part-owner or by sharing in whatever profits she realized.

Typical wages were high enough to attract adventurous or desperate young men. Sailing was part of the culture in maritime communities. Some prospective sailors were running away—from bad situations at home, from the law, or from unemployment and poverty. Some mariners had what we would think of as middle-class aspirations—to save money, learn the skills of an officer, become a merchant-owner. Some used the sea during the agricultural off-season, returning to the farm for work during planting and harvesting.[21]

The skilled mariner was a valuable human commodity in the British Atlantic throughout this period. Because merchant and naval ship technology were so

---

20 Generally speaking, at least, that is so; in recent years, however, we have seen merchant ships plying the Gulf of Aden carrying armed security and repelling Somali pirates with lethal force. Even in the 21st century, lawless "states" with little access to established trading systems foster and support banditry.

21 See Vickers with Walsh, *Young Men and the Sea*.

similar, the skill set of the experienced sailor was transferable from one to the other. As the Royal Navy grew, along with the demands on it, the King's service faced a chronic shortage of skilled men, especially given that merchant sailors' wages were usually higher during wartime. This state of affairs led to the notorious practice of impressment, in which naval officers drafted experienced sailors on the spot from the waterfront or from ships returning to port. Sailors, and sometimes their merchant officers, employed every strategy from subterfuge to violence to escape the press.[22] Many an experienced sailor served on both merchant and King's ships in his career at sea. The Navy sometimes offered signing bounties—like today's enlistment incentives—to recruit skilled men, and after 1692, the government offered sailors wounded or left infirm in naval service the accommodations of Greenwich Hospital, funded by the "seamen's sixpence," a tax on sailors' wages. The severity of naval discipline has been exaggerated somewhat in popular legend, although it is true that insubordination, drunkenness on duty, and unwillingness or inability to carry out orders in what a superior considered a timely manner resulted in what we would consider severe corporal punishment, but what—so long as it was executed with some moderation and in accordance with regulations—most seamen would not have considered excessive. The same was true on merchant ships.[23]

Whatever the type of vessel or service, the ship was uncomfortable, dangerous, and frequently exhausting, and the food provided to the common sailor is now notorious, but it was regularly provided, which was more than could be said for many a laborer on land, and at least as good as what the poorest denizens of the larger cities and towns were likely to get. The ship was cheap to live on, compared to lodging and boarding in port. In fact, the owners and operators of inns, taverns, brothels, and boardinghouses frequently took advantage of a sailor's propensity to live it up while ashore, advancing him money and tempting him with all his favorite ways of spending it, so that he would be forced to sign on to another voyage to pay his debts. Not all such proprietors were unscrupulous. Plenty of sailors were perfectly capable of exhausting their financial resources quickly and completely without any outside encouragement. Their lives aboard alternated between tedium, exhausting work, and terror. When ashore, they wanted everything they had done without at sea—comfort, warmth, decent hot food, plenty of drink, sex, and, for most, female

22 On impressment and its ramifications, see Christopher P. Magra, *Poseidon's Curse: British Naval Impressment and Atlantic Origins of the American Revolution* (Cambridge: Cambridge University Press, 2016).

23 On merchant ship discipline, see Vickers with Walsh, 234–39.

company.[24] For these sailors, life was ephemeral. They lived close to death, they saw death around them, and they saw how suddenly it could come. A slip and fall overboard, a musket ball, a snapped yard, or a fever could take any one of them most any time.

For those with aspirations and talent, and a great deal of luck, a maritime career could eventually prove the means to a station in life and level of comfort far beyond anything promised by the circumstances of a sailor's birth. For those who stood to inherit little or nothing, whose family could not offer them a situation, the sea could prove a path to upward mobility—though a challenging and risky one. That was true for anyone from the son of a laborer to a son of the King.[25] To a more limited extent, it was true for black mariners as well as white, although to be sure, black men had no real prospects of becoming merchant-owners or even the masters of larger vessels.[26] It was common, however, for young black men to go to sea, whether they were free or enslaved. Like their white counterparts, some found sailing to be the best choice among a severely limited set of options. It was common to find them in command of smaller vessels engaged in coastal and inter-island trade, and fishing. Black sailors from Bermuda or Carolina, or one of the Caribbean islands, could spend years on the decks of merchant ships doing the same work as their white counterparts, part of the polyglot nature of so many ships' crews, who hailed from different European-American worlds or African-European-American worlds. In this respect, as in so many others, the merchant ship represented the limitations of mercantilist exclusivist policy.

Those black sailors, or their ancestors, had survived what was routinely regarded by working sailors as the worst of the Atlantic maritime trades—the transport of slaves from Africa to the Americas. Non-natives were very likely

---

24 On women at sea and male sailors' relationships with women, see Lisa Norling, *Captain Ahab Had A Wife: New England Women and the Whalefishery, 1720–1870* (Chapel Hill: University of North Carolina Press, 2000); and Margaret S. Creighton and Lisa Norling, eds., *Iron Men, Wooden Women: Gender and Seafaring in the Atlantic World, 1700–1920* (Baltimore, MD: Johns Hopkins University Press, 1996). On homosexuality among sailors, see Earle, *Sailors*, 102–103.

25 George III' s and Queen Charlotte's son William, the Duke of Clarence and future King William IV, was sent to sea as a boy. He became an officer and was, briefly, Lord High Admiral.

26 On the varied experiences of black mariners, see Bolster, *Black Jacks*; David S. Cecelski, *The Waterman's Song: Slavery and Freedom in Maritime North Carolina* (Chapel Hill: University of North Carolina Press, 2001); and Michael J. Jarvis, *In The Eye Of All Trade: Bermuda, Bermudians, and the Maritime Atlantic World, 1680–1783* (Chapel Hill: University of North Carolina Press, 2010).

to be killed by one of the tropical diseases endemic to Africa while the ship was seeking out or waiting for a cargo of slaves. The intrinsic brutality of the enterprise spilled over onto the upper decks; slave ships were notorious for the abuse not only of the prisoners locked below, but of the sailors paid to operate the vessel and keep those prisoners secured. And yet, almost thirteen million Africans were shipped to the Americas, most of them in the eighteenth century, so countless sailors found themselves on these odious voyages. Once a slaver arrived from Africa, those sailors too sick or too weak to work on the next leg of the voyage—or who found themselves made redundant, as a ship needed less crew for a non-slaving voyage than she did with captives aboard—would be left on shore to fend for themselves. That saved the owners money, as sailors were usually only paid for a voyage once they completed the round trip. It left the abandoned sailors destitute, since they had made the terrible passage for naught. Many were wounded or crippled, and helpless indigent sailors became such a visible fixture in Jamaica that the local government passed an ordinance forbidding masters to abandon members of their crew without means there. It is clear that slaving corrupted the souls of the officers in charge of it, John Newton notwithstanding.

Once the survivors of the Middle Passage had been sold or transferred to pens ashore, the ship's carpenter removed the barricades and platforms with which he had equipped the ship for her slaving voyage, another cargo of local products was loaded, and the ship made her way to her next port of call. The commonly-held notion of a "triangular trade" is an oversimplification of what is better-characterized as a networked or multi-nodal trade. It is clear that many owners ran their vessels on shuttle trades—back and forth from one home port to one destination—but "slave ships" in this period were not a specific type of vessel, and most engaged in other trades as well, seeking to take advantage of the best markets for products available at good prices.

The routine conversion of a general merchant vessel to a slave ship and back again highlights the versatility of this technology. Specialized vessels were the exception rather than the rule. Since our merchant fleet is almost exclusively composed of specialized vessels, we should expect to find marked differences between how maritime trade worked in the Atlantic in this period and how it works now, as indeed we do. We have already addressed the endemic violence and threat of violence as distinguishing features of trade by sea, the hardships of the work, and the most obvious example of violence and hardship, the massive transatlantic slave trade. Other differences, however, are just as instructive.

Those differences are rooted in a much smaller population, which lived in a much smaller total area of the coasts than ours; in cities we would call towns, in even smaller towns, and on farms. Recent studies using GIS to graphically represent the relative sizes of major urban centers make clear how anomalous London was; it dwarfed every other urban center in the British Atlantic.

Smaller markets meant fewer goods and smaller ships; modern visitors to faithful period replicas are commonly surprised by how small they are, given the size of today's merchant ships. As the population grew, and markets grew accordingly, the size of the ordinary merchant ship in the Atlantic grew only modestly; most of the demand for increased capacity was met by expanding the size of the fleet, rather than the size of individual ships. This response is not attributable to technological limitations on the viable size of these ships. Shipwrights were perfectly capable of building much larger vessels than we would have found working in the Atlantic at the time. The upper limit on size was a response to the risk environment.

On top of the natural and man-made hazards endangering the ship and her crew, her owners faced a daunting set of market risks. Even if the ship and crew safely completed a voyage, the owners stood a good chance of receiving the unwelcome news that the voyage had lost rather than made money. Waiting for one's ship to come in did not carry the assumption of eventual good fortune that it does in our contemporary colloquialism.

Although a robust business press developed in the eighteenth century, information could only travel as fast as the fastest ship.[27] Communication by sea from one part of the Atlantic world to another took weeks or months. Without real-time information, as we would call it, the shrewdest merchant took a serious risk of being wrong when he committed himself to a substantial outlay of money or credit, and sent his property and perhaps members of his family to sea, based on the latest information he had about prices and availability of goods. The correspondence of owners and ships' masters is replete with the disappointment of finding higher-than-expected prices for the goods sought, perhaps due to a poorer-than-expected harvest, or recent storm destruction; or lower-than-expected prices for the goods carried, probably resulting from recently-saturated demand for the ship's cargo. If a ship loaded a perishable cargo, and the voyage to her next destination proved overly-long, due to weather or damage, the cargo might fetch a dismal price at the docks; or, worse, prove unsalable.

27 On the business press, see John J. McCusker, "The Demise of Distance: The Business Press and the Origins of the Information Revolution in the Early Modern Atlantic World," *The American Historical Review* 110:2 (April 2005): 295–321.

The distinct possibilities of wreck, capture, disease, and financial loss due to unfavorable market conditions added up to quite a set of risks for ship owners, and the merchant ship was itself one of the responses to those risks, and an attempt to mitigate them. Keeping the maximum size of a merchant vessel in the Atlantic rather modest, relative to large warships and East Indiamen trading to Asia, not only limited the risk to owners from a loss of the vessel, but mitigated the risk of unfavorable market conditions. If a ship could not find a suitable cargo, or an acceptable price for that cargo, the master would have to sail partially or completely empty. A ship with space in the hold was not earning what she was capable of, and costing her owners considerably in spite of that. A ship that was too large for the available trade was more expensive to build, equip, insure, pay duties and taxes on, and man than a smaller vessel—for no gain.

The evidence suggests ways that risk mitigation guided the changes in merchant ships as markets and trades grew. In later chapters, we will explore the evidence for modest size increases in some merchant ships working in the British Atlantic, and how owners and builders tried to effect that change while keeping crew size, and thus cost, down. That will require looking at ton-per-man ratios—the number of tons of ship per man on board—and changes in rig that might be related to labor requirements. Such modest increases could never have kept up with the growth of the British Atlantic. That required building more and more ships, which spread risk among more and more owners as well. Most owners already limited their risk by sharing ownership in individual ships with others; some ships had sixteen owners. A managing owner or two usually ran the ship's operations. The others invested and reaped shares of the profits—or bore shares of the losses. Only the wealthiest shipowners could bear the cost and risk of sole ownership, and they had achieved that status through years of good decisions and good luck. Most owners, in fact, engaged in other business pursuits as well, spreading the inherent risks of shipping even more broadly. Those other pursuits usually figured larger in a merchant's overall portfolio than shipowning.

Owners, then, had to trust specialists with the technical know-how to advise them on what vessels to buy, when and from whom, at what cost, and to supervise the construction, fitting-out, and delivery of newly-acquired vessels. This is clear from their correspondence. Some owners were more knowledgeable than others, and their letters to partners and shipwrights reflect more specific preferences and prejudices than those of, say, the South Carolina magnate Henry Laurens, who reminded his agent in England, Captain Courtin, that he did not know enough about ships to stipulate much beyond basic capacity, trusting his representative to handle the details of ordering a ship and supervising construction.[28] This, then, was yet another set of risks owners had to take on, and their correspondence at times conveys their discomfort with how little control they had over such a major investment.

Sometimes that investment turned out to be too much. If the principal owners could not get enough additional owners to sign on, or some of those owners did not come through with their shares of payment, the ship would be left on the builder's stocks, and, frequently, in litigation. The shipwright, too, bore risks when he agreed to build a vessel. He might start one on speculation, hoping to find buyers before he finished her. Or he might find himself stuck with a vessel, finished or not, that the intended owners could not or would not pay for. Even if the process went smoothly, from laying the keel to delivery, the builder put his reputation on the line every time he launched a ship. To protect that reputation, he had to be cautious about any untested deviation from design or construction. That does not mean he did not make such deviations, but they were incremental adjustments and adaptations, made to suit a client's preferences or in response to the availability of materials. A shipwright who wished to maintain a good reputation could not afford to compromise his core principles, however, to accommodate his clients' desires. Merchants' specifications were usually intended to maximize capacity and minimize cost. The shipwright pushed back against the client if he thought those specifications would unduly compromise seaworthiness or performance, as those aspects of the vessel reflected directly on him. Surviving correspondence indicates that owners deferred to shipwrights on these matters; there was little point in hiring a shipwright and then overriding his professional judgment.

Who were these Atlantic shipwrights? They were artisans—highly-skilled craftsmen who carried a sophisticated set of concepts and a catalog of experience in their heads. That is frustrating to historians, as we want to find pieces of paper on which past people have written down how and why they did what they did, in as much detail as possible. A shipwright designing and building an ordinary merchant ship for Atlantic service in this period did not need to do it on paper, so the evidence we would love to have eludes us, and forces us to combine the relevant documentary evidence we do have, such as building contracts, account books, merchants' correspondence, and a small selection of plans and specifications compiled by the Royal Navy and a few treatise authors, with archaeological evidence, both conventional and experimental. For such an ordinary—even ubiquitous—contrivance, the British Atlantic merchant ship of the seventeenth and eighteenth centuries is a difficult historical quarry.

28 Philip M. Hamer, ed., *The Papers of Henry Laurens*, Volume 3 (Columbia: South Carolina Historical Society, 1968–1976), 366.

From reconnaissance missions at the beginning of the period, carried out by one to three small, stout galleons, establishing and provisioning small, tentative, isolated settlements, to a growing maritime economy based on fishing and on cash crops produced by slave labor, with a growing population and increased movement around the Atlantic basin to and from Europe, Africa, and the Americas, the merchant ship served the Atlantic enterprise as no other technology could. It was adapted to suit changing demands, and its numbers grew along with the population who relied upon it. Exploring it in more detail, as well as its relationship to existing natural resources, human capacities for work, technical aptitude, and suffering, and to its operating environment—both human-made and natural—will reveal the relationship between technology, culture, and nature in this world.

CHAPTER 2

The Ship: a Primer and Field Guide

If we could have walked the docks of a port town on the west coast of England, the east coast of North America, or tucked into a Caribbean bay, the ships we saw tied up or anchored there would have ranged in size, in shape, and in rig, but only within fairly modest parameters. They all shared important core characteristics. This chapter will introduce those, as well as the important variations in vessel type, and their technological ramifications. The variety of craft in use by Europeans in both the Old and New Worlds is well-described in the literature. Here, I have limited the discussion to types most common in transatlantic, British American coastal, and Caribbean use.[1]

The first shared characteristic of all these ships was how they were built: out of sawn lumber, fastened plank-on-frame. We cannot further explore the history of this technology without a basic understanding of how a ship was put together. To a heavy *keel*, or backbone of the skeleton, the shipwright attached the *stem post* at the bow (front) and the *stern post* at the stern (rear), defining the length of the ship and the upward curves at her ends. Those curves would, in turn, largely define how the ship met the water ahead, and what the hull did to that water as it passed behind. To the keel, the shipwright attached ribs, called *frames*, beginning with the frame that would define the ship's maximum *beam* (breadth), usually located just forward of *amidships*, or the midpoint between the front and back ends. Deck structures were formed with horizontal beams, *cambered* (slightly curved) to shed water. Joints between frames and deck beams were reinforced for strength and rigidity with *knees*, bracket-like pieces mounted both horizontally and vertically at these joints. Then, planking was nailed onto the frames, edge-to-edge, to make a smooth outer surface. The seams between planks were stuffed with *oakum*, loose fiber made by untwisting old rope, hammered into the seams with a caulking iron and mallet by a skilled ship's caulker. The seams were then sealed with *pitch*, a heavy, sticky derivative of pine tar.

1 For an overview of European and transatlantic vessel types, see Gardiner, *The Heyday of Sail*. For an introduction to types commonly used in the Americas, see William A. Baker, *Colonial Vessels: Some Seventeenth-Century Sailing Craft* (Barre, MA: Barre Publishing Co., 1962); and Howard I. Chapelle, *American Small Sailing Craft: Their Design, Development and Construction* (New York: Norton, 1951). Inconveniently, this literature, like most, tends to reflect post-eighteenth-century nationalist demarcations.

 | DOI:10.1163/9789004426344_004

The second core characteristic immediately obvious to any observer would have been the *masts*, vertical spars on which the sails were hoisted. These could be as tall as the ship was long. Masts were usually *stepped* on the keel, meaning their bottom ends rested on the central structural member of the skeleton. Wherever the mast passed through a deck, the opening was reinforced by *partners*, box-like structures. This was important, as the masts transferred much of the load from the sails to the hull. To keep the masts in place, lines called *shrouds* ran from the mast *tops* to the side *rails*, and others called *stays* from the tops to the bow and stern, so the masts were secured both side-to-side and fore-and-aft. These served the same function as guy wires on a utility pole or a newly-planted sapling. Between the shrouds, short pieces of line were secured as ladder rungs, called *ratlines*, enabling the crew to climb the shrouds to *go aloft*, or reach the tops of the masts to handle sails.

That brings us, not only to the most important distinction between how the two main types of sailing rigs worked, but to an important intersection of technology and the culture of labor. Climbing the rig to *hoist* and *furl* sails was only necessary on *square-rigged* vessels, so called because their sails were closer to square than triangular in shape (though they were actually trapezoidal in most cases). "Square" sails hung from *yards*, horizontal spars mounted so they could swivel around the mast, and be raised and lowered, at least partially. When not in use, square sails were kept tied to the yards. To deploy them, sailors had to climb aloft, walk or crawl out onto the yards, untie the sails, and let them drop. Other crew on deck used lines tied to the corners of the sails and yards—called *sheets* and *braces*, respectively—to control the sails once they were dropped from the yards.[2] Because square sails were balanced over the centerline of the ship, they were ideal for powerful sailing downwind—reaching and running—as noted earlier. They also projected more sail area aloft than a triangular sail. Air aloft is "cleaner"—freer of surface-friction-induced drag and turbulence from obstacles—than air at deck level.

On the other hand, square sails exact a price for their downwind driving power. To understand that, we should consider the alternative—fore-and-aft sails. The leading (front) edge of a fore-and-aft driving sail is attached to the mast with hoops or slides. To hoist it, crew on deck haul on a line called a *halyard* (from "haul-yard") attached to the top of that leading edge, running up the mast to a pulley, and back down to the deck.[3] There is no need to go aloft to

2 To visualize the sail-handling described here, it will be helpful to watch any of a number of online videos made aboard the working replicas and sail-training ships from around the world. In book form, an indispensable authoritative reference is Harland, *Seamanship*.

3 The halyard on a square-rigged vessel hoists the yard rather than the sail itself.

hoist a fore-and-aft sail, or to *douse* one.[4] The rigging is also simpler, since the crew only has to control one side of the sail instead of both. On most fore-and-aft sails, especially on vessels larger than small boats, the *foot* (bottom edge) of the sail is attached to a spar called a *boom*, whose forward end is attached to the mast and can swing through an arc of up to 180 degrees from left to right. A line, run through pulley *blocks* to create a *tackle* for mechanical advantage, is attached to the after end of the boom to control the angle of the sail to the wind, from tight toward the centerline to all the way out, perpendicular to the center line.[5] This is how the primary driving sails work on modern sailboats.

What is different from modern sailboats is that the latter use triangular sails, whereas most fore-and-aft driving sails of the period were tall trapezoids, with their upper edge attached to a spar called a *gaff*, so that the halyard, the line that hoisted and doused the sail, was attached to the gaff rather than directly to the top corner of the sail, as it is on a triangular sail. This allowed the projection of the same total sail area as a modern triangular sail, but with a shorter mast; the structural limitations of wooden spars precluded their being as tall as modern extruded-aluminum and composite masts.

We would have noticed one other important type of fore-and-aft sail on these ships. Recall that the fixed lines used to hold up the masts longitudinally, or fore-and-aft, were called *stays*. Unsurprisingly, sailors realized that they could attach the upper edge of a triangular sail to a stay. *Staysails* became more and more prevalent, we think, from the late seventeenth through the mid-eighteenth century. One type of staysail in particular stands out as an identifying characteristic of a ship from later in this period rather than earlier. Almost all ships we are concerned with mounted a *bowsprit*—a spar in the bows that protruded out in front of the ship. The bowsprit allowed for the mounting of auxiliary maneuvering sails for the bow, to help turn it, and to help balance the maneuvering sails in the stern (the *mizzen* sail and its auxiliaries). Before

4 *Furl* or *hand* (an older term) refer to securing a square sail by tying it up on its yard. To *douse* means to lower a sail suddenly, as in dousing a fire by throwing water on it, or to furl a fore-and-aft sail. See Harland, *Seamanship*, 17.

5 "Tackle" is traditionally pronounced "TAY-kle." On sailing vessels of this period, this was the only mechanical-advantage technology aboard. The more pulleys the line is run through, the more mechanical advantage the tackle has over the strength of one person hauling on a line un-assisted. The power of a tackle, then, is expressed in a ratio such as 2:1 or 4:1. Pulley blocks usually had from one to three pulleys in them, providing up to 6:1 power; however, 2:1 and 4:1 were more common. (This technology was also used for hoisting heavy cargo.) The trade-off for increased mechanical advantage is increased friction and slower operation. It takes longer to haul in or let out a sail using mechanical advantage, because the line has to run through all those pulleys. That is why warships, with their larger crews, tended to use less mechanical advantage on sail controls, so that they could maneuver faster in battle.

about 1700, these sails were square, hung on yards affixed to the bowsprit, with the primary one hanging below the bowsprit itself. In the eighteenth century, however, these sails were increasingly supplanted by a staysail called the *jib*, which was hoisted on the forwardmost stay—the one that ran out to the end of the bowsprit. The concept was deployed on small craft long before it was found on ocean-going ships. Jibs (larger ships flew more than one, mounted on parallel stays) were driving and steering sails. They were adopted on square-riggers from fore-and-aft-rigged craft, and remained fixtures of square-rigged ships through the end of commercial sail in the twentieth century.

There is one more type of sail important enough to be familiar with, and that is the *lateen* (as in "Latin"). A triangular sail suspended from a long yard all along its upper edge, the lateen rig was ancient Mediterranean technology.[6] See Figure 2, No. 16 (second row, second from left); this small French *tartane* has two lateen sails.

Large lateens were typically used as the primary driving sails on Mediterranean craft, whether singly or in rigs with more than one mast, but on ships we would have found around the British Atlantic, it was only common as a *mizzen* sail—the primary sail found on the after-most mast, called the *mizzen*-mast, of a three-masted ship. As such, its primary function was as a maneuvering sail, not a driving sail. Mounted far aft, it could help steer the ship, and counteract any tendency for the head of the ship to move to one side or the other.[7]

The lateen worked well for this purpose, and it was simple to rig and control, but it was not easy to *tack* or *wear*—move to the other side of the ship when the ship moved through the wind and on to the other side, which was called *tacking* if the maneuver was done by pushing the bow of the ship through the wind, or *wearing* if done by pushing the stern of the ship through the wind.[8]

6 See Julian Whitewright, "The Potential Performance of Ancient Mediterranean Sailing Rigs," *International Journal of Nautical Archaeology* 40:1 (2011): 2–17; and "Technological Continuity and Change: The Lateen Sail in the Medieval Mediterranean," *Al-Masāq* 24:1 (April 2012): 1–19.

7 Masters of period replicas report that, as their ships are equipped with auxiliary engines, they rarely use the lateen mizzen, as its primary function of helping to maneuver the vessel in close quarters, as in coming to anchor in a harbor or docking, is more easily accomplished by using the engines.

8 In modern sailing parlance, "to wear ship" has been replaced by "to jibe" or "to gybe," except on traditional square-riggers, which prefer to wear rather than tack in most cases. Jibing a fore-and-aft-rigged vessel is more dangerous, as without careful control of the sheets, the wind can suddenly catch the back of the sail on the other side, suddenly and violently slamming the sail around and transmitting a strong, perhaps catastrophic, shock to the rig. The heavy boom sweeping suddenly across the deck is also likely to smash the heads of any sailors who do not duck fast enough—perhaps also knocking them overboard.

FIGURE 2 "Several Figures, Representing the different manners of Rigging, which are most in use in the Northern Contrys"
ARCHITECTURAN LXII, FREDRIK HENRIK AF CHAPMAN, SJÖHISTORISKA MUSEET, STOCKHOLM

Half of the long lateen yard projected in front of the mast to which it was attached; that yard pivoted around the mast at its center point. To tack or wear it, however, required the crew to dip the front end of the yard almost vertical, to allow it to pass around the base of the mast onto the other side. Compared to working a gaff-rigged sail, this was more labor-intensive. Gaff-rigged mizzen sails began to replace lateen mizzens in the first half of the eighteenth century, and by the latter half, they were much more common.

As noted earlier, the earliest merchant ships entering service in the British Atlantic were hybrids of Mediterranean and Northern European technological traditions. Now we have added a mix of approaches to deploying sails. We will never be able to pinpoint the precise origins of any of them, as they are all too ancient, but it is clear that they blended techniques long in use in the Mediterranean and Europe, as did their predecessors: the types that made the fifteenth-century voyages of discovery, which combined the type of plank-on-frame hull described above with square driving sails on two or three masts and a lateen mizzen.

A vessel from Columbus' time, however, would have been an archaic curiosity on a seventeenth- or eighteenth-century British Atlantic waterfront. For that matter, some types from the seventeenth century would have appeared old-fashioned to the nautically-savvy of the subsequent century. We need a basic familiarity with types of vessels in common use, in order to investigate the relationship between specific technological features and the pressures put on those features, which sometimes encouraged continuity and sometimes encouraged adaptation.

As a rule of thumb, when a seventeenth-century source refers to a specific type of vessel, that term is associated with the hull form, not the rig. As the eighteenth century progressed, type of rig became more closely associated with vessel names. This is especially important to keep in mind when considering the term "ship." We use that term generically, of course, and so did they. In the eighteenth century, however, it also took on a specific meaning, connoting a mid-size to large vessel with three masts, each with square primary sails. When we see the word "ship," then, in an eighteenth-century source, we cannot assume that such specificity is implied; the source may be using the term generically. It is, in fact, frequently difficult to link a specific type of vessel, with definite characteristics, to a type name used in a period source. Like the language in general, vessel nomenclature became more standardized as time went on, but never completely so in this period. In this book, I use the term in both senses; if I am using it in the specific sense, I will make that clear.

It is also worth keeping in mind that vessels were frequently referred to by the work they did, and that rarely suggested a specific type of hull or rig or combination of the two. We can make no assumptions about a vessel referred to as a "slaver" in this period, except that she was engaged in the slave trade.[9] The Newfoundland "sack ship," engaged in the cod trade, is another such example.[10] Perhaps an easier *entrée* than nomenclature will be a brief discussion of the range of types of hulls, types of rigs, sizes, and employments of the most common vessels, appending common names to specific types as we proceed. We shall take that up shortly.

The vast majority of ocean-going vessels in Atlantic service were between forty and one hundred feet (12.19 and 30.48 m) long on deck. That means they were comparable to the size of most modern yachts or service craft, rather than ocean-going commercial ships. The ratios of their lengths to their breadths was between 2:1 and about 3:1. They were rounder and fuller in the bows (front ends) than more modern sailing craft. Some had fuller lines than others. This means that their hulls were more capacious for their size. Vessels with finer lines sacrificed some capacity for speed, all other things being equal. A useful way to understand this is to borrow a concept from naval architecture called *block coefficient*. Imagine a rectangular block, whose length, breadth, and depth are the same as the ship's.[11] That block has the maximum capacity—volume—that can fit inside those dimensions—a block coefficient of 1:1. A block, however, would present far too much resistance to the water to be an acceptable watercraft—even a very slow one. Ships' hulls must be curved at the ends for them to move through the water, and their lower sides and bottoms must be rounded somewhat for them to have acceptable stability. A hull shape that moves easily through the water is more important to sailing craft than to motor-powered vessels, as motors provide a constant motive power, while the wind does not. A sailing vessel has to be hydrodynamically-efficient enough to move through the water under sail in light winds. So, all ships' hulls have a block coefficient of less than 1:1; their hulls take up less volume than the block.

The less resistance presented to the water, the faster the ship will move under a given amount of sail power. Minimizing the volume of a hull of given dimensions (low block coefficient) achieves that, at the expense of capacity.

9 On slavers, see Rediker, *The Slave Ship*. Rediker notes that we do start to see specialized slavers after 1750, at least in Liverpool; those were further modified in accordance with 1780s changes to legislation on the slave trade (see location 1164). Once the trade was abolished in 1806–1807, craft typically seen in any smuggling enterprise showed up on the slave coast more and more, especially fast schooners such as the "Baltimore clippers."

10 On how sack ships operated, see Olaf Uwe Janzen, "A Scottish Sack Ship in the Newfoundland Trade 1726–27," *Scottish Economic & Social History* 18:1 (May 1998): 1–18.

11 See Chapelle, *Search for Speed*, 43–44.

Maximizing that volume does the opposite. Merchant ships, however, need capacity more than speed, with few exceptions. The whole point is to carry cargo and/or passengers as efficiently as possible to maximize profit versus cost. In general, for merchant ships, the compromise worked like this: merchants wanted a vessel of a certain rough capacity. They wanted that capacity in the smallest possible vessel, because that kept building and operating costs down. Once capacity and size were fixed, they wanted the shipwright to fashion the fastest hull possible—a "fair" hull. The ordinary merchant ship, then, would have had a moderate or moderately-high block coefficient.

The other operational criterion dictating design was *draft* (depth). It should be intuitive that the lower in the water a vessel carries most of its weight, the more stable it should be—and we know that wind in the sails presents a heeling force that must be counteracted by hull stability, or the vessel will capsize. While it is true that a deeper vessel presents more vertical resistance to the water than a shallower one, the deeper vessel can give up some *beam* (breadth) relative to the shallower vessel and retain the same capacity. That narrower beam cuts resistance just as the deeper draft adds it. The deeper draft has other advantages for sailing, perhaps most importantly the increased resistance to leeway (side-slippage). Think of two knife blades. The first is a large chef's knife; the second, a thin boning knife. Imagine pushing both of them forward through water. It will be effortless. Then imagine trying to push each of them sideways through water. The chef's knife is going to offer strong resistance, and your sideways pressure is going to make it slide forward through the water before it slips sideways if you don't restrain it from forward movement. The boning knife will slip sideways fairly easily, because it does not present much vertical resistance to doing so. That is the difference between how a deeper-draft vessel responds to sideways forces from wind and sea and how a shallower-draft vessel does.

Operational considerations, however, can dictate shallow draft. It is natural for sandbars to form at the mouths of rivers and harbors, and ships needed to clear those. Reducing draft requires increasing beam, which helps make up for the reduced stability and reduced capacity at the same time. Shallower, beamier ships were a hallmark of Dutch construction in the seventeenth century, helping to make their ships practical, versatile carriers. Top sailing performers they were not, but we will explore in depth the practical limitations on the importance of sailing performance for merchant vessels. Evidence strongly suggests that, as the period progressed, British Atlantic builders adopted and adapted this basic Dutch trait more and more for ordinary merchant ships.

Now that we have a basic grasp of the considerations influencing hull designs, we must consider rigs—the motive power propelling those hulls through the water.[12] The most important characteristic of a sailing vessel, no matter what specific aspect of design we are talking about, is balance. To operate properly, let alone optimally, the contrivance must balance multiple forces at the same time. In order for a rig to suit a particular hull, it must have the power to overcome the resistance of that hull to the water and the weight of the hull and its cargo. It must also balance the stability characteristics of the hull; a too-powerful rig can overpower the hull and lead to *knockdown* or even capsize.[13]

It would be a simple matter to rig a vessel if weather conditions were constant. Suppose we knew what the constant wind speed was, and from which direction. We could put up a simple rig—perhaps even one sail on one mast, as on a Nile River boat—balanced in its power to the resistance and weight of the hull, and have done with it. Hoist the sail, steer the ship, arrive at the destination at a predicted time.

In the Atlantic, as we have established, weather conditions are anything but constant. Rig versatility, then, was a common trait of Atlantic ships. This introduced complexity, and cost, but it was necessary. A rig composed of multiple sails, usually of more than one type, on multiple masts, was the norm. This allowed the crew to respond to changes in wind force, from calms to gales, and in wind direction. The former imperative explains the division of the sail plan into multiple sails, so that the crew could reduce sail area quickly as the wind built, or add to it if the wind diminished. Some sails were only used off the wind, some sails were only used in light winds, and some sails would be furled at certain wind angles because otherwise they would block a clean flow of air to other sails. Adjustments to the rig were frequent, if not nearly constant, for a significant portion of a typical Atlantic voyage.

That is a stark reminder of the intimate connection between a ship's rig and the labor demands on her crew. Because the manipulation of sails required muscle power, the power of any given sail was limited by the combined strength of the crew; if the sail were too large, and thus too powerful, to be controlled by the available crew, that sail could not be carried, and would need to be split into smaller sails. On larger ships with more complex rigs, it would not do to have a sail so large that all or nearly all of the crew would be required to handle it, as sail adjustments needed to be done simultaneously, or nearly

---

12 For diagrams of the principal sails on fore-and-aft and square-rigged period vessels, see Appendix 1.

13 *Knockdown* does not mean a knocking-down of the rig, but of the entire ship. It means the ship has heeled over so far that she is *on her beam ends*, lying on her side, roughly ninety degrees from upright. This is highly dangerous, of course. A ship could sometimes right herself from a knockdown, and sometimes not. If a substantial amount of her cargo shifted position during the knockdown, she would probably founder.

so, to successfully accomplish ordinary maneuvers. The evidence suggests that the division of the sail plan went along with a general, modest increase in typical merchant vessel size over the course of the period, but it may have as much to do with an attempt at saving labor costs.

To understand how all of this relates more clearly, let us consider two extremes. The first is a Dutch *fluit*, a capacious, economical cargo ship of the seventeenth century.[14] The *fluit* had a high block coefficient, she was full in both the bows and the stern, and she carried a relatively simple, modest-sized rig. That meant that she could carry a large cargo for her size, with a relatively small crew. The trade-off: her speed and handling were no more than adequate. At the other end of the spectrum, consider a Bermuda *sloop*.[15] With a single mast, and an impressively-long bowsprit, this swift vessel carried a huge fore-and-aft main driving sail and a set of large jibs and staysails forward; with her long boom and long bowsprit, she could carry a press of sail larger than she was. Chapman illustrates a typical sloop rig in Figure 2, No. 12 (bottom row, third from left).

The crew of a vessel like this had to be large enough, strong enough, and alert enough to manage such a rig. The sloop was fast and *weatherly* (she could sail close to the wind), but should a *squall* come up, the crew had to jump to it and reduce sail quickly, or they could lose the rig, or even the ship (and with her, their own lives). Dousing those jibs and staysails required going out on that long bowsprit—not a job for the frail or faint-of-heart. The main sail, however, could be doused from the deck; ships could carry somewhat larger fore-and-aft main driving sails, relative to the size of the hull, than they could square ones, and still control them.

It is worth mentioning that crews could reduce sail area without dousing the sail. They could make a sail smaller by *reefing* it—gathering up part of it and tying that part off. This was done to both square and fore-and-aft sails to make more minor adjustments to sail area, and thus driving power, than dousing entire sails. It is also worth remembering here that the height of a rig mattered, since the air aloft exerted more force, and that force acted as more of a lever, on the rig than air lower down. This is why we typically see smaller sails as the rig goes upward. They put less strain on the weaker, upper components of the rig, allowing for a lower *center of effort*—the point at which the force of

14 See A.J. Hoving, *17th Century Dutch Merchant Ships: Text, Photos, and Plans for the Ship Modeler* (Florence, OR: SeaWatch Books, 2014), 62–79.

15 See Michael J. Jarvis, "'The Fastest Vessels in the World': The Origin and Evolution of the Bermuda Sloop, 1620–1800," *Bermuda Journal of Archaeology and Maritime History* 7 (1995): 31–50.

the entire sail plan was concentrated—and allowing less weight aloft, and thus contributing to a lower *center of gravity*. All of that contributed to stability and to the resilience of the rig.[16]

Still, spars broke and lines snapped and sails tore—routinely. The more of those things a ship's rig contained, the more work that rig required to keep it functioning and the more money it cost to rig and re-rig. That fact reinforces the desirability of a split—and thus more complex—rig, in spite of the added complexity and labor it demanded.

The vulnerability to damage of period rigs also points to the importance of materials as decisive factors in how a technology functions. By the end of deep-water commercial sail in the early twentieth century, masts and *standing rigging* (shrouds and stays) were steel. Modern sailing rigs use aluminum alloys developed for aviation, stainless steel rigging wire, and composites such as carbon-fiber-reinforced plastic resins. Some modern composite masts are so strong and light that they do not even need shrouds and stays to support them. Sails have been primarily synthetic since the 1950s. Reducing the weight and increasing the strength of masts, rigging, and sails permits a significant increase in height and sail area over what was possible when spars were wood, rigging was hemp, and sails were natural-fiber textiles. It is of no use, then, to wonder whether shipwrights and riggers would have recognized the advantages of setting up more modern rigs. The materials at hand would never have supported them, and that they did know—not that they shrank from pushing the envelope of what was possible—nor did ship masters, the boldest of whom routinely played a dangerous game with the wind (and with the ship's safety) by "cracking on"—insisting on carrying the tallest rig with the biggest spread of sail they thought the ship could possibly bear, in order to maintain maximum speed. If they miscalculated, rig damage would be the minimum consequence.

The difference in materials accounts in large part for why rigs of this period were so different from later examples. Another reason for that is something we have already discussed: mechanical advantage. Twentieth-century square-riggers would have been impossible to operate without mechanical-advantage contrivances unavailable to the shipping industry before the end of the nineteenth century. Those contrivances require the precise machining of complex metal parts, and they multiply human muscle power manifold, permitting the use of larger sails, provided the materials used in the rig can stand up to the increased pressures and are cost-effective. That said, *blocks*, utilizing wood species harder than some metals, were a key technology aboard all ships of the

16 These terms were not in use at the time, but mariners were well-aware of the physical phenomena to which they refer.

period, and they allowed the use of tackles for limited, but highly-effective, mechanical advantages of 2 or 3 to 1.

Splitting the rig and deploying it lower-down, then, made it possible to work with limited mechanical advantage and modest-sized crews. It also relieved wooden spars and natural-fiber sails and rigging of shocks and strains they could not have borne, while allowing for a sufficient sail area to power the vessel satisfactorily.

It would be easy to infer, given our deeply-held prejudices about technology, that this comparative treatment of period and modern rigs suggests a superiority of later over earlier technology; if only they had had what we have, they could have done so much better. That sort of thinking must be recognized and expunged. Our focus instead must be on different ways of accomplishing a task, given that almost all the variables acting upon such accomplishment are different. The wind and sea are constants, as are the physics of hydro- and aerodynamics. Most other influences on how to go about building and operating a successful commercial cargo vessel—or sailing vessel—then and now are different. Since we have already begun a discussion of period materials, and some ways in which they influenced rig design and operation, we should continue that before exploring others.

By far the dominant material used in ship construction, including rigging, was wood. We have replaced wood with so many other materials in so many applications in our world that it would be easy to relegate this natural material to the status of quaint and primitive precursor to much more effective alternatives. One could add that any objection to such a dismissal based on the useful properties of wood as a material would be irrelevant, since wood was the only material available at that time, in that place. Surely, such an argument might continue, if wood had been the best alternative, it would not have been so thoroughly superseded by iron and then steel in later periods.

A central challenge to doing the history of technology is overcoming a simplistic conceptual structure in which the invention and diffusion of a new alternative drive technological choice independently of a historically-unique combination of wants and needs. In our case, that structure would have us believe that wood was used to build ships because it was the only material available for the purpose. Once someone invented a way to produce rolled iron plate economically, ships were made of iron, because it was "better" than wood. Once someone figured out something similar with steel, the latter replaced iron because it was "better." This is ahistorical; implicit in it is the fallacy that the wants and needs of one group at one time are the same as those of a group in another time, and that blinds us to the need to understand all of the ways in which that is not the case. Such a structure also strongly suggests inevitability and a linear technological determinism—the assumption that the availability of a certain technology drives the adoption of that technology. These too are fallacies; historians of technology have shown us cases in which available technology was not utilized in some particular time and place because external factors pressured for the use of an alternative.[17] We could certainly explicate the advantages of iron over wood, or steel over iron—or wood over either—but we can only do that synchronically, not diachronically. When we have managed to grasp the historical context of a given technology, then we are able to explain why, even if an alternative had been available, it might well have been ignored. Our task here, then, is to understand wood as a shipbuilding material in the British Atlantic between 1600 and 1800.

Wood has several advantages as a shipbuilding material, provided it is the right wood; it stands to reason that, with the variety of tree species found on Earth, properties would vary significantly. In general, however, wood has a high strength-to-weight ratio. It is buoyant. It is not brittle; it can flex and return to shape without fatigue damage. It can be harvested and worked with human-powered tools; complex and expensive infrastructure is not required to use it, so wooden shipbuilding could take place most anywhere there was access to water, access to trees, and a shipwright—fortunately for the castaways from *Sea Venture* in Bermuda. Because trees grow relatively straight and tall, large straight boards may be milled from their trunks (though it would have been more convenient had tree trunks been square instead of round). Because trees grow branches, mature specimens presented naturally-bent sections that shipwrights could harvest for the curved structural pieces they needed to fashion floor sections, knees, stem, and stern pieces. These were called *compass timbers*, and they were highly-prized; in the Royal forests, reserved for naval construction, trees from which shipwrights intended to eventually harvest compass timber would be marked accordingly.

As more and more mature trees of suitable species were harvested around shipbuilding centers, the sources of compass timber grew scarcer. Shipwrights then had to fashion their curved pieces out of multiple smaller straight pieces. Fortunately, wood is easy to join together with simple fastening techniques: nailing, through-bolting, and *treenailing* or *trunnelling*. This last technique involved driving a wooden dowel through holes bored in both pieces—holes just slightly smaller than the dowel. Because wood will give a little when compressed without splitting (how much depends on the species), the *trunnel* formed a tight fastening. After the protruding end of the trunnel was cut off flush with the surface, the connection was made even more secure by driving

17 See Lane, and the discussion of his argument in the Introduction.

a small wedge into the head of the trunnel to expand it. Wood, then, was used to fasten wood. Using the same material meant no issues with incompatible dissimilar materials, the bane of many fastening techniques. The acids in some woods tended to corrode nails and bolts—as did, of course, salt water. Dissimilar materials do not expand and contract at the same rate with changes in temperature and moisture, which can help fasteners work loose over time.

This does not mean that period shipwrights could do without iron fastenings—nails and bolts. Trunnels must be fairly large to be adequately strong, so they are used to join larger pieces together. Planking was nailed on, and the keel assembly generally utilized long bolts. The limitations of wood trunnels meant that the disadvantages of iron—its more difficult manufacture, its expense, its limited availability, and its corrosion—had to be borne.

Wood does not rust, but it does rot. It also burns. Fire can and does occur on ships made of any material, but rot was the constant enemy of the wooden ship, and it could (and did) negate the skills of the best shipwright and the highest initial quality of the materials he used. By the opening of this period, shipwrights had long known which available species of wood offered the best combinations of strength, workability, and rot resistance. As the Americas and Asia were explored and colonized, they discovered and used more such species.[18] While the primary structural wood was, and remained, white oak, New World species, such as live oak from the southern Eastern Seaboard colonies and Bermuda cedar, gained currency over time for their own desirable properties. Each presented advantages and challenges that influenced its use. Live oak was strong, and even more rot-resistant than the traditional white oak, but it was so hard when seasoned that it was difficult to work with hand tools, limiting its use before the end of the period, when more powerful sawmills were set up in New England. Bermuda cedar became renowned for its high strength-to-weight ratio and rot resistance, and the aforementioned Bermuda sloop gained a reputation as the finest, swiftest small vessel type available in the British Atlantic for much of this period.

Given adequate availability, the specific properties of certain species led shipwrights to use between two and four different types of wood in a single vessel. They knew which available type would best serve for a plank, a knee, a frame, or a sternpost. More knowledgeable owners knew this, too, and shipbuilding agreements typically specified woods to be used in construction.

Untrammeled extractive exploitation of natural resources is a hallmark of modern western civilization, and trees for shipbuilding were certainly no

18 See Amy Mitchell, "A Comparison of Wood Use in Eighteenth-Century Vessels" (Master's thesis, East Carolina University, 1994).

exception. Mature white oak was already a protected commodity in Britain during the reign of Charles II (1660–85). To European explorers and settlers, the forests of the Americas were limitless. Relative to what was available in Britain and Western Europe, they were—but access to them was not. While poor survivability makes archaeological remains rare, we should examine those we do have for any correlation between date and place of construction and evidence of any adaptation of construction techniques indicative of scarcer materials, such as fashioned pieces where we might expect to find compass timber, or less-than-ideal wood species used in certain applications, which might point to the depletion of timber around centers of shipbuilding such as Massachusetts Bay and the Delaware Valley.

What we would call the "science" of wood as an engineering material was a matter of ongoing concern, given its importance. The mitigation of rot was a special priority among both shipwrights and chemists. The former experimented with the placement of floor timbers and associated structures in an attempt to improve water and air circulation in the *bilges* (lowest parts of the hold, where water collected), and the latter eventually included no less a light than Michael Faraday.[19]

Wood was not the only primary shipbuilding material subject to rot. Rigging a ship required a web of natural-fiber rope and a suit of sails made from cotton or flax. The preferred fiber for rope was hemp, and the preferred source for manufactured hemp rope was the Baltic. In fact, the Baltic was also the preferred source for sail canvas (called *duck*) and for *naval stores*, the pine products necessary to seal wood and rope for marine use—pitch, tar, and turpentine. Given those established preferred sources, and the generally-high cost of skilled labor in the Americas (due primarily to a scarcity of it for most of this period), the development of British American manufacture of these goods was challenged indeed—and we can add to that challenge the mercantilist principle that colonial manufactures were to be discouraged in favor of the home country's. Nevertheless, those industries did grow up in British America.

---

19 Michael Faraday, "On the practical prevention of dry rot in timber," published account of a lecture by Faraday at the RI on 22 February 1833, (London: John Weale, High Holborn, 1838), Pamphlet Vol. 249 No. 18, Library of the American Philosophical Society. The APS Library's pamphlet collection contains two other items on the subject from our period: Edmund Saunders, "An Address to the Owners of Ships, &c., in particular, and in general to all who have an Interest in preserving Timber under Water" (Plymouth: M. Haydon & Son, 1790), Pamphlet Vol. 116, No. 3; and Henry Guest, "Observations on sheathing vessels: seasoning timber; the proper time to fall timber; the nature and what force it is that causes the sap to rise; with a number of other valuable observations" (New Brunswick, NJ: A. Blauvelt, 1805), Pamphlet Vol. 1162, No. 7.

Ropemaking and sail-making were local concerns in New England; and, although the materials may have been imported from the Baltic, they were likely to have been shipped in vessels built and owned in British America, especially in the latter half of the eighteenth century, when the British American economy fully matured. A naval stores industry developed earlier, assisted by a bounty from the Government for supplying the Royal Navy, in effect from 1705 to 1725 and again from 1729 to the Revolution, although at a reduced rate and only on tar and pitch. The encouragement of American pine product production was prompted by the restriction of Baltic supplies by war. The American products, however, especially those from Carolina, were widely regarded as markedly inferior to those from the Baltic—so much so that the Navy protested the bounty, although mercantile interests on both sides supported it for their own reasons.[20]

Prejudice against the products of a certain group in a certain place plays an important role in the fortunes of markets and technologies. There are frequently some grounds for the prejudice, beyond simple rivalry for the same markets, but the prejudice typically survives long after the grounds for it have shifted. It is not clear that the American pine product industry in this period ever caught up in quality to its better-established and better-regarded Baltic counterpart. The case for home-built versus American-built British ships is one in which an early prejudice against the quality and longevity of American-built ships had a long life, surviving into the period when the British American shipbuilding industry was a century old. Data on longevity yield mixed results.[21] Lord Sheffield, who consistently and publicly decried American competition with home-based shipping and shipbuilding, claimed American ships would only last five or six years, but John McCusker says this was almost surely an exaggeration, and that the real average was probably more like ten.[22] In 1799, Lloyd's of London rated southern-American-built ships their A-1 rating at the time—twelve years.[23] So, at least in their probably dispassionate judgment,

20 Justin Williams, "English Mercantilism and Carolina Naval Stores, 1705–1776," *Journal of Southern History* 1:2 (May 1935): 169–85.

21 A Hull shipwright testified that London-built ships were "unquestionably better" than ships built in the English outports, as London builders got better prices on materials and thus could afford to build to a higher standard. This testimony supported Lloyd's policy of automatically rating Thames-built ships higher than any others. *Annals of Lloyd's Register* (London, 1884), 15.

22 John J. McCusker, "The Pennsylvania Shipping Industry in the Eighteenth Century" (unpublished bound typescript, Historical Society of Pennsylvania, Philadelphia, 1973), 121, 125.

23 McCusker, "Pennsylvania Shipping Industry," 126. Lloyd's rating determined the terms and price of ship insurance. The higher the rating, the longer the ship was expected to

these were fine vessels, and we should not paint British prejudice with too broad a brush. Southern live oak had managed to gain recognition for its outstanding combination of rot-resistance and strength by that point, obviously. It had taken time, however, and the southern shipbuilding industry remained small.[24] Live oak is too hard to be worked easily with hand tools, once it has seasoned. In the South, the wood could be worked green, but by the time it could be transported to New England, it would have hardened. It would be the 19th century before New England yards, with their new industrial tools, could make full use of Southern live oak in their yards.[25]

Marshall Smelser and William Davisson complicate the matter, writing in the same year as McCusker. Examining New York Navy Office records from 1715 to 1764, they found that the average age of the ships listed there was 4.7 years, and that only 9.4% were over ten years old. What was most striking, however, is that the British-built ships tended to be twice as old as those built in America. They attribute this to construction methods—thus casting an aspersion on British-American shipwrights.[26]

Christopher French followed up on this in 1991, and his findings remind us of the inherent limitations of data from one source—a linchpin of most scholarly disagreements on shipping productivity over the past 50 years, as we have seen. French looked at ships trading between London and Virginia or Jamaica, and says that, while their age was always relatively low, it was "also noticeably higher than the mean age of vessels trading between London and New York." The former averaged 7.2–7.7 years.[27] French, however, supports the Smelser-Davisson conclusion that British-built ships lasted about twice as long, and he also attributes that to construction methods. He points out that the British-built ships were bigger, and that bigger ships were more expensive

---

last—hence the year range assigned to it. For perspective, consider that, while in 1799 an A-1-rated ship was expected to stay in service for twelve years with her original integrity, in 1911 Lloyd's rated the German steel bark *Peking* A-1, with a projected service life of 100 years. She is no longer in service, but she is still afloat.

24 See Converse D. Clowse, "Shipowning and Shipbuilding in Colonial South Carolina: An Overview," *American Neptune* 46 (1984): 221–44.

25 Warren Riess with Sheli O. Smith, *The Ship That Held Up Wall Street* (College Station: Texas A&M University Press, 2015), 54.

26 Marshall Smelser and William L. Davisson, "The Longevity of Colonial Ships," *American Neptune* 30 (1973): 16–19. It was much more common at the time, though, for the prejudice to attribute the relative inferiority to the quality of timber used; the belief was, simply, that English oak was better.

27 Christopher J. French, "The Longevity of Ships in Colonial Trade: Some Further Evidence," *International Journal of Maritime History* 3:1 (June 1991): 158. Thanks to Prof. French for suggesting related sources.

to build—thus it was more compelling to build them to last—especially since bigger ships were also likely to be more profitable if properly employed.[28]

The listed age of a ship in a document, however, is not the same thing as how long she lasted. These studies tell us that, on these routes at these times, there were more old British-built ships than old American ones. That does support their conclusions, to be sure, but it does not prove that British-American building methods resulted in lower durability. Were some of these ships pulled off these routes after a time and employed in other service? Were ships on other routes, or at other times, older or younger, on average? In her small data sample of wrecks found on this side of the Atlantic, Kellie VanHorn found evidence for economizing on timber in the British-built, not American-built, vessels. Those were smaller vessels, however, than those considered by Smelser, Davisson and French.

Merchants' and shipbuilders' papers provide additional information. When Brown, Benson, & Ives purchased the brigantine *Maria* in 1799, she was six years old.[29] The sloop *Success* was seven years old when considered for purchase by Brown and Starbuck.[30] That same year, 1777, they either bought or thought about buying the sloop *Bonetta*, estimating her age at eleven.[31] The Crowninshield ship *America* was in that family's service in 1796, and sold by them in 1803, after which indications are she was to continue in Atlantic service.[32] The Derby ship *Astrea*, already in West Indies service in 1784, was deemed fit enough for a voyage to the Far East in 1789–90.[33] The Browns' papers first mention their brig *Harmony* in 1774–1775.[34] Further records indicate she was still in their service in 1794.[35] The American-built ships in service to these merchants in this small sampling were between six and twenty years old.

---

28 French, "Longevity," 161.

29 Bill of sale, brigantine *Maria*, 1799, box 676, folder 3, Records of Brown & Ives 1796–1914, Brown Family Business Records, John Carter Brown Library, Brown University, Providence, Rhode Island (hereafter JCB).

30 Memorandum of sale, sloop *Success*, 1777, box 682, folder 4, Records of Nicholas Brown and Company, 1762–1782, Brown Family Business Records, JCB.

31 Memorandum of sale, sloop *Bonetta*, 1777, box 682, folder 5, Nicholas Brown and Company, JCB.

32 John Crowninshield to George and Richard Crowninshield, 31 March 1803, Letter Book No. 7, Letters from France, 1803, box 4, folder 8, Subseries A. Correspondence, Series IV. John Crowninshield (1771–1842) Papers, Crowninshield Family Papers MH 15, The Phillips Library of the Peabody-Essex Museum, Peabody, Massachusetts (hereafter PEM-Phillips).

33 Papers of ship *Astrea*, box 1, folders 5–8, Derby Family Papers MSS 37, PEM-Phillips.

34 Cargo manifest, brig *Harmony*, 1774 or 1775, box 682, folder 3, Nicholas Brown and Company, JCB.

35 Papers of brig *Harmony*, 29 March–9 April 1794, box 683, folder 8, The Records of Brown, Benson, & Ives, 1792–1796, Brown Family Business Records, JCB. Note: the finding aid lists this box's dates reversed: 9 April 1794–29 March 1794.

A correspondent to the British Society for the Improvement of Naval Architecture pointed out the importance of environment to this issue when he wrote, in 1791, that of two identical ships, one on station in a hot climate would be useless in 12 years, while one on station in a cold climate might last 30 or even 40 years. He cited British colliers as an example of the latter.[36]

An 1831 anecdote from Alexis de Tocqueville's *Democracy in America* further muddies the waters:

> I accost an American sailor, and I inquire why the ships of his country are built so as to last but for a short time; he answers without hesitation that the art of navigation is every day making such rapid progress, that the finest vessel would become almost useless if it lasted beyond a certain number of years.[37]

Whether American and British shipwrightry differed in such a way as to make American-built ships less durable is a question for which we need much more archaeological evidence to answer, although more studies along the same lines as Smelser's and Davisson's and French's would help too. VanHorn's evidence suggests that, if so, it is unlikely to be attributable to disparities in know-how, as she found no "clear distinction between regions of build or even between the two nationalities." Variations "appeared to relate more to vessel type or perhaps the individual practices of shipwrights than location of construction."[38]

The opinions codified in *Lloyd's Register*, however, indicate that by the end of the period, some industry leaders in London considered the best American-built ships almost on par with Thames-built vessels—provided they were built with Southern live oak; the prejudice against New England white oak versus British oak persisted.[39] One advantage this prejudice might have offered

36 "A Collection of Papers on Naval Architecture," Part 1, Society for the Improvement of Naval Architecture, reprinted in London by *European Magazine*, Pamphlet Vol. 744, No. 3, Library of the American Philosophical Society, 57. Colliers were also stoutly built, which is why Cook's first long-distance voyager was a converted collier.

37 Alexis de Tocqueville, *Democracy in America*, Book 2, Section 1, Chapter Eight, trans. Henry Reeve. The Gutenberg Project, accessed 24 Jan. 2019 at https://www.gutenberg.org/files/816/816-h/816-h.htm#link2HCH0008. Thanks to Larrie Ferreiro for pointing this out to me.

38 Kellie M. VanHorn, "Eighteenth Century Colonial American Merchant Ship Construction" (Master's thesis, Texas A&M University, 2004), 216.

39 To preclude confusion, we should note that, as of 1799, there were two rival shipping registers, as one group within Lloyd's parted ways with their colleagues over a disagreement over the very sort of classification system we are discussing. This group seems to have been composed entirely of shipowners, whereas Lloyd's was dominated by underwriters.

to American shipwrights was to keep the prices of their vessels low enough to attract more English and Scottish buyers, but whether this offset whatever business was turned away by the American reputation is unclear. Regardless, merchants in Britain began buying American-built ships by the mid-1600s, and that trend only increased, prompting some loud complaints by the London shipbuilding lobby. The classic mercantilist model here would have had cut timber from America shipped to England for ship construction, but that was not cost-effective. In the eighteenth century, most commercial vessels operating on the American coast and from there to the Caribbean islands—and a great many of those on transatlantic routes—were American-built.

The last material essential for ship construction was iron, for fasteners. Besides nails and bolts, iron was used for rigging and deck hardware and for the rudder's *pintles* and *gudgeons*, the hinge assemblies that attached it to the ship and allowed it to pivot. British Americans faced a set of challenges in setting up their own iron foundries. In addition to hostile mercantilist policy from London, scarce and thus expensive American skilled labor made it difficult to recruit and retain the necessary workforce. The initial capital investment was significant, and while credit was available, foundries went bankrupt when they could not turn a profit and repay start-up debt.[40] They never gave up, and we find foundries, for example, in Tidewater Virginia as part of the plantation elite's campaign to diversify their economy in the face of sustained low tobacco prices in the later eighteenth century, but any shipwright's supply of nails for a given vessel were likely to have come from England. In fact, shipbuilding agreements with British American shipwrights tend to specify that the owner will supply some materials, and iron work is typically among those.

The ship, then, was an intersection of myriad influences and considerations. It was successful when it managed to balance all of those: the demands of the sea and the demands of the market, capacity for cargo and ability to sail. We will delve more deeply into those influences and considerations, but at this point we have laid out the gist of building and operating a wooden sailing ship. We have also seen hints of how the ship was a confluence of skills

The original *Register* became known as the "Green Book" and the rival publication the "Red Book." The "Green Book" allowed for the twelve-year longevity classing of live oak-built ships, just behind the highest classification—thirteen years—awarded to the best Thames-built vessels. Other American-built vessels, though, were classed at only six years. The "Red Book" rules maintained the strong prejudice for British oak. *Annals of Lloyd's Register*, 19–21.

40 This was also true of attempts to operate copper mines, such as the one in Connecticut that struggled through the first half of the eighteenth century before being used as Connecticut's first prison from 1773 to 1827.

and materials from a broader world than the British Atlantic: building traditions from Europe, Britain, and the Mediterranean; materials from England, America, and the Baltic. The next step is to explore in more detail the ways in which specific variants of the technology attempted to meet certain combinations of imperatives, and what influences may explain why those specific variants exhibited their defining characteristics. We need to get familiar with the most common types of merchant ships working in the British Atlantic.

Contemporary sketches and paintings of commercial port waterfronts show us a collection of vessels exhibiting a variety of differences, within the parameters just laid out. By considering the differences between vessels contemporary to each other—perhaps anchored near each other—we are exempt from worrying about teleological presuppositions creeping into our thinking, or about traveling the comfortable and well-worn rut of the linear-progress model, in which one form is improved to create the next form, *et cetera.*

The smallest vessels in harbor used for voyages over open water were built and operated for short-distance coastal use. Only rarely would they sail out of sight of land. Such vessels could also use estuaries and rivers, sounds and even creeks. They could thus directly serve small settlements or engage in small-scale on-the-spot trade, perhaps with an isolated farmstead or with a single indigenous trader offering a bundle of pelts or deerskins. The two important sources for such craft were, first, the ships' boats that early explorers and settlers brought over with them; and second, the craft operated by indigenous Americans. It will come as no surprise that a blending of the two traditions occurred in short order. The earliest of the former type was called the *shallop* (pronounced *sha-LOPE*), the English analogue to the French *chaloupe*; the similarity is a reminder that traditional vessel types developed across borders.[41] It is not worth delving deeply into the etymology of terms describing two-masted smaller utility craft like the shallop; the terms were loosely and broadly employed, and such craft operating in the same time and place shared fundamental similarities. An example relevant to the early British Atlantic is the *pinnace*, a term that could mean anything from a two-masted boat like the shallop to a small, relatively lightly-built ship.[42] An illustration from Chapman which he labeled "pinnace" will serve as a general idea of these two-masted boats, used throughout the period for coastal and inland work—see Figure 2, No. 21 (second row, far left).

---

41 On small boats like the shallop used along the British American coast, see William A. Baker, *Sloops and Shallops* (Barre, MA: Barre Publishing Co., 1966).

42 For a discussion of the confusing use of the term "pinnace" in period nautical etymology, see Baker, *Colonial Vessels*, 52–77.

John Smith used a shallop to explore the Chesapeake during the establishment of the Jamestown settlement. The shallop was what we would think of as a large boat, stoutly-built, with one or two short masts, with one *loose-footed* fore-and-aft sail on each mast. "Loose-footed" means the bottom edges of the sails were not attached to booms. On a small craft, booms threatened the crew with a potentially-fatal blow to the head; dispensing with them was much safer, and was feasible with small sails such as these. While it may seem that such a craft could have gotten by with a single-masted rig, recall that splitting the sail plan allows for smaller, more manageable sails for the same total area—and thus power—and allows for that area to be carried lower, increasing the vessel's stability. The shallop was also frequently decked over, at least partially, providing it with greater sea-keeping ability. It was safe and useful on open bays and along the coast, and could ride out a squall. It could be carried, whole or in pieces, on a ship. It was also small enough to row, should the wind die or turn *foul*—turn into an unfavorable direction.

Such versatility was of immense value in a world where water transport was the rule and land transport the exception. While 90% of our trade goods still move by water, they do so in great bulk; vastly more small-scale transportation is accomplished by land now than it was in this period, provided navigable water was available. On the Atlantic coasts, the relatively few roads were poor, and road-building was slow and labor-intensive. The coasts, however, are laced with inlets, sounds, rivers, and creeks. Settlements clustered on or near the water. Small craft moved people and goods, for everything from personal visits to provisioning the homestead to small-scale trade.

The first such trade, of course—which quickly grew into large-scale trade—was with indigenous Americans, whose commodity trades were long-established. From them, settlers learned to use indigenous materials and techniques to fashion their own watercraft. An important example, analogous to the shallop and used for similar purposes, was the *periauger*—derived from a Spanish word; yet another reminder that the colors of colonial maps bled more freely than policymakers would have preferred. This coastal craft was made by hollowing out a large tree trunk, using the *dugout* technique, in which burning coals begin the process, which is finished with *adzes*—like axes, but with the blade turned perpendicular to, rather than parallel to, the handle.[43] In the southern colonies, the bald cypress was a preferred tree. The cypress, being a

43 It is worth noting here that dugout techniques developed independently in Europe, the Americas, and Africa. See Joseph T. Butler, ed., *The European Origins of the Small Watercraft of the United States and Canada* (Thibodeaux, LA: Center for Traditional Louisiana Boatbuilding and the Museum Small Craft Association, 1992).

swamp tree, is highly rot-resistant, and it can live for thousands of years, reaching diameters of several feet. As time went on, settlers needed somewhat larger periaugers to carry more goods and people as settlement grew, but at the same time, the available stands of prime ancient cypress were exhausted, and they had only smaller specimens with which to build. Their solution combined the dugout technique, adopted from indigenous tradition, and the imported tradition of assembling a craft from milled timber: the largest available trunk was dug out, the sides of it flattened, and "wing" pieces attached to the sides with trunnels, effectively re-creating the older, larger periauger with three pieces.[44]

Shallops and periaugers were the sorts of vessels that would bring goods and people to and from the principal port towns where ocean-going ships would be anchored or docked. They might bring tobacco, deerskins, tropical produce, or salted meat to ships that could transport those goods to more distant markets. They might transport captured native prisoners to be sold into slavery and transported to West Indies plantations. They would then transport people and goods arriving on ships to smaller markets, retail outlets, and settlements.

The smallest vessels used for ocean crossings were only slightly larger. The main differences were in construction. These vessels had to be at least somewhat more heavily built, and fully-decked. They needed more powerful rigs to move their greater weight. In the seventeenth century, these vessels were more likely to be square-rigged than they would be in the eighteenth. One of the earliest such vessels from this period known to history, the pinnace *Sparrow-Hawk*, left London for the Jamestown Colony in Virginia in 1626. She was blown far off her intended course by storms, and wrecked at Cape Cod in Massachusetts. Her remains were discovered in 1863. She was only about 40 feet (12.19 m) long—small, even for the time, for a trans-Atlantic crossing. The word "pinnace" (Dutch *pinas*, from French *pinasse*), as mentioned in the Introduction, was broadly and loosely employed and does not denote a narrow set of characteristics.[45] When applied to a small ship, rather than a boat,

44 Chapelle relates the North American coastal periauger to a West Indian dugout type—see *American Small Sailing Craft*, 19. The rarity of ancient bald cypress trees along the southern eastern seaboard bears witness to the scale of their eighteenth-century harvest. A few survive in a very shallow swamp in southeastern North Carolina. They are believed to be the oldest trees east of the Mississippi River. These are the cypress trees in the Three Sisters Swamp off the Black River, which contains a tree proven to date to at least 364 CE, although the scientists who discovered it are convinced that other trees near it are older—perhaps dating to the beginning of the Common Era. Most of the ancient trees, though, have hollow cores now, making them impossible to date accurately.

45 On the Dutch *pinas*, lavishly and expertly illustrated, see Hoving, *17th Century Dutch Merchant Ships*, 46–61.

the term refers to a vessel built relatively narrow, sharp, and light enough to be good sailers, at the expense of maximum capacity and structural strength. They were commonly used for exploring, reconnoitering, and light trading. Naval versions served as small warships. The pinnace could be two- or three-masted, with square driving sails and lateen mizzens typical. A larger pinnace, *Kalmar Nyckel*, brought the first colonists to New Sweden (now in Delaware) in 1638 (see Figure 15).

The even-smaller *Discovery*, built in England in 1602, was called a “fly-boat,” analogous to the Dutch *vlieboot*. She was an explorers’ vessel, making the first European voyage into what became known as Hudson Strait before joining the 1607 expedition that founded Jamestown. Shortly after, she served Henry Hudson in the search for the Northwest Passage. Like “pinnace,” “fly-boat” was a loose appellation given to smaller, lighter ocean-going vessels combining good speed, endurance, and shallow draft—ideal for reconnaissance. It was also at times interchanged with “flute,” in English, which is confusing because a Dutch *fluit* or *fluyt* was a capacious and relatively slow vessel, as we have noted. This is a good reminder that, absent specific context, names of ship types can be misleading, as their meanings were so fluid over relatively short spans of time.

While these early seventeenth-century vessels were small when judged solely by their basic dimensions, they would strike us more as little ships than as boats, were we to see one up close—which, to a certain extent, we can; *Sparrow-Hawk*’s remains are on display in Massachusetts, and reconstructions of her have been made; a carefully-research speculative reconstruction of *Discovery* is currently operational at Jamestown, Virginia.

We would also have noted the simplicity of these vessels’ rigs—both are believed to have been two-masted, and the sails could be few because they were small. Throughout the period, more two-masted types of smaller vessels were in service than single-masted. In fact, one of the most important types of eighteenth-century coastal and island traders, the sloop, is rather an anomaly in this respect, with its single mast. We have discussed this type already, and noted that its large fore-and-aft main sail and jibs helped make it fast and weatherly, but also required diligence and fortitude on the part of its crew. Sloops, however, were frequently considerably larger than *Sparrow-Hawk* or *Discovery*, which serves as an interesting exception to the general rule that larger vessels would have more, not less, divided rigs. We are safe to hypothesize that the sloop made labor demands that similar-sized craft did not, and that their speed and weatherliness could justify that in certain applications—such as carrying dispatches (“advice boats”), fresh fish, or contraband.

Later in the eighteenth century, at least in British America, the sloop gave up some of that popularity to a two-masted alternative, the *schooner*.[46] (Etymologically, *sloop* and *schooner* both derive from Dutch; both vessels' precursors may be found among older Dutch types.)[47] Like the sloop, the schooner's hull was generally built to be at least moderately fast while still offering useful capacity. Extreme versions were certainly built, well into the nineteenth century, and the rig was still common to yachts in the twentieth. A schooner's driving sails are fore-and-aft, and its main mast, the after one, is only slightly taller than its foremast. In that sense, its rig is similar to the two-masted rig on small boats like the shallop and the periauger, which preceded it. See Figure 2, No. 6 (third row, second from left). The schooner's success in British America may well have to do with its deserved reputation as a fast, handy New England fishing vessel.

In the seventeenth century, fishermen made good use of another two-masted modest-sized vessel called the *ketch* (related to *catch*, appropriately enough). The ketch, however, carried its taller main mast in front of a much shorter mizzen mast, and was square-rigged on that main mast. See Figure 2, No. 3 (top row, second from right). It would seem that the ketch's popularity gave way to the schooner's in the eighteenth century. This may have to do with the latter's fore-and-aft rig, given larger trends in that direction, but that hypothesis needs to be investigated.

In the eighteenth century, especially during and after mid-century, the dominant mid-sized ocean-going types listed in shipping records were referred to as *brigs* (or *brigantines*), *snows*, and ships. The brig or brigantine (the word was used interchangeably) denoted a square-rigged vessel with two masts, the main mast forward of the mizzen mast, and slightly, but not dramatically, taller. See Figure 2, No. 4 (top row, far right). The snow was similar, and the difference is not immediately obvious from a small drawing, but the snow hoisted its mizzen sail on a separate, small *trymast*, set just behind the mizzen mast (in most period art, it is difficult to tell that they are two separate spars). See Figure 2, No. 2 (top row, second from left). This allowed the mizzen sail to be operated without interfering with whatever sails were being operated on the

46 Joseph A. Goldenberg, *Shipbuilding in Colonial America* (Charlottesville: University of Virginia Press, 1976), 78. Goldenberg writes that the schooner needed a smaller crew than a comparable sloop. We should make a serious attempt to put hard numbers to that claim.

47 See Karl Heinz Marquardt, *The Global Schooner: Origins, Development, Design and Construction, 1695–1845* (Annapolis, MD: Naval Institute Press, 2003); and David MacGregor, *The Schooner: Its Design and Development from 1600 to the Present* (Barnsley, UK: Pen & Sword Books, 2003).

mizzen mast. Note that the snow's driver is cut differently than the brig's—its after side is cut more vertically, so it does not project as far astern, and it does not have a *boom* (the spar to which the foot—lower edge—of the sail attaches). The two would seem to be variants of the same idea, but the evidence suggests that they were actually adapted from different antecedents, becoming more alike as those adaptations accumulated. This reminds us to avoid making unsubstantiated assumptions about how and why this technology developed as it did—assumptions such as that of a simple linear progression from one type to another.[48] The history is so long and convoluted that we are just as likely to be wrong as right, even if our assumptions "make sense."

The snow was the largest two-masted square-rigger in service in the later eighteenth century, but the largest vessels of all, throughout the entire period, were ships, in the specific sense of the term. *Ship*, as noted, could generically refer to any type, or it could more specifically refer to a three-masted square-rigger—consistently the most common ocean-crossing merchant ship of both centuries. In the earliest years of the period, the mizzen (aftermost) mast frequently carried only a lateen mizzen sail. That was true of Columbus's *Santa Maria* and of Christopher Newport's *Susan Constant*. Later in the seventeenth century, and on larger vessels, the mizzen mast could also carry a mizzen topsail, as in Figure 3, *Cadiz Merchant* from 1682.

The small galleons that founded and supplied the first tentative English colonies carried this rig.

The later ship rig split its sail plan into more sails, and carried triangular staysails and jibs terminating at the bowsprit.[49] See Figure 2, No. 1 (top row, far left). An earlier version of an eighteenth-century ship rig would have carried a lateen mizzen rather than the gaff spanker we see here. This vessel also has spritsail booms under her bowsprit; the snow to the right is shown flying those sails.

Ships were not always larger than brigs and snows, but the largest vessels in service were ships. They were the earliest vessels to cross the Atlantic in the early modern period, and the type most consistently in service throughout

48 For a *tour de force* of the point I am making here, see E.P. Morris, *The Fore-and-Aft Rig in America* (New Haven, CT: Yale University Press, 1927). Morris attempts to trace the origins and transmission of fore-and-aft rigs from Europe, to the European American settlements, and through the subsequent history of North America. The complexity—and conjecture—intrinsic to his subject makes for a challenging read, despite the care and clarity of his writing.

49 A study of why that is, combining experimental archaeology and maritime economic history's data, would be worthwhile.

FIGURE 3 The *Cadiz Merchant* underway, 1682, from Barlow's Journal, 1659–1703

the period. The brig and snow, however, were increasingly popular in the mid-size class of vessels beginning around 1750 but showing up prominently in the records well before that date. There were no hard-and-fast divisions between size "classes" in this period, and plenty of overlap between types of vessels and their sizes. For example, a brig and a schooner could be the same size, or a brig and a snow, or any of those and a ship.

While it is clear from the foregoing that rig type became a standard basis for vessel nomenclature in the eighteenth century, the earlier period's penchant for describing vessels based on hull characteristics did not die out. In fact, in 1768, the Swedish-English naval constructor Fredrik Henrik af Chapman included a list of hull types with their defining characteristics in his famous *Architectura Navalis Mercatoria*. Compiling that information in a simple table draws our attention to the significant aspects of hull design that we should be familiar with going forward—aspects certainly familiar to any nautically-savvy person at the time. The terms used here may be found in the glossary, and I will explain the most important ones shortly, but full acquaintance with them is not necessary to glean insights from the table. Keep in mind that, since the information here was published in 1768, we cannot apply it to, say, 1668, even though most if not all of these terms were in use then too.

The Max Size column lists length, breadth, and tons in burden (we will discuss the significance of tonnage measurements later; for now, we will use it merely as a comparative). The Bows column basically notes whether the vessel had a "fancy" bow or a plain one. A plain bow was like that of a boat—the sides curved inward to the stem, and that was that. A fancy bow, like the frigate-built hull's, was more complex and ornate, like a warship's, including a figure-head, which was purely ornamental. On a British warship, the requisite figurehead was a lion. On a merchant vessel, it was typically a woman. The *head* was an additional structure just below the end of the upper deck, where the crew's toilets were; they were oblong holes on boxes, as we would find in a wooden latrine, and they opened directly to the water below.

Type of stern was one of the most common ways of referring to a vessel. In letters or logs or depositions, one will frequently see something like "a square-sterned brig" or "a pink-sterned vessel ..." The stern was a prominent feature and fairly easy to make out from a distance. "Square stern" means that the hull terminated in a flat surface or surfaces. A "pink stern," in fact, could refer to a flat-built stern, but it would denote a relatively narrow one. (The term is etymologically-related to "pinch.")

A square stern was easier to build, as it is always easier to build a wooden structure with straight pieces than curved pieces. It was relatively weak, however.[50] The "round stern" was not fully round, but it allowed for a rounded tuck of the bottom of the hull where it curved up into the stern. It allowed a smoother water flow past the stern, and that was known at the time, which probably explains why the round-tuck stern became ubiquitous, despite the added complexity. The guiding principle of what we call hydrodynamics was the pursuit of a clean, uninterrupted flow of water from *entrance* (where the bow first contacts the water), past the midsection, to the *run*, where the hull rises and narrows toward the stern. While compelled to build merchant hulls for capacity, and thus rather full in general, shipwrights strove to avoid bumps, abrupt curves, asymmetry, and hollows, to encourage the water to flow unimpeded and without turbulence to the ship's rudder, so that the steering would have maximum effect in "clean" water.

Another advantage of the round tuck, aside from hydrodynamics, was that it presented a stronger and less-resistant surface for storm waves to slam into, offering a less-vulnerable transition from stern to hull for the sea (or an enemy's guns) to loosen by force. The Royal Navy recognized this, and it is apparent in the sterns of their warships in the eighteenth century.

Wales are thicker planks that serve as longitudinal reinforcement. The best-known is the *gun-wale*, or *gunnel*, the uppermost such piece, and a term that has become ubiquitous over time, which now refers to the top edge of the side of any craft; if you own a canoe, you step over the gunnel to get in or out. The wales (and regular planking) terminate at the stern—either at the transom or the tuck, depending on how high or low along the hull they run.

In most of these types, which had a flat transom overhanging the tuck, the head of the rudder was inside that overhanging *counter* below that structure. In the flute (from the Dutch *fluit* or *fluyt*), with no such structure, all of the rudder was exposed ("outside"). In addition to the basic structural difference—square or round tuck—we see here another basic distinction, as we did with the bow, between fancier and plainer construction.

The frigate-built vessel, then, was the most ornate and complex here, given the full treatment at the bows and at the stern—including quarter-galleries, the wrap-around protrusions set with windows. She had a square tuck, however, unlike the *pink*, the other most common hull form in ocean-going British merchant service at the time. As noted in our discussion of earlier hull forms, the word *pink* referred to the narrow ("pinched") stern, the most obvious aspect of these generally finer-lined hulls.

50 Brian Lavery, *The Colonial Merchantman* Susan Constant *1605* (Annapolis, MD: Naval Institute Press, 1988), 15. The face of the stern was framed by three or four light timbers running almost parallel to the stern timbers.

TABLE 1 Hull types from Chapman, via MacGregor (abridged)[a]

Class	Max size	Bows	Stern	Mid-section	Waterlines
frigate-built	160–0 ft. 41–5 ft. 1,275 tons burden	full head, cheeks, rails, figure-head	square tuck, quarter-galleries, wales go to wing transom or tuck	deadrise varies from small to big, slack bilges, round sides	moderately fine entrance and run
bark	150–9 ft. 38–0 ft. 1,250 tons burden	plain stem, no head	square tuck, wales go to wing transom or tuck	flat floors, hard bilges, vertical sides	full entrance and run
flute/ flyboat[b]	128–3 ft. 29–7 ft. 700 tons burden	plain stem, no head	round stern, square taffrail, wales go to sternpost, outside rudder	small dead-rise, slack bilges, big tumblehome all round	full entrance, parallel sides, finer run
pink	109–2 ft. 29–6 ft. 434 tons burden	full head, cheeks, rails, figure-head	round stern, square taffrail or narrow lute, wales go to sternpost	hollow garboards, big deadrise, slack bilges, round sides	fine entrance and run

a MacGregor, *Merchant Sailing Ships*, 20. Note that MacGregor converted Chapman's tons-burthen figures to English tons, so these will not match the figures in the original.

b Chapman uses "flyboat," MacGregor uses "flute;" as noted, both terms were used loosely, but both denote Dutch design, or at least Dutch antecedents.

The mid-section, or *midship bend*, defined the shape of the vessel as much as any other component. Recall that the midship frame was the first of the "ribs" to be fashioned onto the keel. It defined the shape of the midsection, and thus much of both the capacity and sailing qualities of the finished vessel. In the Mid-section column, we find the basic terms used to describe hull shape—then and now. *Deadrise* refers to how much the bottom of the ship slopes up to where it meets the side of the ship, expressed as an angle or, as is the case here, in relative general terms ("small" or "big").

Zero deadrise means a completely flat-bottomed hull. Where the bottom meets the side is called the *turn of the bilge*, or in short form *bilge* (recall that "bilge" also refers to the lowest part of the inside of the hull, where the water collects). *Slack bilges* mean that the turn of the bilge is a relatively gentle curve. *Hard bilges* mean that the turn of the bilge is an abrupt bend, almost a square

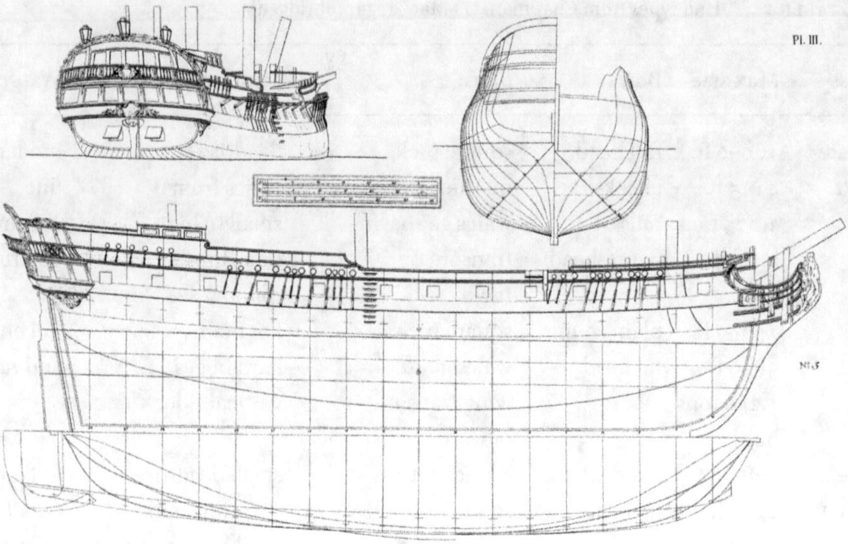

FIGURE 4 Frigate-built ship
ARCHITECTURAN III, FREDRIK HENRIK AF CHAPMAN, SJÖHISTORISKA MUSEET, STOCKHOLM

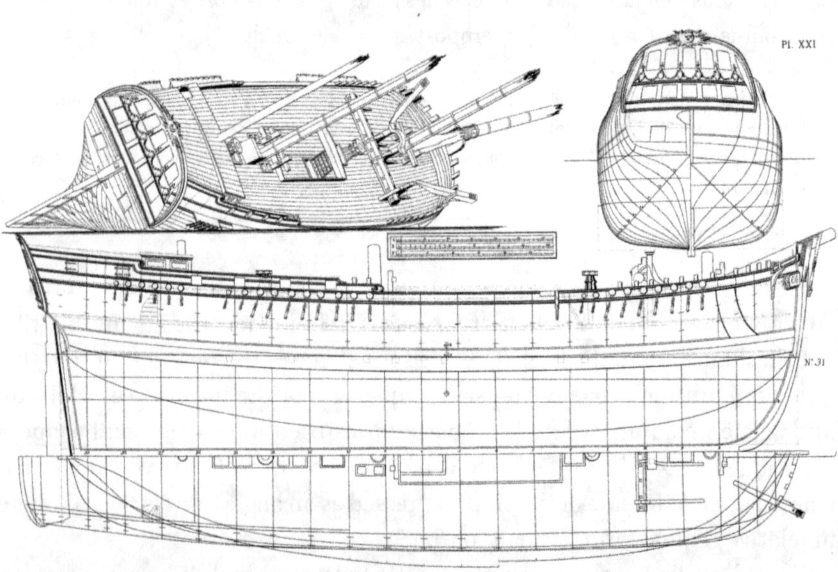

FIGURE 5 Bark
ARCHITECTURAN XXI, FREDRIK HENRIK AF CHAPMAN, SJÖHISTORISKA MUSEET, STOCKHOLM

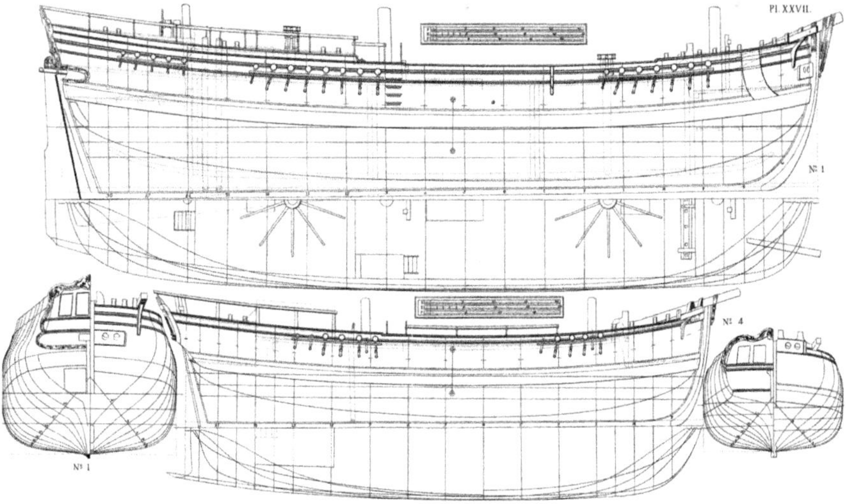

FIGURE 6 Flute ("flyboat")
ARCHITECTURAN XXVII, FREDRIK HENRIK AF CHAPMAN, SJÖHISTORISKA MUSEET, STOCKHOLM

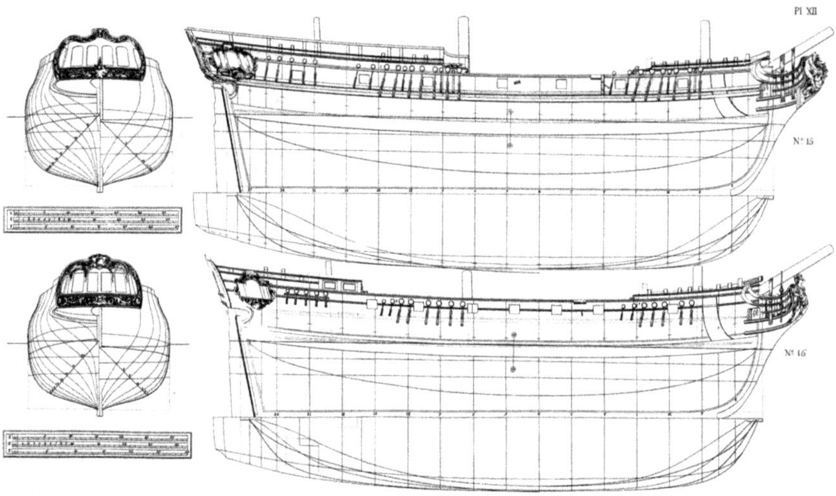

FIGURE 7 Pink
ARCHITECTURAN XII, FREDRIK HENRIK AF CHAPMAN, SJÖHISTORISKA MUSEET, STOCKHOLM

one. Consider the *bark* hull form. This is the closest to a box shape we will find in these ships (the bark is best-known as an eighteenth-century collier—coal carrier). The flute, on the other hand, is the roundest.

Tumblehome refers to the curving-inward of the ship's sides as they rise from the turn of the bilge toward the gunwale. Pronounced tumblehome was one of the most prominent traits of seventeenth-century ships, and the shift away from that over the course of the period bears further discussion.

The frigate-built type had the widest variance in midship form, although Chapman does list slack bilges and round sides as constants. With varying deadrise, the type could be built with deeper or shallower draft, depending on intended purpose and operating area. This type is also listed with the maximum size. In Atlantic service in this period, a ship of 1,000 tons was unheard-of; these sizes are only applicable to the largest East Indiamen. For our purposes, 200 tons was a mid-sized ocean-going ship, while 400 tons was a large one—even at the end of the period.

The bark and the flute, then, were the plain workhorse types here, and the frigate- and pink-built types were the fancier and faster ones. They were also more common for most of the period in Atlantic service, with the other two more often employed in coastal and Baltic trades in the home islands and northern Europe. That has to do with the perennial difficulty of safely employing the most capacity-efficient, but slowest and least defensible, carriers in the Atlantic, with its consistently high levels of war and maritime predation.

While I have tried to keep this as brief as possible, it may still convey the impression that this is a most complex subject. It is, but it is also the case that there was standardization of ship design and construction in the British Atlantic, at least to an extent that we can pick out clear and consistent patterns. We would not have found an infinite variety of vessel types plying their trades in a given Atlantic port. The types discussed in this chapter, and their defining characteristics, would have accounted for the overwhelming majority of craft entering or clearing or anchored or tied up at the pier.

Deadrise, the turn of the bilge, and tumblehome are defining characteristics of a ship's hull, and some of the most important areas in which a shipwright had choices to make in designing and building that hull. Changes in one or more of them would affect how the ship behaved in a seaway, affecting speed, stability, the ability to carry sail without excessive heel, and the ship's motion in the water. For example, in general, a shallower draft dictates a wider beam, for stability. Big deadrise will present more resistance to side-slippage through the water than small deadrise, allowing the ship to sail more closely to her intended track, especially to weather.

With a basic understanding of types and features, we can consider questions related to design choices, construction techniques and their meanings, stasis in ship design, and the innovations that we do observe in hulls, rigs, and prevalent vessel types throughout the period. As an *entrée*, we will consider the creation of an ordinary merchant ship, from the document setting out the agreement between principal owners and the shipwright, to the launch and fitting-out of the finished vessel.

CHAPTER 3

From the Stocks to the Ways: Building a Ship from Contract to Launch

On 17 January, 1733, two Bristol merchants, John Winne and Walter Hawksworth, wrote a letter to another Bristol merchant, William Jones.[1] In it, they laid out a detailed set of specifications for a ship they wanted built. Jones was to forward the letter to a shipwright he knew in Boston, in New England. The letter specified the principal dimensions, *scantlings* (how stout the construction was to be), some stated preferences for hull shape, and details of equipment and accommodation. Unusually, it also specified a name: *Bristol Merchant*. At four handwritten pages, it is the most detailed such letter I have found—though what it leaves out is as instructive as what it includes.

There has been little historical scholarship published on British Atlantic merchant shipbuilding in this period. The focus of what literature there is has been on shipbuilders and their business more than on the technical details of what they built, which is typically the preserve of archaeology. For this discussion, it will be important to focus on those details, and their connection to the priorities of both builder and client.[2]

Unpacking the Winne & Hawksworth letter allows us a detailed, first-hand introduction to how an ordinary merchant ship was built, and knowing how the letter was written and in what order will give the reader a sense of what was important to the prospective owners as well as what practical matters had to be spelled out in order for the ship to be built.[3] For clarity, I have used

---

1 Winne and Hawksworth to William Jones, 17 January 1733, box 1, folder 8, Joshua Coffin Papers MSS 457, 1647–1862, PEM-Phillips.

2 For an overview of the British shipbuilding industry, see Davis, 41–76. The standard historical monograph on shipbuilding in British America is still Goldenberg's *Shipbuilding in Colonial America*. For a social history of New England shipbuilders, see Joseph Cullon, "Colonial Shipwrights and Their World: Men, Women and Markets in Early New England" (PhD dissertation, University of Wisconsin, 2003). On shipbuilding in the southern colonies, see Clowse, "Shipowning and Shipbuilding in Colonial South Carolina." There is a clear need for new scholarship in this area by both historians and archaeologists well-informed by the other discipline.

3 For a fully-illustrated (with color photographs) account of the construction of the replica *Susan Constant*, providing full visual detail of the process, see Peter H. Spectre and David Larkin, *Wooden Ship: The Art, History, and Revival of Wooden Boatbuilding* (Boston: Houghton Mifflin, 1991).

 | DOI:10.1163/9789004426344_005

a transcription rendered into modern English. A literal transcription may be found in Appendix 2.

> Bristol, England
> 17 January 1733
>
> Dimensions for a new ship, to be built at Boston in New England, of 58 feet [(17.68 m)] on the keel, 21 feet [(6.4 m)] in the beam, 12 feet [(3.66 m)] rake by the stem, 5 feet [(1.52 m)] rake by the post, 9 ½ feet [(2.90 m)] in the hold, from under the lower deck beams to the ceiling; 3 feet 10 inches [(1.17 m)] between decks, from under the upper deck beams to the lower deck; and to carry her breadth well forward, even to the after part of the fore-channel. The length of keel and beam are to be measured as customary. The breadth of floor, from sirmark to sirmark, is to be 11 feet [(3.35 m)], with 7 or 8 inches [(17.78–20.32 cm)] of deadrise. The floor timbers are to be 12 inches [(30.48 cm)] square.

The first specification in the Winne & Hawksworth letter is for the length of the keel. As the principal structural member of the ship, the keel length roughly defined the length of the ship, although the manner of her bow and stern construction would dictate how much longer she was on deck than at the keel. *Bristol Merchant* was to have a keel of 58 feet (17.68 m), which was ordinary for a smaller-to-middling Atlantic merchantman. The keel would be the first piece of the ship to be fashioned and set in place on blocks. If a suitable log were available, it could be fashioned in one piece, and for this size of vessel, that was common. For keels longer than the longest log available, pieces would be *scarfed* together, at an angled, interlocking joint. One of the chief skills of the shipwright was the ability to form precise, complex joints in large pieces of timber using hand tools, such as the adze mentioned earlier, and the crosscut saw. The keel was the backbone of the ship. The frames were the ribs, and their centers were attached to the keel.

The second primary specification is the beam. At 21 feet (6.4 m), that gives the ship a keel-length-to-beam ratio of 2.76, which is moderate, but not tubby. The maximum beam will be defined by the midship frame, which will be the first to be fashioned, assembled, and fastened to the keel.

The next two dimensions—rake of stem and rake of stern post—may be added to keel length to get the approximate overall length of the ship. *Rake* is the inclination of the stem line from the vertical, and determines how much length would be added to the horizontal if one were to drop a plumb bob from the end of the stem or stern post to the ground—the *overhang*. In this case, adding both stem and stern rake to the keel gives us an approximate length on deck of 75 feet (22.86 m)—on the small side of middling for an ocean-going merchantman of this period, but in no way unusual.

Those dimensions, then, gave the shipwright the desired length, breadth, and amount of *overhang*, bow and stern. The overhang of the bow and stern determines how much longer the deck will be than the keel. The next specifications are for height—or depth really; a ship's hull had height when sitting dry on the stocks, but in the water, we speak of depth rather than height. Much of that vertical dimension was determined by these interior-space stipulations. *Depth in hold* means the vertical space between the lower deck beams to the *ceiling* (*ceiling* in ship construction means something completely different than it does in house construction; it is the surface covering over the frames, which means over the sides and bottom of the interior space, not the *overhead*, which is the term used for what we would call the ceiling of a building). The hold is the lowest space in the ship, where most of the cargo—and certainly the heaviest of it—would be stowed. If you were, say, 5′ 8″ (172.72 cm), you would find this space commodious, with about four feet (1.22 m) of space above your head.

This ship was to have two decks, and the space between them would not have been commodious at all: 3′ 10″ (1.17 m). The crew would have had to crawl in this space, but it was not intended for crew quarters. It was intended for cargo and stores for the crew, and added structural reinforcement for the vessel.

While some aspects of the ship's exact shape were left to the shipwright, as was customary, the basic concept of the *fish-form hull* was specified here; the "[b]readth [was] to be carried well forward …" means that the ship's maximum beam (breadth, width) was to be "well forward" of a point midway between the tip of the bow and the middle of the stern. The "midship bend," then, was not to be literally amidships. The practice of carrying maximum breadth forward of amidships was ubiquitous throughout the period, as a glance at any of the vessel illustrations in this book will tell.

That "midship bend," as we have noted, was the principal determinant of the basic hull shape. All other frames would be derived from it, using incremental modifications and the builder's eye. This method of design and construction is called *whole moulding*, and it too was ubiquitous throughout the period. As important as that first frame was, it is fitting that primary dimensions for it come early in the letter. The floor was to be 11 feet (3.35 m) *sirmark* to sirmark, which means the flat section was to be 11 feet (3.35 m); sirmarks were marks to note where a flat section was to meet a curve, or two curves were to meet.

"Flat" does not mean completely horizontal, though; the letter specifies 7 or 8 inches (17.78–20.32 cm) of deadrise (builder's discretion)—the number of inches from the horizontal that the floor would rise on each side over the span of its flat breadth—11 feet (3.35 m). That is small deadrise, another indication—along with her length-to-beam ratio—that *Bristol Merchant*, like the vast majority of other merchant ships, was to be built more for capacity and sea-kindliness than outright speed.

> The stem is to be a good crooked piece, and one that will tread well in proportion. The stern post is likewise to be a good piece, and both pieces not less than 10 inches [(25.4 cm)]. The keel is to be 12 inches [(30.48 cm)] deep and 10 inches [(25.4 cm)] thick, with a 3-inch [(7.62 cm)] false keel, nailed on well.

Recall that "compass timbers" (naturally curved, or "crooked") were prized for making strong, one-piece curved members, such as stem and stern posts, which are not straight. The owners specify compass timbers for these posts, as well as a minimum thickness of 10 inches (25.4 cm). The keel is to be 10 inches (25.4 cm) thick as well, and 12 inches (30.48 cm) deep. These are substantial timbers.

The false keel was a sort of shoe—in this case, a 3-inch-thick (7.62 cm) piece, the width of the keel, nailed on to the bottom of the keel. This piece would take the brunt of damage in the event of a hard grounding or shipworm attack. A damaged false keel would of course be far easier and cheaper to replace than the keel itself. The false keel also provided three extra inches (7.62 cm) of depth to the keel assembly, helping the ship track straighter through the water and resist leeway (side-slippage).

> The transom is to be 15 ½ feet [(4.72 m)] high, 16 feet [(4.88 m)] broad, and 12 inches [(30.48 cm)] square. The height of the *harpin* is to be at builder's discretion, to give the ship a good sheer—not too much, nor too little, but middling in proportion to most they build.

The *harpin* was a *wale*—a plank thicker than the others, contributing to structural rigidity—around the bow, and its height relative to the other specifications, as well as its curve, would help define the *sheer*—the overall curve of the top rail from bow to stern.[4] Its placement is yet another aspect of the design

4 According to Richard Steffy, both harpins and hawse-pieces were associated with "large, round-bowed vessels." While we know that *Bristol Merchant* was not large, this lends credence to the assumption that she would have been round-bowed, typical of ordinary merchant ships of the time. See J. Richard Steffy, "Illustrated Glossary of Ship and Boat Terms,"

that was left to the builder's discretion, and thus something we would have to guess at in a reconstruction. However, it is clear that the owners wanted something typical here. It is not anybody's guess where the harpin would have gone and what sort of sheer *Bristol Merchant* would have had at her railings. Those characteristics existed within a fairly narrow range of common practice.

Transom is used here as a general reference to the section of the stern formed by the *transoms*, which were individual athwartships timbers, fastened to the *sternpost*, that defined the shape and size of the stern, and strengthened it.[5] Its specified breadth of 16 feet (4.88 m) tells us that the ship would have been five feet (1.52 m) narrower at the stern than at her point of maximum breadth. That would have been visually noticeable—and typical.

> She is to be 12 inches [(30.48 cm)] high in the waist, with a rise for a forecastle, and another rise abaft at the main mast for a half deck, both to be of 10 or 12 inches [(25.4 or 30.48 cm)] with 2 hances proper for each rise, with a plank rail from hance to hance set on 4 timbers left for that purpose. The two bulkheads for the forecastle and half deck are to be 4 inches [(10.16 cm)] thick, at least, and timbers are to be left up at each hance with crutches of a proper height to put the spare topmast in.

These specifications speak to accommodations. A ship's waist was her midsection, where the main deck was exposed and the crew worked sail controls that did not require going aloft—hence the term "waisters" for the less-experienced seamen who were not qualified to go aloft as the able seamen were. The rises for the *forecastle* and half deck allowed for construction of the standard accommodations for crew—forward—and master and officers—aft. The half deck mentioned here would form the after cabin, to which the letter will refer again. The term *forecastle* (shortened and pronounced *fo'c'sle*, (FOKE-səl)) harked back to the late Middle Ages, when this structure truly resembled a castle, and was used as a fighting platform—along with its after counterpart, the *stern castle*. These structures shrank during the sixteenth and especially seventeenth centuries, and by this time, they would barely provide standing headroom inside on a ship this size. The *hances* mentioned here were curved

---

in Ben Ford et al., eds., *The Oxford Handbook of Maritime Archaeology* (Oxford: Oxford University Press), retrieved 8 Jan. 2019, from http://www.oxfordhandbooks.com/view/10.1093/oxfordhb/9780199336005.001.0001/oxfordhb-9780199336005-e-48).

5 The term is still used thus; like so many terms that originated in plank-on-frame wooden construction, and referred to specific structural members, "transom" is now a more generic term referring to the flat part of a stern that has such a flat part, whether of composite, metal, or traditional wooden construction.

pieces to connect the higher structures to the waist railing. An aesthetic sensibility is almost always present in us, and the general sensibility governing attitudes about what looked proper on a ship emphasized smooth curves and unbroken lines. Sharp discontinuities and abrupt transitions did not look right. That is analogous to our common aesthetic sensibility regarding cars. Bulges and disproportionate design elements will cause even a casual observer to deem a car's design ugly, and such was the case with a ship. Ugly mattered; no owner wanted an ugly ship, and the lowliest sailors routinely commented on a ship's appearance as well as her performance. It was a point of pride for everyone associated with the vessel that she be pleasing to the eye.[6] It was also indicative of the builder's overall care in her design and construction. In June of 1799, George Crowninshield wrote to his sons John and Richard about a brigantine they had recently purchased. "You ought to Put a Ringtale Saile to her," he suggested, as "She will look the Better for it…."[7] John Odiorne wrote to James Locke in July 1805 about the latter's ship, on which he was apparently serving as a supercargo. "We are allowed to be the handsomest ship here by everyone," he claimed, "outwardly but not so inwardly."[8] Elsewhere in the letter, he described the inadequacy of the ship's interior accommodations, which he claimed were poorly-arranged and full of disorganized stores and cargo, which put off a potential buyer who had been interested in purchasing the vessel until he went below. Accommodations and proper stowage therein conveyed a sense of order and sea-readiness.

A *bulkhead* is another basic structural component of a ship; it is the equivalent of a wall in a building. Bulkheads divide compartments on the ship, but

6 Even as thoroughgoing an economic history as Davis' *Rise of the English Shipping Industry* ends with the author musing on a hypothetical shipowner's aesthetic appreciation for his vessel as she heads out to sea.

7 George Crowninshield to John and Richard Crowninshield, 29 June 1799, box 4, folder 6, John Crowninshield Papers, PEM-Phillips. "RING-TAIL", a small triangular sail, extended on a little mast, which is occasionally erected for that purpose on the top of a ship's stern. The lower part of this sail is stretched out by a boom, which projects from the stern horizontally. This sail is only used in light and favourable winds, particularly in the Atlantic ocean. RING-TAIL is also a name given to a sort of scudding-sail, hoisted beyond the after-edge or skirt of those main-sails which are extended by a boom and gaff; as in all sloops, brigs, and schooners: this ring-tail is accordingly of the same depth with that part of the main-sail upon which it borders. See "SAIL" in William Falconer, *An universal dictionary of the marine: or, a copious explanation of the technical terms and phrases employed in the construction, of a ship. Illustrated with variety of original designs of shiping The second edition, corrected* (London: 1771; Gale Ecco, Print Editions, 2010).

8 John Odiorne to James Locke, 15 August 1805, box 1, folder 6, Shipping Papers (a topical collection from different sources, 1707–1886) (1989–112), NHHS.

they also provide necessary structural reinforcement, helping to more fully connect decks to each other or the lower deck to the floor. A *crutch*, as used here, is a fork, in two of which the spare topmast can be carried. Spare upper spars were required for ocean work, as it was highly likely at least one would break in heavier weather. Crutches were also mounted for booms to rest in when their sails were furled.[9]

> Eleven flat timbers of identical height and length are to be placed on the apron of the keel, 10 inches [(25.4 cm)] apart and with the same thickness as above, 12 inches [(30.48 cm)] and 8 inches [(20.32 cm)] at the rung heads. The rest of the floor timbers are to be of the same thickness both before and abaft, and to carry their length well, so as to give a long floor and to rise in proportion from the middle of the keel aft on, and from the foremost flat forward to what is satisfactory, so as to make a clean sail and entrance.

This section is an illustrative mix of specificity and vagueness. In keeping with how whole-moulded construction worked, it stipulated exactly how the owners wanted the midship section of frames—the largest ones—to be fashioned and installed, while the remaining instructions clearly leave the means to the end up to the builder's discretion. The midsection floors (11 flat timbers) are to be identical and flat. Ten-inch (25.4 cm) spacing of such heavy timbers gives a good sense of how stout ship construction was. (In eighteenth-century naval construction, there was almost no space between frames—hence the term "wooden wall"—a form of armor.) The *rung heads* were the tops of the floor timbers. They were "compassed"—curved. Frames were composite members, not one-piece. They consisted of the floor timber at the bottom, then one or more *futtocks*, then *top timbers*, the uppermost pieces of the frame, above the waterline. These would be trunnelled or bolted together.

Consider the next stipulation:

> The rest of the floor timbers are to be of the same thickness both before and abaft, and to carry their length well, so as to give a long floor and to rise in proportion from the middle of the keel aft on, and from the

9 In another usage, *crutch* referred to a structural member: "A bracing timber used to prevent a mast step from shifting laterally; also, a curved or angular timber, similar to a breast hook and used for a similar purpose in the lower part of the stern" (Steffy, "Illustrated Glossary of Ship and Boat Terms").

> foremost flat forward to what is satisfactory, so as to make a clean sail and entrance.

There is much more here to unpack about how ship design and construction worked than at first meets the eye. It is obvious that this contains none of the specificity of the instructions immediately preceding it. That is because the remainder of the design of a whole-moulded vessel was almost entirely up to the builder's discretion; the key to the successful execution of it was the builder's eye, based on his experience and grasp of the accepted, time-tested principles of shipbuilding. The thickness, and thus stoutness, of the remaining floor timbers is specified, but not their placement or their shape. "Rise in proportion" means that, as the hull shape rises and narrows from its fullest section—the "11 flat timbers"—it is to do so proportionally, with balance fore and aft. There was nothing more central to the shipwright's sensibility than proportion. Period treatises express every piece of the ship in a proportional relationship to other pieces. We could defend the claim that what we really mean by "the time-tested principles of ship design" is a complex set of accepted proportional relationships. Those evolved somewhat, and there was some wiggle-room there, but only within fairly narrow parameters. When we reconstruct a period vessel, and are forced to interpret a likely approach the shipwright might have taken, we do so using the proportions we know were in common use at the time.

To form the forward section, the shipwright was to rise and narrow the hull from the foremost flat to what was satisfactory, so as to make a clean sail and entrance. There is much room to maneuver there. Given how standard the forward and bow sections of this sort of vessel were, what the owners were saying as much as anything was "don't mess it up." They did not know how to design the forward section of a ship. They were trusting that the shipwright did.

> The lower futtocks should be 9 inches [(22.86 cm)] broad and 8 inches [(20.32 cm)] thick at the rung heads, and grooved rounding fit for their work. The upper futtocks should be 8 inches [(20.32 cm)] broad and 7 inches [(17.78 cm)] thick, and not less than 6 inches [(15.24 cm)] at the bend. The top timbers are to be 7 inches [(17.78 cm)] thick at the foot, and not less than 4 ½ inches [(11.43 cm)] on the top; all these timbers aforementioned to be of white oak, and every timber and futtock to butt and butt with a square butt and no less scarf than 4 ½ feet [(1.37 m)] (or 4 [(1.22 m)] at least).

As for the rest of the framing, we see here that the frames were to have two futtocks, an upper and a lower, and one top timber, for a total of four component pieces, with each component a little smaller than the one below. The lower the member, the more weight and strain it carried, and the higher the member, the more advantageous it was to make it weigh as little as possible given the strength it needed, so as to keep the weight of the vessel concentrated lower down for stability. White oak was by far the most widely-accepted wood for ship framing, and the stipulations on butt and scarf were to ensure that the framing joints were as strong as possible.

> A keelson is to run from stem to stern, and a knee to be fixed against the stern post; the feet of the stern and bow timbers are to be well-chocked and bolted, and a broad apron placed on the stem, not less than 6 ½ inches [(16.51 cm)] thick, with two breast hooks in the hold, one pretty low down, the other to lodge the lower deck on, and to come well around the bows.
>
> Two breast hooks are to be placed between decks, one in the middle, the other to lodge the forecastle deck on, to come well around the bows and armed [?—original "arm'd"] so it may take the knuckle and some other timbers in.

This next section is all about structural reinforcement. The *keelson* was an inner longitudinal member, running the length of the centerline on top of the floors; it served to clamp the floor timbers between itself and the keel below. *Knees*, *chocks*, and *breast-hooks* served a similar function to brackets and braces; they filled and reinforced ninety-degree joints to prevent flex, or what the shipwright would have called "working"—as in "working loose."

> The quarter and stern timbers are to rake according to builder's pleasure, with a transom athwart them, and be well-kneed. Two hawsepieces are to be placed forward, long enough to run low down in the bows, so that they may be firm [?], and the catheads are to turn from the hawsepieces and not to lodge on the deck; the lower deck is to lodge on the transom abaft, and the beams are to be 11 inches [(27.94 cm)] thick and 12 inches [(30.48 cm)] broad to round 7 [(17.78 cm)] or 8 inches [(20.32 cm)], all double-kneed, with knees of compass timber, and the beams 4 ½ or 5 feet [(1.37 or 1.52 m)] apart, with ledges and carlings in proportion. The beams of the upper deck are to round according to builder's discretion, and are to be 7 ½ inches [(19.05 cm)] thick and 10 inches [(25.4 cm)] broad, all double-kneed with lodging knees of compass timber, except a hanging knee on each side at the bulkhead of the forecastle and half deck abaft the beams, 4 ½ or 5 feet [(1.37 or 1.52 m)] apart, with ledges

> and carlings in proportion. The half deck and forecastle deck beams are to be 4 ½ inches [(11.43 cm)] thick and 7 inches [(17.78 cm)] broad, to round according to builder's discretion, all double-kneed with knees of compass timber.

Hawsepieces and *catheads* were associated with anchoring. Hawsepieces were bow timbers through which *hawseholes*—the holes through which the anchor cables or chains ran—were cut.[10] Catheads were protruding beams from which the anchors themselves were handled. Traditionally, they had cats' faces carved into their ends. The owners were describing pieces of the ship that would be added later in construction, once the framing was in place and the bow and stern were being built up. The rest of this section specified scantlings and reinforcements for strength. *Ledges* and *carlings* were reinforcing members for the deck beams. *Lodging* knees were horizontal; *hanging* knees were vertical. Note that, again, the owners specified compass timber for these pieces. In compass timber, the grain of the wood was curved; a curved piece would have grain running all the way through it. A curved piece cut out of a straight-grained larger piece would have much shorter grain running through it. Since the grain is the cellular structure of the wood, pieces with short lengths of grain from edge to edge were much weaker than those with longer grains. Wood was always cut along the grain rather than against the grain wherever possible.

Once the ship was fully-framed, with bow and stern structures built, deck beams in, and all reinforcing pieces—knees, chocks, breast-hooks, carlings—fastened in, she could be planked.

> The ship is to be planked with white oak, from the keel to the bend of 2 ½ inches [(6.35 cm)] thick, except two strakes of 3 inches [(7.62 cm)] and one of 3 ½ inches [(8.89 cm)] at the point of the bilge, one strake of 3 inches [(7.62 cm)] and one of 3 ½ inches [(8.89 cm)] under the bend, two bends one on top of the other, 9 inches [(22.86 cm)] broad and 5 ½ inches [(13.97 cm)] thick, one strake of 3 inch [(7.62 cm)] plank on top of the bends and one on top of that, of 2 ½ inches [(6.35 cm)], the rest to be 2-inch [(5.08 cm)] plank to the gunwale, except a channel bend of 8 inches [(20.32 cm)] broad and 3 ½ inches [(8.89 cm)] thick is to carry around the side and to tumble in at the channel bend, and then turn off to the gunwale "so as to make a handsome side." Every strake and

10 According to Steffy, hawsepieces (or hawse-pieces) were associated with round-bowed vessels. See Steffy, "Illustrated Glossary."

plank is to be worked fair, and none broader than 9 or 10 inches [(22.86 or 25.4 cm)] at most.

The ceiling in the hold and between the decks is to be 2-inch [(5.08 cm)] white oak plank, except 2 strakes of 2 ½-inch [(6.35 cm)] and one of 3-inch [(7.62 cm)] plank on the rung heads, or point of the bilge, and one of 2 ½ inches [(6.35 cm)], the other 3 inches [(7.62 cm)] on top of that, for rising under the lower deck; one of 2 ½ inches [(6.35 cm)] for rising under the upper deck, and one of 2 ½ inches [(6.35 cm)] for spirket rising on the lower deck, all the planks both within and without to be free of sap. All the decks are to be laid with full 2- or 2 ¼-inch [(5.08 or 5.72 cm)] oak deal plank, free from sap and at least 30 feet [(9.14 m)] long, except a 2- or 2 ¼-inch [(5.08 or 5.72 cm)] oak plank for the waterway, and 4-inch [(10.16 cm)] oak for partners on the lower deck. The upper deck waterway must be thick oak and [two words partially obscured] so as to make a good seam above the deck.

The partners shall be of oak plank 2 ½ inches [(6.35 cm)] thick, both before and abaft.

We are accustomed to thinking of "planks" as about 1″ [(2.54 cm)] thick. These planks, at 2–3″ [(5.08 to 7.62 cm)] thick, were much more substantial. The craft standard for shipbuilding called for using only seasoned wood—wood that had been stored long enough to dry out, so that there was no sap—hence the stipulation that the planking be free of sap. This was for rot resistance. *Strakes* were planks (or joined planks) that ran the entire length of the hull. The details of the *channel bend* (channels in this sense were small platforms on the outside of the hull near the gunwales, around which the shrouds, the standing rigging that held up the masts side to side, were bent) carry an aesthetic component: by tumbling in (as in tumblehome—curving inward) and then turning off—curving back out—the resulting curvature would, to the owners, "make a handsome side." Slab-sided or wall-sided vessels were not considered as pleasing to the eye as those with some curvature to them. The *waterway* was a "timber or gutter along the side of a deck whose purpose was to prevent the deck water from running down between the frames and to divert it to the scuppers [openings in the rail for water to run overboard]."[11] "*Deal*," as used here, probably means planks sawn to standard dimensions.[12]

11 Steffy, "Illustrated Glossary."

12 In more modern usage, deal planks are softwood, usually pine or fir. The owners here specify oak deal. Like most terms having to do with shipbuilding, this one has obviously morphed in meaning over time. The consistency is the idea of uniform sawn dimensions.

Partners were the structural reinforcements built around holes in decks through which masts passed. Given that the owners wrote "both before and abaft," we can reasonably assume that this ship was to be two-masted, making her a brig or a snow. Given her size, that would be no surprise. It might seem odd that the owners did not mention a rig anywhere in the letter, but that was common.

> The half deck is to run aft with a fair sheer, so that it may be 5 feet 9 inches [(1.75 m)] in the cabin, under the beam; and the quick work on top of the deck is to be 6 or 7 inches [(15.24 or 17.78 cm)] high and 7 or 8 inches [(17.78 or 20.32 cm)] on the forecastle, and all the upper deck planking, both for main and half deck, is to run the full length of the deck, so that there are no butts.

Here, the owners came back to accommodations; the half deck, which should run aft with a fair sheer—another aesthetic stipulation—was to have 5′ 9″ [(1.75 m)] of headroom in the cabin underneath. That does not strike us as generous, but few ships, especially ones this small, had vertical interior spaces that would strike us as generous. *Quick work* refers to the part of the hull above the loaded waterline, while *dead work* refers to the part below it. Note the stipulation that there be no butt joints in the upper deck planking. Butt joints are simple, straight, end-to-end joints, as in a hardwood floor in a house, and they lessened strength and increased the likelihood of leaks.

> A false stern post and rudder, and the counter, are to be planked with 2 ½-inch [(6.35 cm)] oak plank, and the stern with 2 ½-inch [(6.35 cm)] deal plank. A drum-headed capstan for 6 or 8 bars is to be fixed in its place, and a substantial windlass and bitts, of 6 inches [(15.24 cm)] thick and 18 or 19 inches [(45.72 or 48.26 cm)] broad, is to be fixed high up enough from the deck so that it will not hinder the loop holes in the bulkhead of the forecastle under.

The false stern post served a similar function as the false keel, protecting the key structural member (the stern post) from damage should the ship be struck in the tail, or should she ground in the tail. In addition to helping the rudder provide steerage, in the same way the false keel helped the keel provide good tracking and resist leeway, the false stern post would take the brunt of the damage of a twisted rudder should the rudder be heavily damaged by grounding or striking an object.

The *capstan* was the rotating drum that wound up the anchor cable so the ship could *weigh* (retrieve and raise) *anchor*. The bars mentioned were heavy wooden bars placed in notches in the capstan, by which the crew turned the capstan by walking around in circles, pushing on the bars. (There grew up a large body of *sea chanteys* composed and repeated specifically for this work, as was common for most repetitive and physically-demanding tasks involving human muscle power—swinging a scythe, picking cotton, etc.) A *windlass* was the same sort of winch, only the drum was oriented horizontally. We are now getting into the equipment the shipwright was to install, or have installed, after the ship was mostly complete.

> She is to have a head cutwater and badges with topsail sheet bitts afore and abaft, those before to go down to the lower deck, and be so substantial and strong as to make a gallows to step the bowsprit in, which must be well-chocked both above and below. Builder is to find crosstrees and trestle trees, anchor stocks, and cleats for the yards and masts—as many as may be required.

A *head*, *cutwater*, and *badges*—which were ornamental—suggest that *Bristol Merchant* was to be frigate-built, with more than just a plain stem. As noted, that was common in the British Atlantic even for modest-sized merchantmen. (Badges were ornamental pieces placed on the outside of the hull near the stern, either containing a window or "a representation of it," and decorated with ornament.)[13] The owners were ordering a smart-looking ship rather than one that would cost them the least amount of money. A cutwater, as explained by Mainwaring in the *Sea-man's dictionary*, was

> the sharpness of the ship before, which doth (as it were) cut the water and divide it before it comes to the bow, so that it may come by degrees and not too suddenly to the breadth of the ship, otherwise the ship would beat so full against the water that she would make but little way; and therefore many times when a ship is too bluff we put-to a false stem, and as it were lengthen her forward on; and this we call a cut-water, which will not only make her sail better but also make her keep a better wind and not to beat so much against a head sea.[14]

13 See Falconer, *An universal dictionary*.

14 Henry Mainwaring, *The Sea-mans Dictionary: or, an exposition and demonstration of all the parts and things belonging to a shippe* (1644), in *The Life and Works of Sir Henry Mainwaring*, Volume 2, edited by G.E. Mainwaring (London: The Navy Records Society, 1920), e-reader edition published by The Internet Archive, 2007, location 2043.

Bitts were short stout posts, around which a line could be wrapped to secure it. We see them today on piers—heavy short steel double posts for securing the large mooring lines of ships. In this case, they are for securing the topsail sheets (a primary sail control line), and also to serve as the posts within which the bowsprit would be set (or *stepped*). *Crosstrees* and *trestle trees* were straight wooden pieces attached to the mast. Crosstrees were attached athwartships to help spread the shrouds for better mast support. Trestle trees were mounted at the tops of masts to support the *top* (a small platform used for observation) and accept the heel of the topmast (masts on these vessels were not all one piece). The *anchor stock* was the heavy wooden bar, bound with iron, at the top of the anchor, and *cleats* are t-shaped or horn-shaped pieces for tying off lines, like bitts but smaller. All of these pieces were to be made or procured by the shipwright; in some agreements, the owners agreed to supply this sort of thing already manufactured elsewhere, as well as other pieces of equipment and even fasteners.

> The ship shall have a main hatchway with hatches, an after hatchway with hatches, and grating for the half deck, and likewise a grating with high coamings for the forecastle, a small scuttle to go down out of the forecastle, and another abaft, in the cabin, to go down to the powder room. Two ports will be cut out and hung through the counter for stern chasers, one on each side of the cabin, and two on each side in the steerage. These ports are to be of equal distance from each other, as nearly as possible, and between 20 and 24 inches [(50.8 and 60.96 cm)] from the lower deck. All are to be of the same height, so that the gun carriages may serve either one, all to be hung well and lined, so they shut tight, and contrived so that they do not cut into the planks below them on the outside, so that the two strakes above the bends may run fair along. One port is to be cut out, hung, and lined on each side of the forecastle, and the lower sill made so that it will be clear of the two planks above the bends.

Hatches, *gratings*, and *scuttles* provided access, air, and light belowdecks. They allowed for the loading and unloading of cargo and stores. The mention of a scuttle (opening with a lid) in the aft cabin, to go down into the powder room, gives us our first indication that *Bristol Merchant*, like most Atlantic merchant ships, would be armed. The letter specifies ports for ten guns, including "stern chasers," backward-pointing guns mounted low in the stern, ubiquitous on both armed merchant and warships of the period. The letter reiterates the importance of making sure the ports sealed tight, and that their cutouts did not compromise the strength of the planking. Perhaps the owners were thinking of economizing on armament, however, by loading her with only enough guns for one side at a time, since they stipulate that the gun ports be the same height so that the gun carriages may serve either one.

It is worth noting that, as Winne & Hawksworth were ordering this ship in 1733, the British Empire was at peace, and had been for four years. The next war affecting the Atlantic world would not break out until 1739, although one of the chief provocations of that war, aggressive boarding and seizures by the Spanish *garda costas* in the Caribbean, was already going on. The owners' willingness to incur the expense of armament and its accoutrements reflect their sense of the state of security of *Bristol Merchant*'s intended operating area. Anywhere in the Atlantic *Bristol Merchant* might find herself, she could not expect the kind of protection the Royal Navy would later be able to offer. She would be on her own, and her guns would reflect that reality.

> A tiller is to be fitted to the rudder, and a well to be built in the hold. Builder is to furnish oakum, and the ship is to be well-caulked all over. Builder is to find pitch and a coat of white stuff for her bottom. She is to be trunnelled with dry white oak well-seasoned trunnels. All the timber and planking are to be well-seasoned, and the planking above the bends must be thoroughly dry.

The *tiller* is the handle used to turn the rudder to steer the ship. When the rudder was turned, all the force of the water flowing under the ship was applied to one side of the rudder, encouraging the ship to turn in that direction. Under normal circumstances, there was no mechanical advantage—labor-assistance from pulley systems—for the tiller; it served as a lever, so the longer it was, the easier it was for the helmsman to steer.[15] On ships of this size, it was practicable to make tillers of sufficient length to provide adequate leverage. On larger ships, that was more difficult, as a larger rudder required a longer tiller for the same leverage. That problem probably contributed much to the impetus for perfecting and implementing wheel steering systems, which were in common use by this time on larger vessels, but wheel steering systems are complex, with more parts vulnerable to failure or malfunction.

15 In heavy weather, the crew could rig a block-and-tackle from the rails or deck on each side to the head of the tiller (called "lashing the tiller") to assist with steering. It was also common in heavy seas to assign more than one man to the helm, combining their muscle power.

A well in the hold collected bilge water, making it easier to pump out. We have already discussed the use of oakum and pitch to seal the seams of the planks.

White stuff was a mixture of whale oil, rosin, and brimstone (sulfur). It was commonly used for *paying*—coating the ship's bottom with something to retard weed growth and repel shipworms. Weed growth happens quickly in warmer waters and can reduce a ship's speed by half, which concomitantly reduces her maneuverability and makes her decidedly less safe. Shipworms destroy the ship's timbers. The use of expensive rolled copper plating for sheathing did not come into common use until the end of the period, and it presented dissimilar-metal corrosion problems between the copper and iron fastenings that had to be worked out. For the vast majority of Atlantic merchant vessels, paying was done as specified here.

By specifying that the trunnels used by the builder were to be seasoned white oak, just as the framing and planking were, the owners were ensuring quality fastenings and avoiding the problem of using dissimilar woods together, which expand and contract differently, have different strengths, and different rot resistance qualities.

> Sufficient timbers are to be left to come through the gunwales of both sides, fit to belay to with stanchions and rails for a breast work at the bulkhead of the half deck. Double timbers are to be fixed through the gunwales on each side of the half deck, so as to make round ports, and two in the stern, through the taffrail, and the timbers on each side must be strong enough to bear a cable if one is put out through the stern, as it will sometimes be. Substantial rails are to be fixed on them, breast-high. Two chesstrees, and various other things we may not have thought of here, and may be required, shall be provided, and all work is to be done in a workmanlike manner and the ship built to our satisfaction.

Stanchions were upright posts to which ropes or rails could be attached, forming a railing at the edge of a deck. By stipulating that the half-deck rails have round ports in them, the owners may have intended the ship to have more guns mounted on that deck; guns mounted on the upper decks were common. The ports in the *taffrail*—the stern rail—were stipulated to be strong enough to pass mooring cables through, to tie the ship to a pier or wharf or mooring buoy.

As the owners concluded their attempt to list all the items necessary to finish out the ship, they transferred the obligation to do that to the shipwright, telling him to attend to anything they forgot to mention, and to do a good job.

By insisting that the vessel be built to their satisfaction, they were reserving the right to refuse to pay for her if she did not. Whether or not that scenario would have played out in the owners' favor in court—such cases were not rare—we do not know, but the owners were covering their bases as best they could.

> We would have her built as a floaty ship, not to draw more than 11 ½ feet [(3.51 m)] of water when laden—if less, the better. She should sail well; in order to do that she must have a clean tail, carried fair, without hollows, and a good entrance, with a round harpin to keep her up, and a round taut [original 'taught'] body so she may carry sail; and be stiff, not only when laden but also at her light water mark, so she may shift without ballast if possible. For we shall want her to ship in wild roads and with as little ballast as possible; so it will suit us to have a stiff-shifting ship which will be no hindrance to her sailing. Recall that our galleys in the last war were so, and sailed as well as most ships in England; we should be glad if this ship did too.

This paragraph is valuable, as it gives us several clues as to how the owners intended to employ *Bristol Merchant*. By wanting her built as a "floaty ship," they were asking that she have as shallow a draft as possible. They immediately added, however, that they wanted her to sail well, and to be stiff. The builder could satisfy all of the stipulated criteria only by building in moderately, not drastically, shallow draft. In fact, the only way to achieve a satisfactory design was moderation in all respects. By asking for "a clean tail, carried fair, without hollows, and a good entrance," they were repeating the commonly-held preference for as smooth a flow of water and as little resistance as possible in the hull shape. The reference to "a round harpin to keep her up," however, reminds us of the rounded, bluff bows of these ships—a shape that provided good buoyancy forward, to "keep her up"—as in, keep her bow up. That is not the most hydrodynamic shape for a bow; wanting "a good entrance" and a round bow were somewhat contradictory desires, but such is the very nature of ship design. It is an exercise in the balancing of complex compromises. Good buoyancy in the ends—bow and stern—not only helped keep the ship from burying herself in the waves by the bow, or being swamped by them at the stern, but helped protect her from *hogging*—drooping at the ends from insufficient longitudinal rigidity.

The other inherent weakness of the wooden ship, besides her vulnerability to rot and leaking, was structural rigidity—particularly, longitudinal structural rigidity. With so many joints and so many pieces, fashioning an interlocking structure that would hold its shape under the stresses of motion in a seaway over a period of years was quite a challenge. Failure to maintain longitudinal rigidity resulted in one of two conditions, both of which were seriously detrimental to the ship's hydrodynamic performance, let alone her integrity. The first was *sagging*, in which the middle section of the ship drooped relative to her ends. The second was *hogging*, in which the ends drooped relative to the middle. William Hutchinson, in his well-known treatise, recounts his first job as a supervising shipwright, building a merchantman for the Jamaica trade, 90 feet [(27.43 m)] on the keel. He built her on concave stocks as a prophylaxis against hogging, and recommended building all vessels on stocks with a concavity of two inches (5.08 cm) for every 30 feet [(9.14 m)] of length.[16] The thought was that, as the ship settled with age and use, her originally slightly convex keel would become horizontal, not concave. There is a rough analogy to the modern flatbed semi-trailer, built with a curve so that it will not sag under load. Hutchinson did not believe that a slightly convex keel was detrimental to performance.[17]

Hogging was closely connected to the bearing of armament. As the seventeenth century went on, there was a trend in English construction to increase displacement in the ends as a countermeasure.[18] That means broader, not sharper, bows and sterns (a trend that, as we shall see, continued in the eighteenth century, although probably for a different primary reason). The structural problems of wooden construction have much to do with why we would never see the knife-like prows of modern steel warships on vessels in this period.

A "round taut body so she may carry sail; and be stiff" means a hull form that will resist heel and roll when carrying enough sail to move at an acceptable speed. A "tender" ship heeled and rolled easily. Such ships frustrated their masters and owners, as they could not carry as much sail as a similar vessel without such a flaw, and thus were slower. They also contributed to excessive crew fatigue and discomfort. A hull stiff enough to resist rolling even when unladen was a stiff hull indeed; as noted, most ships had to have some weight aboard to be stable even at anchor. The owners may have been overly-ambitious here, which they were perhaps acknowledging when they added "if possible" to that stipulation. At this point, we get our best clue as to where these instructions

16 William Hutchinson, *A Treatise on Naval Architecture*, Fourth Edition (London: Conway Maritime Press, 1970), v–vi. Originally published 1794.

17 The major drawback of wood not addressed here is its flammability. That is because, aside from providing effective pumps, there was nothing the shipwright could do to mitigate that risk.

18 William A. Baker, *The New Mayflower: Her Design and Construction* (Barre, MA: Barre Gazette, 1958), 33–34.

were pointing. The owners want their vessel "to ship in wild roads and with as little *ballast* as possible …" "Wild roads" meant unprotected *roadsteads*. Shallow draft and a stiff hull would permit the ship to get over sandbars at the mouths of rivers and in coastal shallows. The ability to remain upright and operable without ballast in unprotected roadsteads meant the ship could ride at anchor off an open beach or off a river mouth with no maritime infrastructure whatsoever. This suggests the owners had in mind sending the ship to sparsely-settled, undeveloped parts of the Americas or, just as likely, to the coast of West Africa. Bristol was a principal slaving port. Ships put into the "Guinea trade" or "Africa trade" would need to spend considerable time close in to the coastline, with no developed harbors to shelter them. It took weeks or months to negotiate trade terms with local African principals, and might require making several stops, remaining in each for some time.

Galleys as used here meant small warships; the "last war" would have been the War of the Spanish Succession, which was a continuation of the War of the League of Augsburg against France. The British-French maritime imperial rivalry was in full swing, with only pauses, never a lasting peace. This remark reminds us of the connection between smaller warships and smaller Atlantic merchantmen; the very term "frigate-built," with its connotations of borrowing important design elements from those small warships, suggests the connection, and we have noted that *Bristol Merchant* was likely to be frigate-built.

> Do not forget that she should rise at the middle of the keel, aft on gradually and fair, not rise quickly up all at once abaft.
>
> As we shall send a ship over with all materials, we shall not lack all the masts and yards, as is common, but you may agree to have them as is customary, and what we don't take, oblige the builder to provide for.
>
> Take care to agree that two feet (0.61 m) of the 58-foot [(17.68 m)] keel is to be given in as if it were out of the rake, which is common, so we may pay back for 56 feet [(17.07 m)], and that the ship must be finished and launched in August at the latest, but sooner if possible. She should be named the *Bristol Merchant.*
>
> If there is anything amiss in these dimensions, which may be mended to our interest, please do so, and we will be obliged.
>
> [signed Winne and Hawksworth]
>
> [to William Jones in Bristol, to be forwarded to Boston as soon as possible]

Still focused on re-emphasizing their preferences for hull form, the owners indicated once again that *Bristol Merchant* should have a rather full, and thus capacious, hull—no pink-stern here. As for masts and yards, the prospective owners agree to supply them, but if the builder wants or needs more or different spars than what they provide, that will be his responsibility.

Before closing, the prospective owners tried to save a little money. The shipwright's fee was typically based on a tonnage calculation called *carpenter's measure*, and this calculation was based on the agreed-upon length of the straight part of the keel. The owners contended that it was customary to consider two feet (60.96 cm) of the 58-foot (17.68 m) keel as "out of the rake," or as if it were part of the rake, and thus not part of the straight section. This would make a difference of 4.6 tons or so, Massachusetts carpenter's measure, and could save the owners perhaps £12–15.[19] Peter Earle writes that the wages of an ordinary foremastman, which we can take as representative of a sailor's pay for the purposes of this comparison, "showed remarkable stability" throughout the period "at around twenty-five shillings per month, except in wartime, when they rose to fifty shillings a month or even more."[20] The Empire was at peace in 1733, so at the lower wage rate, and assuming the lower cost-per-ton estimate, Winne and Hawksworth could have paid 9.6 sailors a month's wages each out of this savings on builder's tonnage.[21]

Finally, they give us an idea of the time frame for ordering, building, and launching of an ordinary merchantman of this size. The letter was written in January. Jones has to forward it to Boston; that could take two months. A shipwright in Boston has to agree to terms and send word of such back to Bristol, as the letter does not suggest that Jones was planning to go to Boston himself. That could take another two months, as could the reply. That would leave the summer—June through August—to build and launch the ship. That was certainly possible. David Lowell, a shipwright working in Amesbury, Massachusetts, built a schooner in three months during the New England winter of 1792. In a period of less than two years, from January 1792 to September 1793, he built or helped build five vessels.[22]

It was also possible that favorable winds and fast sailing passages would shorten the time needed to come to terms by several weeks. On the other hand, bad weather, sailing delays, or disagreements could drag it out for many

19 Building costs per ton around the British Atlantic fluctuated significantly over time. They differed from region to region and even within regions, from large urban centers to outlying towns (see Goldenberg, 94–95). I base this rough estimate of per-ton savings on the aggregate of shipbuilding contracts I examined.

20 Earle, *Sailors*, 33–34.

21 One pound equals twenty shillings of twelve pence each.

22 David Lowell Account Book, MSS 1229.1, PEM-Phillips.

months longer. The owners were being optimistic here, but not overly so; such uncertainties were the fact of business life in the eighteenth century.

It is also worth noting that ships were not finished when they were launched. After they slid down the slipways into the water, they still had to be rigged, and much of the interior and deck fitting-out would be done post-launch. If *Bristol Merchant* was actually launched in August, perhaps she could have been ready to cross the Atlantic in the fall, before the winter gales set in. In the meantime, several tradesmen were likely employed to finish her. The blacksmith and the caulker would have done their work prior to launch, although the former may have had more hardware to fashion for the post-launch fitting-out. A glazier would be needed to make and install the windows in the stern galleries. A mason would be hired to build the ship's brick galley stove. A cooper would fashion the ship's water casks. Riggers would install the masts, yards, shrouds, stays, and ratlines. A sailmaker would fill the order for the ship's suit of sails, including spares, which were most necessary, given the ubiquity of damage and fast wear. In some surviving business records of shipowners, the pay chits of all these tradesmen may be found—small scraps of paper on which they wrote, briefly, what service they had provided or items they had sold, and what they were owed. All these costs were in addition to the price-per-ton paid to the shipwright.[23]

Once everything was done to the owners' satisfaction, the owners would make final settlement with the shipwright and other tradesmen. Most of the time, at least in British America, the shipwright's substantial bills would be paid with a combination of currency and goods, as currency—especially the most-prized pounds sterling—was always in short supply. "Goods" were most often "West India goods"—sugar, molasses, and rum. This would be specified in the actual shipbuilding agreement. (Note that the Winne & Hawksworth letter was actually a letter to a merchant correspondent, not directly to a shipwright.) Sugar products were in such high demand that they served this world as a form of currency in themselves. New England's principal industry in the eighteenth century was the distilling of rum, both for domestic consumption and for export. Sugar products, among other goods, were used to purchase slaves in Africa, most of whom were shipped to the West Indies to make more sugar products—and die early deaths in the process. Many of the vessels built in New England would immediately be put into the West Indies trade, beginning the 1640s, and perhaps also into the Africa trade. It was all connected, and sugar products were the linchpin of the connections.

---

23 Plentiful examples may be found in the Brown Family Business Records, JCB.

Some merchant owners, such as we may assume about John Winne, had direct experience running vessels at sea. Some did not. Henry Laurens was a planter and merchant with no direct maritime experience, and that is reflected in his typically self-effacing order for a ship from England. Laurens does not specify much technical detail in his order:

> You know that I do not take upon me to be a good Judge of Shiping, therefore let it suffice that i say I would have a Ship of about 700 Barrels burthen as well built as possible, having an eye to profit & shall have no objection to a good Cabin & some little expence to decorate the whole.[24]

Such a brief and vague commission as this turned out to be the exception, but even this one reminds us of every owner's chief priorities. The first stipulation is capacity, and that is typical. Since owners primarily saw ships as conveyances of their goods to market, the characteristic of a ship they were most concerned with was how much of what cargo she could carry for her size and cost. Capacity cannot be considered alone, however. Larger ships cost more per ton to build than smaller ships.[25] So, merchant owners had strong incentive to maximize the capacity-for-size of their vessels. (Shipwrights, on the other hand, had strong incentive to build larger vessels, if they had the means and could get the contracts.)

Next, Laurens wants her as well-built as possible without costing him so much that he cannot make a profit with her. He is willing to provide a nice cabin for her master, which should help him entice someone worthy to serve in that capacity, and to make the ship look nice, which will reflect well upon her owner in foreign ports. We find in merchants' papers that gilders and carvers were routinely on the payroll when a ship was ordered or refitted. Such aesthetics mattered enough to owners that they were willing to incur costs for them. It is clear, however, that Laurens is perfectly willing to trust whatever builder he commissions—and, as it turns out, his agent in England, Captain Courtin, whom he charges with overseeing her construction and delivery—entirely when it comes to matters of design and performance.

While researching colonial shipbuilding in the late 1960s, Joseph Goldenberg found an unusual correspondence in the Coates and Reynell Papers at the Historical Society of Pennsylvania.[26] Having spent a month looking for

---

24 Hamer, ed., *The Papers of Henry Laurens*, Volume 3, 366.

25 For illustrative data, see McCusker, "Pennsylvania Shipping Industry," 130.

26 Goldenberg's archival sources were the starting point for the archival research for this project. I thank Prof. Goldenberg for personally providing advice.

something similar in the same repository, I can attest to how unusual it was. It is found in an exchange of letters between John Reynell in Philadelphia and Daniel Flexney in London. Reynell is acting as agent for Flexney, having a ship built for him by the successful Philadelphia builder Charles West. This was a typical role American merchants played for their better-capitalized counterparts in England, especially before 1750. The first letter, from Flexney, dated 24 May 1740, is primarily concerned with the usual business of goods and prices, but two lines of it read: "… if thou hast no allready begun I desire thee build a vessel about L 120 Tun Square Sternd beleive to be a Brig will sail with least expence & lett her be finished with all possible expedition …"[27] A 120-ton brig was a typical mid-sized Atlantic merchantman of this period, as we have noted, and Flexney tells us why.

We find a follow-up reference in a letter of 19 June 1740, unusual because it describes the details of the ship and expresses a technical opinion on nautical matters, if only in passing. Reynell writes back to his client:

> Have agreed with Charles West to Build a Square Stern Vessel of 55 foot [(16.76 m)] Keel, 21 foot [(6.40 m)] Beam, 10 foot [(3.05 m)[Hold, 4 foot [(1.22 m)] betwixt Decks, 12 foot [(3.66 m)] Rake 3 [(0.91 m)] of which to be in the Keel and 9 [?] Inches [?] [(22.86 cm?)] Dead Rising for thy [Custom?] to be lanch'd the 1st of May next. Could not get her done sooner by any good Carpenter besides am informed, it would be a great Disadvantage to the Vessel to plank her up in Winter am to give him £4 6 June. Would not advise thee to make a brigantine of her their large Booms soon beat [?] 'em to Pieces, in my Opinion a Snow is a much Handyer vessel.[28]

The specifications listed here do not just indicate size. As we know, the specification for "Rake" and for "Dead Rising" (deadrise) indicate aspects of the shape of the hull and thus the design of the vessel. It is not clear whether Reynell or West came up with those specifications. Reynell politely suggests to Flexney that he exercise patience, as building a vessel out of wood takes time to do

27 Daniel Flexney to John Reynell, 24 May 1740, box 2, folder 8, Series 1. John Reynell a. Incoming Correspondence, 1729–1764, Coates and Reynell Family Papers, 1677–1930, Collection 140, Historical Society of Pennsylvania, Philadelphia (hereafter HSP). The "thee" lets us know that Reynell and Flexney are Quakers; Quakers formed their own business networks around the Atlantic World, as did other groups with ties of religion or political leanings or extended kinship.

28 Reynell to Flexney, 19 June 1740, Series 1. John Reynell b. Outgoing correspondence, Letter Book Vol. 4, October 1738–October 1741.

properly and the construction schedule must take into account how the seasons of the year affect the exposed timbers. The final sentence here, advising Flexney to specify his new vessel as a snow rather than a brigantine, is one of only two such explicit opinions found in merchants' papers in two months of work in five archives with extensive collections of such papers from this period. Obviously, a fair number of owners—or at least builders and masters—had to believe that the snow offered advantages over the slightly simpler brig, but it would be worthwhile to explore, experimentally, what those advantages were. The line drawings of the two rigs in the previous chapter may indicate support for Reynell's opinion. Note how far the brig's mizzen boom extends past the stern compared to the loose foot of the snow's driver. Contemporary line drawings by Chapman and Falconer, reproduced by David MacGregor, also support the idea that the snow was a "Handyer vessel," at least in this respect.[29]

On 6 July, Reynell includes the following in another letter:

> As the vessel I had agreed with Mr. Charles West to build for thee is not yet put up have desired him to get a keel 3 foot [(0.91 m)] longer, that it may do for Capt. Stephenson w[hi]ch he has promised he will, and certainly I think 21 foot [(6.40 m)] Breadth will be full narrow for that length.[30]

We can read into this that Captain Stephenson, who, later letters reveal, Flexney uses as his man-in-the-yard and probably senior captain at sea, has voiced the opinion that the keel should be extended, and thus the vessel made longer for her breadth, and that Reynell believes that at that length, she needs to be 21 feet [(6.40 m)] in the beam. Too narrow a ship will not stand up well to her canvas. Apparently, there was some discussion of this—perhaps Flexney was concerned that broadening her would increase his cost even more, which it probably did, but Flexney's side of that correspondence was not located, if indeed it still exists. The next letter mentioning this ship is dated 14 December, from Reynell:

> The new ship is raised, we are obliged to make her 21 foot [(6.40 m)] Broad. Agreeable to my former agreement, or the Carpenter would not

29 MacGregor, *Merchant Sailing Ships*, 73–75. Richard Unger opines that "the snow could carry more canvas with greater reliability." Richard W. Unger, "Early Modern Vessels," subentry under "Trading Vessels," in John B. Hattendorf, ed., *The Oxford Encyclopedia of Maritime History*, Vol. 4 (Oxford: Oxford University Press, 2007), 171. Also see Harland, *Seamanship*, 78.

30 Reynell to Flexney, 6 July 1740.

> build her. I cannot help being of the Opinion that she will be abundantly the better for it, much more than the Cost of it will be. If she had been made narrower, it would have been a greater Disadvantage to thee, than [then] the Carpenter, and he was Unwilling to make her any Deeper then [than] our former agreement which I thought was full deep for warr Time. However we got over that.[31]

Reynell and West, the builder, were probably in cahoots here. We do not know how much of the insistence on changing the beam was Reynell's and how much was West's. We do know that a builder of West's stature would have to uphold his own standards for the sake of his reputation. If he was convinced a narrower beam would make for a bad ship, it is completely believable that he would refuse to build her so; West was an established shipwright. The builder's eye was a powerful guide, and still is in wooden shipbuilding. This passage also reminds us that, while Flexney may have had the money, his control over what was happening in the yard in Philadelphia was limited. He had to trust Reynell and West. That was another owner's risk he incurred.

Finally, it would be instructive to know what Flexney meant by "full deep for war Time." It is certainly an acknowledgement that the shipping business was a different business in wartime than in peacetime, and that difference, in his mind, reached all the way back to the design of the ships themselves. I do not know whether Flexney was thinking more of commercial considerations here—different markets in terms of geography or commodity—or more of defensive matters—speed, maneuverability, the ability to carry guns.

Formal shipbuilding agreements, as we might expect, tend to be more detailed the larger the vessel—and thus the more money and work being risked on her. These agreements are clearly as much about "risk management," as we might call it today, as sailing instructions and insurance policies. A short one, for a small vessel, reads much like the excerpt from Reynell's letter to Flexney in 1740. Twenty years after that, William Peirce contracted with Joshua Coffin and Samuel Hale, all of Newbury, Massachusetts, on 5 September for "The Hull of a Small Vessel Designed for a Sloop …," to be 40 feet (12.19 m) on the keel, 6 feet 4 inches (1.93 m) in the hold, 17 feet (5.18 m) in the beam, planked outside with 2 ½ inch (6.35 cm) thick sound white seasoned oak, with ceiling plank (inside lining) 1 ½ inches (3.81 cm) thick. Deck planks were to be 2 ½ inches (6.35 cm) thick and free of sap and rot, timbered with good sound white oak, top timbers to be all white oak, short quarter deck "as Customary for such Vessels …"

31 Reynell to Flexney, 14 December 1740.

a long floor, but short floor timbers, about 7 inches (17.78 cm) deadrise—the same specified for *Bristol Merchant*. The builder was to find the spars—mast, boom, bowsprit and gaff—and was to launch her, caulk the treenails, and find a rudder, tiller and windlass, grave her and stop the worm holes, and launch her before 8 December. The owners were to contribute all iron nails, pitch, turpentine, and oakum. The price per ton was three pounds one shilling and four pence, to be paid in West Indies goods, corn and/or money.[32] This was more typically brief than the Winne & Hawksworth letter of 1733. It is likely that Peirce either had prior dealings with Coffin and Hale, or trusted their reputation, whereas for Winne & Hawksworth this was probably not the case.

Again, we see design details specified here, and we do not know whether those were specified by the builder or by the owners. Perhaps the owners, if they were knowledgeable and experienced, expressed design preferences to the builder, who translated those into quantifiable specifications. Note that there is no capacity specified here. That omission was more common for these small vessels than for larger ships. Any good builder, however, would have been able to translate a roughly specified capacity to the dimensions we see here.

With some familiarity with how a shipwright was to build a vessel from a letter of agreement, we also have a sense of what important aspects of design were left up to his experience and judgment. In the next chapter, we will delve more deeply into those, as we attempt to answer questions the archives will not answer for us. For all the details about *Bristol Merchant* conveyed by the Winne & Hawksworth letter, it does not tell us the shape of the hull, except in the most general terms. It does not tell us the sheer—curve of the top rail—from amidships to the bow. If we were to build an operational reconstruction of *Bristol Merchant*, we would have to commit to judgment calls about important aspects of her design, based on our best understanding of similar vessels. In that sense, our reconstruction would necessarily convey a false sense of certainty. A virtual reconstruction, on the other hand, could retain the representation of those uncertainties, by presenting them in a contrasting color, for example. Models, too, can present areas of uncertainty by using dissimilar materials. This is now common in hull models made by archaeologists, in which the actual remains are represented faithfully, and missing parts are added in somewhat abstract form, as with metal. Figures 15 and 16 are photographs of just such a model, created from the remains of the "Indigo wreck" discovered at Alexandria, Virginia, in 2017.

32 Shipbuilding contract, William Peirce, 1731–1778, MSS 0.549, PEM-Phillips.

FIGURE 8 Indigo wreck, Alexandria, Virginia, model, profile view
PHOTOGRAPH BY CHRIS DOSTAL

FIGURE 9 Indigo wreck, Alexandria, Virginia, model, bow-on view
PHOTOGRAPH BY CHRIS DOSTAL

This technique has also been used with actual remains. Perhaps the most famous examples of this are the remains of the Viking ships of Roskilde, Denmark, which are on display with plain metal structures indicating what is missing.

We can make an analogy between the Winne & Hawksworth letter and the partial remains of a hull like the Indigo wreck. Both leave gaps in our comprehension of the ship as a whole. We can work to fill in those gaps, but to do so, we have to consider all the evidence, from all the sources available, together. Only then do we stand a chance of understanding the mind and craft of the shipwright.

CHAPTER 4

The Mysterious Art of the Shipwright: Deciphering Merchant Ship Design

In the previous chapter, we used a detailed shipbuilding order to walk through the process of building an ordinary merchantman. The prospective owners specified some aspects of design, but much of that, as always, was left to the judgment and skill of the shipwright. In this chapter, we will explore how the shipwright applied that judgment and experience to design the physical characteristics of the vessel, which affected how she would perform at sea. We will see how these characteristics related to each other. We will consider how some of them changed over time, and why that might have been.

Artisanal shipwrightry will be considered here in the context of the relationship between artisanal craft in general and the Baconian project expressed in the works of what we call the Scientific Revolution and the Enlightenment. This relationship was, to an important extent, characterized by a divide, but that divide was not a clean one, as scholarship on the subject has pointed out.[1] Artisans were instrumental in furthering that Baconian project, and some of them, literate and ambitious enough to get their treatises published, were quick to deride others for, as they saw it, stubbornly refusing to do so.

Given the dearth of reliable documentary sources, this inquiry will take advantage of a mix of evidence, including that derived from maritime archaeology, experimental archaeology, and digital modeling, a young and promising approach. A brief methodological introduction to all three, specific to the subject at hand, is in order.

In the Atlantic and Caribbean, interpreting physical ship remains as indicators of overall hull design almost always requires a degree of conjecture on the part of the archaeologist, as what remains of the hull rarely includes anything above the floor timbers. The archaeologist estimates the sizes and shapes of the upper structure based on the corpus of knowledge extracted from all relevant remains yet discovered and analyzed, as well as any documentary sources that may exist. Conjecture from partial hull remains employs an advanced level of educated guessing. Any archaeologist will take pains to point out to laypersons the degree of conjecture involved in a hypothetical hull reconstruction.

1 See for example Pamela O. Long, *Artisan/Practitioners and the Rise of the New Sciences, 1400–1800* (Corvallis: Oregon State University Press, 2011).

 | DOI:10.1163/9789004426344_006

While traditional archaeology investigates what past objects were, experimental archaeology attempts to discover how they worked. Of course, in order to do that, researchers first need some confidence in what those objects were; they need the results of traditional archaeology. From those, investigators can reconstruct careful copies capable of performing their originally-intended function—or as close to that as current understanding can get. In ship archaeology, the approach was first used successfully in demonstrating how ancient craft worked, and their voyaging capabilities.[2] Using the same techniques for early-modern vessels has received less attention, although it has been underway since the *Mayflower II* in 1957.[3] In addition to whatever conjecture is required to effect a reconstruction, the building and operation of replica ships introduces another set of potential pitfalls that must be taken into account before interpreting the results of any experiences or experiments: any differences between original and modern materials, construction techniques, operating conditions, and human abilities. These do not, however, present intractable difficulties, but rather adjustment factors that must be applied to results for accurate interpretation. Construction techniques for early modern vessels are for the most part discernible from the archaeological record, and typically followed as closely as possible in replica construction. The original wood species used in construction are frequently replaced with more durable alternatives, given the expense of building a replica, but the weight of the original must be—and can be—closely approximated, and this is unlikely to affect performance characteristics to an appreciable degree. Natural-fiber sails and rope rigging are in common use, so we know their characteristics. We can also compare them closely to their synthetic counterparts, and that information can be used as a corrective for experiments on vessels using synthetics.

By "operating conditions," I am not referring to wind and sea conditions; it is a rather straightforward matter to match wind and sea conditions for a given set of experiments or experiences to those recorded by, or likely experienced by, the original mariners. I am referring to the condition of the vessel, and the two most important aspects of condition are ballasting and fouling. Modern safety regulations require vessels to be ballasted according to a set of written standards. This ballast must, in most cases, be fixed. Period merchant ships,

2 For a detailed and well-illustrated account of building and using replicas of ancient and medieval craft based on archaeological remains, see Steffy, *Wooden Ship Building and the Interpretation of Shipwrecks* (College Station: Texas A&M University Press, 1994).

3 On *Mayflower II*, see Baker, *The New* Mayflower; Alan Villiers, "How We Sailed the New *Mayflower* to America," *National Geographic* 92:5 (November 1957): 627–74; and David Thorpe, "The *Mayflower II* Story—Reflections of a Sailor," Parts 1–6, *The Mayflower Quarterly* 78:2–79:3 (June 2012–September 2013).

however, did not carry fixed ballast; they used cargo, carefully-arranged, for ballast; or, if lightly-laden, they used ballast stone or shingle they picked up for that purpose. To conduct any experiments on performance characteristics, we have to establish a correlation between how the replica is ballasted and how the original might have been.

The other major consideration is fouling. We have established that speed and maneuverability could be drastically affected by a foul bottom. Modern vessels generally sail with clean bottoms, as they use highly-effective anti-microbial bottom coatings. Most of the time, then, we can expect a replica to be somewhat faster than an original would have been, unless the latter were fresh from the stocks or from a careening. We cannot make direct performance comparisons between replica and original without taking into account the state of fouling. Fortunately, the documentary record does give us an idea of what fouling was likely to be under certain conditions, and ships' logs also mention the state of fouling and record bottom cleanings.

As for human abilities, we are still learning the skills and methods of period sailors, as will become clear in Chapter Seven. We know from the documentary record how many crew were aboard a certain vessel for a certain voyage, and remarks on under-manning, and the problems it caused, are not uncommon. We also have enough aggregate experience operating replicas to know what is required to accomplish most important tasks aboard on a given vessel, with a given rig, under given conditions. It is important to keep in mind not just what a sailor knows how to do, however, but what that sailor is willing to do (or permitted by modern law to do, for that matter). When considering specific labor demands aboard, it is worth remembering the differences in "risk management" and "risk tolerance."

As we use, and consider the results of, experimental archaeology on ships, then, these are the potential caveats to keep in mind when interpreting the results of experiments or when attempting to compare modern experiences to those of period sailors on period ships. There is an element of conjecture inherent to using any type of historical evidence. That does not mean we refuse to use it. It means we make ourselves aware of the potential for error and take that potential into account.

This chapter will draw on another relatively new research method—digital or virtual modeling—that promises to help get us unstuck in our bid to understand as much as we can about how this technology worked and allow us to make better guesses as to why the people who used it made the choices about it that they did. The use of the same software used by naval architects and marine engineers can do much more than provide a digitized analogue to paper blueprints. It can simulate dynamic forces and then analyze the results of applying those forces to a virtual vessel. This permits the exploration of what-if scenarios that we cannot practicably create in the physical world. It costs millions of dollars to build and rig an operational replica of one of these ships. Digital modeling costs are minor in comparison. Note that here, I am using "digital modeling" to mean the analysis of dynamic forces on the vessel, rather than the detailed virtual reconstruction of the structure, which, thanks to 3D printing, can be realized in tangible form, as discussed in the previous chapter. The important caution to be exercised with dynamic-forces modeling is the set of limitations of the applications being used, and that is why collaboration with a well-trained and experienced professional is essential for any researchers attempting to exploit this approach.[4]

By collaborating with replica operators, naval architects, and marine engineers, we can overcome some of the uncertainties left to us by the historical record. While it would be easy to assume that, because that record includes several treatises on shipbuilding, we should be able to fill in whatever gaps we find in the shipbuilding agreements between owners and shipwrights that we have discussed, that is not the case, and we need to understand why.

Besides shipbuilding agreements and a smattering of eighteenth-century plans, the most important written source for information on seventeenth and eighteenth-century ship design and construction is a small number of treatises on shipbuilding—some published at the time, some of those obscure and some well-known, and others never published in their own time but re-discovered and published in ours.[5] For the most part, these were written by shipwrights or

4 We do not yet possess the computing power to analyze complex natural dynamic forces, such as weather patterns and the physical behavior of the ocean, with exactitude. We are also still learning the mathematical principles behind "randomness."

5 The treatises I refer to are: Edmund Bushnell, *The Compleat Shipwright*, Sixth edition (London: For Richard Mount, 1699); Fredrik H. af Chapman, *Architectura Navalis Mercatoria: The Classic of Eighteenth Century Naval Architecture* (Mineola, NY: Dover Publications, 2006); John Charnock, *An History of Marine Architecture*, Volume II, ECCO facsimile (London, 1801); F. Dassié, *L'Architecture Navale, contenant la maniere de construire des navires …* Google e-book via archive.org (Paris: Chez Jean de la Caille, 1677); Leonhard Euler, *A compleat theory of the construction and properties of vessels: with practical conclusions for the management of ships made easy to navigators*, trans. Henry Watson (London: J. Sewell, 1790); William Hutchinson, *A Treatise on Naval Architecture*; Brian Lavery, ed., *Deane's Doctrine of Naval Architecture, 1670* (Annapolis, MD: Naval Institute Press, 1981); Mungo Murray, *Treatise on Shipbuilding, 1765, Includes translation of Du Hamel du Monceau's treatise of the same year* https://archive.org/stream/treatiseonshipbuoomurr#page/no/mode/2up; accessed 24 Jan. 2019; Ollivier, *18th Century Shipbuilding*; W. Salisbury and R.C. Anderson, eds., *A Treatise on Shipbuilding* and *A Treatise on Rigging Written About 1620–1625*, The Society for Nautical Research Occasional Publications Number 6 (London: The Society for Nautical Research, 1958); Marmaduke Stalkartt, *Naval Architecture or the Rudiments and Rules of Ship*

public officials involved in naval construction, either writing for an audience assumed to be well-versed in the craft or dismissive of what later authors saw as the outmoded and arcane ways of the traditional artisan and enamored of the new "science," with its promise of universal principles universally available and the opportunity to impose centralized, standardized control on ship design.[6] However, finding material from a treatise copied into Massachusetts shipwright David Lowell's journal proves that, at least to some extent, the material in these treatises was something useful to a practical tradesman far from the naval dockyards.[7]

Lowell writes that to find the height of the mainmast of a ship, "Multipli the Wedth and Depth together of the Ship then Multiply by 3 and divide by 5 gives the Length the Ship's Main Mast in Yards-----" So it goes for the rest of the journal. The length of every spar is a proportion of the length of something else. For the mainmast, take the length of the lower deck, and the extreme breadth, and adding them together, take half—that is the length of the mainmast.[8] So

Building, Google Books facsimile (London, 1781); and William Sutherland, *The Ship-builder's Assistant: Or Some Essays Toward Compleating the Art of Marine Architecture* (London, for Richard Mount, 1711).

6 For an introduction to this subject, see Larrie Ferreiro's *Ships and Science*. The earliest English treatise relevant to this period is called "Fragments of Ancient English Shipwrightry," attributed to Matthew Baker c. 1570, in the Samuel Pepys Library, Magdalen College, University of Cambridge. For such a long period, there are not many, reminding us that ship design and construction was overwhelmingly artisanal. They are listed in the bibliography as published primary sources, along with two important French treatises and one in German. William Baker discusses the English treatises in *The New Mayflower*, and allows the non-specialist some insight into how he used them to help him design *Mayflower II*. (A treatise discovered later, and published in 1994, long after Baker had done his work, provided some guidelines and dimensions for merchant ships not available to Baker and not available to Brian Lavery when he worked on the design for the 1991 *Susan Constant* replica, but according to Peter Wrike, both designs fell within the parameters of what is specified in the tables published in 1994. For the manuscript, see Richard Barker, "A Manuscript On Shipbuilding, Circa 1600, Copied by Newton," *Mariner's Mirror* 80:1 (February 1994): 16–29; for the 1991 *Susan Constant*, see Lavery, Susan Constant; for Wrike's comments, see Peter Wrike, "The Jamestown Replicas," in Bennett, *Sailing Into The Past*, 120–33.

7 Hocker pointed out to me that, with the exception of Sir Anthony Deane, who wrote his treatise for Samuel Pepys, treatises were written for those who already knew how to build ships, and that is why they leave out so much of the technical detail we would love to have (Hocker, personal e-mail, 18–20 August 2015—hereafter Hocker corr. 8/15). I was able to match the original source of most of Lowell's treatise notes as a later edition of William Sutherland's *Ship-builder's Assistant*, a treatise first published in 1711 and re-printed, with much editorial license, throughout the eighteenth century. See Phillip Reid, "Notes from a published treatise in an ordinary eighteenth-century shipwright's journal," *Mariner's Mirror* 104:1 (January 2018), 79–83.

8 Lowell Account Book.

it goes in every treatise on shipbuilding surviving from this period. The answer to the question *how did they design these ships?* is, ultimately, *by proportion*, as indicated in the previous chapter. The beam is in certain proportion to the keel, the depth in hold in proportion to both, and so on. The actual proportions vary some—but usually not much—from author to author. The clear overall impression is that, within a rather narrow range of variance, the proportions for building certain types of vessels were generally agreed-upon. Some of the questions left us by the shipbuilding agreements can be answered in the treatises: "Put a Sloop's mast halfway between one Third & one Quarter of the Keel from forward or from the Rake forward----"[9] For historians, however, this is still frustrating. We want to know why. Why do you put a sloop's mast there? Why is the fore yard 7/8 of the main yard? If David Lowell already knew the answers to those questions, why would he have bothered writing this down so carefully?

Scholars studying old ships today want to see the kind of proof—the kind of supporting evidence—that comes out of a modern laboratory. We want to see test results, computer analysis computations—or their equivalents. In the eighteenth century, however, there were no such equivalents. There was experience—inherited and adapted—and there was the eye of the artisan. The proportions so carefully listed in the treatises, and the calculations necessary to come up with specific dimensions that remained faithful to those prescribed proportions—offered an accessible means of transmitting information about building ships to those who already knew how to build ships. They do not give us the information we need, because their intended recipients did not need that information.

British naval historian Brian Lavery, principal researcher for the 1990–91 replica (currently operational) of the small English galleon *Susan Constant* at Jamestown, Virginia, edited a modern edition of the *Doctrine of Naval Architecture* by Sir Anthony Deane, Master Shipwright of two naval dockyards under Charles II. Lavery sums up Deane's *Doctrine* by saying that Deane's calculations, while "impressive" in number,

> are purely geometric, giving a means of forming the shape of the ship. With the exception of the calculation of the draught of water, they give no indication of the likely performance. He also attempts to impose a series of proportions on his design.... This is, of course, not a truly scientific system, for it depends entirely on the experience or prejudices of the builder, and often it would have to be modified in practice.[10]

9 Lowell Account Book.

10 Lavery, *Deane's Doctrine*, 22.

Artisanal shipwrightry did indeed depend "entirely on the experiences or prejudices of the builder." The easy trap for a modern student to fall into is the tacit assumption that it was therefore inferior to what was "scientific."

For the seventeenth century especially, proportion is much bigger than ship design. It gets at the cultural centrality of understanding natural reality geometrically. So much of what we would call "scientific," was understood by "natural philosophers" in terms of geometry and proportion, including the cosmos itself. This was the age of Kepler, who tried again and again to figure out how to reconcile the heavens to a set of interlocking geometric forms. This is the larger cultural milieu, a direct tie of technology to ontology.[11]

This is not to suggest that an aesthetic sensibility predisposed seventeenth-century shipwrights to design their ships to an abstract concept rather than practical considerations. Jonathan Adams turns that on its head when he rejects the idea that "hulls were designed by methods of some geometrical elegance which had little to do with performance…." He counters "that the geometric procedures developed to generate the complex underwater body was elegantly simple."[12] All that we know about this culture and its shipwrights would suggest that, to them, they were designing for both—they would have assumed that "geometrical elegance" and "performance" were to be found in the same place. Aesthetic considerations cannot be divorced from "functional" considerations in design.[13] A ship had to look proper to those who built it, owned it, and sailed it. Aesthetic sensibility tends to slow down the rate of change, providing there are no urgent short-term pressures for that change, as incremental changes in prominent visual elements are less jarring to that sensibility than drastic ones. While this is likely impossible to prove, such a brake might have acted on the rate of decrease in *castling*—superstructure above the main deck—which progressed through the seventeenth century and into the eighteenth.

One of the easiest visual cues distinguishing the earliest ships of this period to the later ones is the high *castles*—especially stern castles—of the earlier vessels. The historical record does allow us to trace a fairly linear devolution of these from medieval ships through galleons and into later seventeenth-century vessels. The castles tended to lower until, by the mid-eighteenth century, they

11 For a work placing maritime technology in its cultural context with exceptional depth and breadth, see Robert Hicks, *Voyage to Jamestown: Practical Navigation in the Age of Discovery* (Annapolis, MD: Naval Institute Press, 2011).

12 Adams, *A Maritime Archaeology of Ships: Innovation and Social Change in Medieval and Early Modern Europe* (Oxford: Oxbow, 2013), 179.

13 David Pye, *The Nature of Design* (New York: Reinhold, 1964).

were gone. It is worth pointing out that later seventeenth-century ships did not look like *Mayflower II, Susan Constant, Sea Venture*, and the other small English galleons of the early exploration and settlement period.[14] They had much lower stern castles and their forecastles did not protrude much above the waist—the lowest upper deck, amidships.

Illustrations of merchant ships from the seventeenth century are uncommon. We have the detailed, expert sketches of the Van de Veldes, father and son, in the major maritime museums of the UK and the Netherlands, and they are a singular, priceless resource. We also have the drawings made by mariners, whether sketches or graffiti on walls and stones. Notable among those is the work of Edward Barlow, who started his career afloat as a seaman and ended it as a master. Barlow made a point of drawing every ship he served on, as well as harbors and islands and other features he knew would be useful to other mariners. Figure 3 in Chapter 2 is Barlow's drawing of a ship called *Cadiz Merchant*, which he sketched in 1682. This would have been an ordinary English-built Atlantic merchant ship of the time.[15]

Conventional wisdom has it that later castles were vestiges of the fighting platforms that gave taller ships an edge in combat when ship-to-ship fighting was less about cannon volleys and more about small-arms fire at close range, boarding, and hand-to-hand combat. As that situation reversed itself, so this hypothesis goes, there was selective pressure to reduce castling, due to its weight, cost, and *windage*—vertical area presented to the wind, against which it can push. As with any de-escalating arms race, however, letting this guard down took time and caution.

Usually, however, selective pressures to retain or discard a technological characteristic are multiplex, not simplex. So, even if, to some extent, the decrease in castling was a de-escalating arms race, we should also assume that other pressures existed and that if we are to understand this design aspect, we need to find those, and how they might be tied to each other.

There is no necessity for the stern works to look just as they do on these ships, even if we assume that they did not detract from performance. Note the almost-delicate point to which the stern castle comes on the *Duyfken* replica, pictured in Figure 10 on the following page.

---

14 For an illustration of the basic changes in ships from the Middle Ages to the 19th century from a leading expert, see Harland, *Seamanship*, 36–37.

15 Barlow, *Barlow's journal*, 281. Barlow's original journal is now held by The National Maritime Museum, Greenwich (JOD/4), and it has been partially digitized—see http://images.rmg.co.uk/en/search/do_quick_search.html?q=Edward+Barlow&qw=&srchType=on, accessed 13 December 2019.

FIGURE 10 *Duyfken* replica of *VOC jacht*, Australia
PHOTOGRAPH COURTESY OF THE *DUYFKEN* 1606 REPLICA FOUNDATION

To some extent, this is an aesthetic flourish—among other things, of course. Speaking of aesthetics: Lavery cautions that artists may have at times exaggerated stern heights, citing drawings of both Matthew Baker and the Admiralty, which show lower stern heights than some artistic representations of the period.[16]

When William Baker designed the *Mayflower II* in 1955–1957, he took pains to reduce weight in the scantlings of the stern castle, and he expected significant windage from it. David Thorpe, one of her crew on her transatlantic crossing in 1957, remembers that "the sterncastle acted like a weathervane but only enough for the ship to take the seas forward of amidships such that she lay comfortably."[17] Her master, Alan Villiers, learned in heavy Atlantic storms

16 Lavery, Susan Constant, 13.

17 David Thorpe, "The *Mayflower II* Story—Reflections of a Sailor," Part 6, *The Mayflower Quarterly* 79:3 (September 2013): 287.

that the stern castle and the other characteristics of the ship's hull design allowed her to *lie a-hull* easily. Lying a-hull is stopping at sea by putting the helm over with no sail up, so that the vessel makes minimal forward or sideways progress. It is a way to ride out storms when continuing to sail becomes too dangerous. Most modern sailing vessels will not do that. They must *heave-to*, which requires setting some sail against the rudder. That is potentially dangerous to crew, sails, and rig when the weather is violent.[18] All evidence touching on the subject notes the ease with which one of these ships could lie a-hull in a storm, to ride just like a duck with her head tucked under her wing, as the old simile goes.[19]

Captain William "Chip" Reynolds, master of the *Half Moon*, an operating replica intended to represent Henry Hudson's vessel *Halve Maen* (see Figure 11, p. 118), recounted his first experience of sailing that vessel out of a protected harbor and into high seas.[20] The swells outside were high enough that he feared burying the bow—an anxiety heightened by the height of the *poop deck*—the deck built on top of the stern castle. As he stood on its steep downward slope, he felt almost vertical going down the waves. He was worried about losing the forecastle. The bluff bow, however, was so buoyant that she rose up the other side of the swell without "one splash of water on the deck of the ship." There is no question, as far as Reynolds is concerned, that the bluff bow slows the ship down relative to a sharper one, but "it has such buoyancy that it's nearly impossible to bury that bow in a sea." The speed sacrifice is "immaterial" in comparison, and it would likely only be a speed sacrifice in calm seas when a sharp, and thus less-buoyant, entry would not be at a disadvantage.

In 36 hours, they had "one splash of water" come aboard. With high rails and dry decks, he permitted the crew to move about without jacklines.[21] "It was almost inconceivable to think of getting knocked off."[22] Harland writes that it was "said that the high sterns of the Elizabethan era reflected the fear of the contemporary seaman of taking a sea over the stern, having it rake the decks,

18 Villiers, "How We Sailed the New *Mayflower* to America," 667.

19 Capt. William T. "Chip" Reynolds corroborates that, based on his experiences handling the replica *Half Moon* in frightening weather, including Hurricane Floyd in 1997. Reynolds, Skype interview, 8 January 2015—hereafter Reynolds interview.

20 As of this writing, the *Halve Maen/Half Moon* replica has been transferred from the U.S. to the Netherlands.

21 Jacklines are ropes modern sailors attach to the decks, to which sailors clip their harness tethers so they cannot be washed overboard.

22 Reynolds interview.

FIGURE 11 *Half Moon* replica of *VOC vlieboot Halve Maen*
PHOTOGRAPH BY JOHN W. MANGRUM

and sweep away all in its path."[23] What Reynolds discovered was perhaps one of the reasons why the high stern castle persisted as it did.

Reynolds' anecdote reminds us again that no design aspect exists independently of others, and changing one means the list of pros and cons of the others will change too. The bluff bow provided enough buoyancy to keep the forecastle protected from boarding seas. The stern, too, would need to be buoyant to support the weight of the stern castle, and the hull would need to be stable enough to resist the higher center of gravity and the windage of the protruding upper works. So, moving toward a leaner, sleeker hull would create an imperative to reduce that superstructure. Was there any such trend in hull design in the period? Yes, and it would be worthwhile to experiment with the relationship between that and the reduction of superstructure.

A second obvious—and related—feature of early hulls with higher castles was pronounced *tumblehome*. Tumblehome is the curving-inward of the topsides toward the rail, as opposed to the topsides rising straight up or flaring out. Since both castling and tumblehome decreased over the course of the period in these vessels, we should look for any possible connections between the two.

Ship historians have identified the following benefits of tumblehome:

- the inward-sloping topsides made it easier and safer to come alongside the vessel for boarding or loading and unloading, especially if she were rolling at all;
- the *shrouds* and *backstays*—the standing rigging holding up the masts, attached at the sides and stern of the vessel and to the masts—could be tensioned in a more direct line;
- it was easier for the shipwright to find, and pay for, shorter deck beams;
- it would make it much easier to careen the ship;
- it would reduce wave stress on the hull and decks compared to flared topsides.[24]

To all this we must add the observation that small open boats tend to have flared topsides, especially in the bow, and this helps keep them dry inside. It also increases buoyancy as the flare resists immersion. Open boats, however, do not have decks, and it would seem from the foregoing that tumblehome's advantages have much to do with decks—keeping them strong, keeping them light, and keeping them dry. Small boats do not weigh much—neither do their structural members, nor their cargoes.

---

23 Harland, *Seamanship*, 214.

24 Richard Barker, "Why tumblehome?" *Mariner's Mirror* 84:1 (1998): 95–97, which refers to previous entries in the discussion as well as other works; J.V. Bartlett, "Why tumblehome?", *Mariner's Mirror* 83:4 (1997): 478.

We should also consider the relationship between tumblehome and stability. As the ship heels, or leans laterally in response to the pressure of wind on the sails and waves on the hull, more and more of her side is immersed. Tumblehome will not offer as much resistance to that heel as straight or flared sides would. At normal angles of heel, this was probably not a concern.[25] Tumblehome also keeps the edge of the deck out of the water longer if the ship heels drastically; on the other hand, it decreases *righting moment*—the tendency of the ship to resist heeling and right herself.[26]

Tumblehome could bring heavy guns closer to the centerline. Did that contribute to stability? John Smith—of Virginia fame—seems to have thought so, writing in his *Sea-man's Grammar* that "it is certain this makes her wholsome in the Sea without rowling [rolling], because the weight of her Ordnance doth counterpoise her breadth under water."[27] Ship archaeologist Jonathan Adams, however, refers to this as a fallacy, well-known as such by the eighteenth century. He does point out that tumblehome did have an "indirect benefit, as by lightening the deck structures above the waterline it lowered the centre of gravity and this may have been part of the thinking behind reducing castling."[28] Harland, on the other hand, seems to accept Smith's older assumption, writing that an inward-curving hull's ability to keep the upper deck guns "inside the immersed waterline plane" was more stable than having a wall-sided vessel's guns "'hanging' over the sea, as it were."[29] It would be worthwhile to conduct engineering-laboratory experiments with tumblehome.

Stability, then, has to do both with hull form and weight distribution—of deck structures, guns, cargo, and the hull structure itself. Adequate stability is as important as any design criterion in any vessel, and in a seagoing vessel, stability must take a more directive role in the set of compromises making up the design than it does for vessels engaged in sheltered-water operation. This is especially true of *ultimate stability*—the ability of the hull to recover from being rolled by waves. We know that stability was a problem to be worked out by English shipwrights at the beginning of this period. We know, for example, that it was not uncommon for seventeenth-century English merchant ships to

---

25 Barker, "Why tumblehome?" 95–97.

26 Barker, "Why tumblehome?" 97.

27 John Smith, *The Sea-man's Grammar and Dictionary, Explaining all the difficult Terms in Navigation, and the Practical Navigator and Gunner: In Two Parts* (London: 1691 edition, Google e-book facsimile), 52. Smith refers the reader to a set of principles of ship design, which he endorses, written by Sir Walter Raleigh in 1612. Those may be found in Baker, *The New Mayflower*, 37–38.

28 Adams, *A Maritime Archaeology of Ships*, 179.

29 Harland, *Seamanship*, 44.

be *girdled* after construction to increase stability. *Girdling* (also called *furring*) involved nailing an extra layer or two of planking just above, at, and just below the waterline.

Captain John Narbrough, in command of the flagship *Prince* in the Third Anglo-Dutch War in 1672, commented on her:

> I do believe the ship will bear a good sail, for she stands at her bearing. Girdling the ship would make her one of the finest ships in the whole universe, for it would make her much more floatier and carry her guns higher, and she would bear the better sail and be a better and securer ship to receive shot, and I believe it will not prejudice her sailing.[30]

Smith's *Sea-man's Grammar* mentions the technique as a solution for a "crank-sided ship"—a ship that will bear no sail.[31] William Baker, the naval architect and ship historian who designed *Mayflower II* and other seventeenth-century replicas, found from his research that "during the period in question more English ships were furred than those of any other nation."[32] Mainwaring had said the same thing 300 years or so earlier, but he was not so detached about it:

> I think in all the world there are not so many ships furred as are in England, and it is pity that there is no order taken either for the punishing of those who build such ships or the preventing of it, for it is an infinite loss to the owners and an utter spoiling and disgrace to all ships that are so handled.[33]

Ferreiro says builders did not know why the technique worked, but they knew that it did.[34] One does not have to know why something works to know that it does. Much of what we know about the aerodynamics of sailing, we learned from aviation—we learned well after the fact.

Girdling was a drastic solution to poor stability. The preferred first solution was adding ballast. Recall, however, the note about load waterline. If the ship needed so much ballast that she floated too deep, especially if that meant her point of maximum breadth was below the waterline, that carried negative consequences for stability and performance. Ferreiro believes that

30 R.C. Anderson, ed., *Journals and Narratives of the Third Dutch War, Publications of the Naval Records Society, Vol. LXXXVI* (London: Navy Records Society, 1946), 71.

31 Smith, *Sea-man's Grammar*, 53.

32 Baker, *The New Mayflower*, 6.

33 Mainwaring, *The Seamans Dictionary*, location 2291.

34 Ferreiro, 191.

stability problems were not usually inherent to the design, but due to loading and seamanship issues.[35] It is certainly true that proper loading and ballasting were critical to the ship's stability. The evidence, however, will not allow us to have done with this by citing this generalization. We have to deal with the early seventeenth-century English method of determining the midship bend, and with the possibility that a 1694 tonnage rule enacted by Parliament had a deleterious effect on the stability of later seventeenth- and early eighteenth-century vessels.

To review: the *midship bend* refers to the principal frame of the ship's skeleton—the frame that determines her maximum beam (breadth), and thus the overall shape of the hull—since breadth and length are related by rules of proportion. After the keel was laid, this frame would be the next piece of the ship to be raised and attached. How shipwrights arrived at the shape of the midship bend is a principal interest of period ship archaeology.

The midship bend is a compound curve. No matter what precise shape the shipwright had in mind, the goal was to produce a "fair" curve—with no bumps or angles. The trick to that is to join the smaller, single curves that make up the entire shape in a way that ensures a smooth flow of one into the other. There were basically three ways the shipwright could do that. They were not mutually exclusive.

First, he could simply join together the component pieces of the frame—five would be a typical number—to form the approximate shape and dimensions he had in mind, and then fair (smooth and round) the joint areas with tools. Second, he could use flexible battens—long strips of wood—to define the curves. Neither of these methods required drawing anything or using mathematics or geometry. It was entirely done by hand and by eye, based on experience and rules of thumb. This is how Ab Hoving presents early seventeenth-century Dutch building techniques.

> The Dutch method was ad hoc, executed by men who used old rules of thumb and never used compasses to draw a circle, only to measure (take a foot in a pair of compasses and run it point over point along a long plank to count the feet). There is no real trustworthy quote in Dutch literature even suggesting predefined main frame shapes existed in the Dutch method.[36]

---

35 Ferreiro, 191.

36 Hoving, personal e-mail, forwarded by Paul Fontenoy, 2 April 2015. Also see Hoving, "Early Shipbuilding Literature in the Dutch Republic," *Nautical Research Journal* 58:3 (Autumn 2013): 163–64. Hoving most recently laid out his understanding of Dutch artisanal building

That brings us to the third method—defining the midship bend by drawing arcs of circles with a drawing compass—either full-size on the ground or to scale, on paper. This method allows the translation of prescribed proportions and dimensions to an actual shape before building the frame. The mathematics required to do that are rudimentary, provided one has access to a treatise or at least tables of dimensions and proportions, and tutelage by someone who knows the method.[37] In the literature, and in the ongoing discussion based on it, we refer to this method as the "tangent-arc method," as it relies on arcs of circles and lines tangent to those to define the fair compound curve. There is, however, more than one tangent-arc method, and the resulting shape is determined by which one the designer uses. The debate is over who used this method, who did not use it, and which specific method they used, if they used it at all. That cannot be resolved here, as it is still under investigation, but that investigation is yielding clues. We can consider a summary of what experts know about it, and tie the issue back to that of the stability of seventeenth-century English ships.

We start with a three-sweep tangent-arc system described by Matthew Baker (1530–1613) in the earliest known English treatise on ship design and construction, "Fragments of Ancient English Shipwrightry." The provenance of parts of it is somewhat cloudy, but it is clear that Baker did write much of it, in the late 16th century. William Baker used the earlier Baker's method for drawing the midship bend for *Mayflower II*, as his research indicated that is how it would have been done originally.[38] Indeed, it was long assumed that, in England at least, Matthew Baker's prescriptions for ship design were influential for at least a century after his death.[39] The problem is that either a three- or four-sweep tangent-arc method produces a hull form featuring "a midsection with narrow floors and little initial stability—the arcs will not reconcile except when the floors are very narrow ..."[40] Burningham points out that such a hull, extant today in the replica *Mayflower II*, could not "stand up without ballast and could

---

(in which design and construction are inextricable) in *17th Century Dutch Merchant Ships*, 36–40.

37 Following Donald Johnson's instructions for drawing a four-sweep tangent-arc midship bend, which we will discuss presently, yielded the result it was supposed to, minus proper scale, which was not attempted. It required no mathematics other than simple arithmetic, and it seems to be something that could be taught in under an hour. See Donald Johnson, *Charting the Sea of Darkness: The Four Voyages of Henry Hudson* (Camden, ME: International Marine, 1993).

38 Baker, *The New Mayflower*, 77.

39 Accordingly, they have been influential in the design of every thoughtful replica of an early seventeenth century English ship in our own time—raising issues we will take up.

40 Burningham, "Early Seventeenth-Century Ships," in Bennett, *Sailing Into The Past*, 111.

not safely take ground in drying harbours …"[41] The replica *Mayflower* "… [was] so heavily ballasted that the main deck was at the waterline and the gun ports were too close to the waterline to be opened at sea."[42] While acknowledging that the tangent-arc system as he describes it led to some successful warship designs, he finds "its application to English merchant ships … puzzling."[43]

William Baker's diagrammatic representation of a three-arc method based on Matthew Baker supports Burningham's observation—about fairly narrow floors, at least.[44] Lavery's midship-bend diagram of the current *Susan Constant* replica also depicts a hull with narrow floors, but not shaped the same as *Mayflower II*'s or Burningham's diagram of the Matthew Baker three-arc method.[45] By the time we get to Dassié's *L'Architecture Navale* of 1677, we see fourteen diagrams of midship forms, varying significantly, and those with similar construction lines to Baker's show hulls with much broader floors and reduced height relative to beam.[46] *Deane's Doctrine* also shows a tangent-arc-derived midship bend with broader floors than what Burningham drew.[47] Like Dassié, however, Deane wrote in the 1670s.

Did English merchant ships use Matthew Baker's three-arc system to form their midship bends in the early 1600s? If so, did that produce hulls whose initial stability left something to be desired, leading to girdling and over-ballasting, which in turn compromised other aspects of performance and utility? If so, did shipwrights move away from such a midship bend by the 1670s?

It is clear that one can produce a variety of midship bends using different tangent-arc methods. The tangent-arc method *per se* does not inevitably produce Matthew Baker's midship bend, and thus does not necessarily introduce an inherent instability—if indeed Baker's method does carry that fault with it, which the evidence strongly suggests it did.

The way out of this thicket combines a thorough familiarity with the period manuscripts on ship design and construction, some archaeological remains, the training and experience to analyze and interpret them, and access to a marine engineering computer lab. Jonathan Adams worked on the remains of *Sea*

41 Burningham in Bennett, 111.
42 Burningham in Bennett, 111.
43 Burningham in Bennett, 111. Matthew Baker did not exclusively describe a three-arc method—see Adams, *A Maritime Archaeology of Ships*, 132–36.
44 Baker, *The New Mayflower*, 77.
45 Lavery, Susan Constant, 61. Lavery also used Thomas Harriot's notes on shipbuilding "to the extent possible." See Jacqueline Stedall, "Notes made by Thomas Harriot (1560–1621) on ships and shipbuilding," *Mariner's Mirror* 99:3 (2013): 325–27.
46 Dassié, *L'Architecture Navale*, 78.
47 Lavery, ed., *Deane's Doctrine*, 69.

Venture, the English galleon wrecked on the Bermudan reefs in 1609, and then set out to reconstruct her hull—digitally—from all the available evidence. Then he would be able to analyze that virtual hull.[48]

Adams took data from all the English manuscripts from 1545 to 1670. Because *Sea Venture* was a typical Atlantic wreck—only her bottom timbers survived—the only data from her wreck that Adams had access to were her *flat of floor*—the flat, lowest section of the floor timber, attached to the keel—and the *floor sweep radius*, a measurement of the first curved section of the floor timber, from the outer end of the flat section to the beginning of the first futtock that forms the lower side of the hull. Adams needed to look at the relationship of the wreck's flat of floor and radius of floor sweep and compare that relationship to the relationship between those dimensions prescribed or described in the period manuscripts, and see how close a match he could find. Then, because of the centrality of proportion, he would be able to extrapolate the other important relationships and reconstruct the major lines of the hull.

Adams drew four midship sections to a common scale and superimposed them. Each was based on a different source—two from Baker—one three-arc and one four-arc—one from Wells, c. 1620, and one from Deane, 1670. The chronology shows a "progressive drift in the principal design criteria."[49] It also shows that "the curvature they produce is not dramatically different."[50]

Adams graphs the two dimensions from 1545 to 1670 and shows a fairly linear increase in the flat of floor as a percentage of breadth, and a corresponding, also mostly linear decrease in the radius of floor sweep as a percentage of breadth over the same period. So, judging by the manuscripts, English floors grew wider and thus the percentage of the total breadth made up by floor sweep radius decreased concomitantly. The overall shape of the curvature of the midship bend otherwise remained quite consistent, the most obvious change being from more of a wedge shape early on to a more rounded shape later. We see an increase in that round-shape tendency as we move into the eighteenth century, although Adams points out that the change in the flat of floor to floor sweep radius trend leveled off after Deane (1670).[51]

Locating *Sea Venture* on that time continuum, Adams then extrapolated all the dimensions needed to come up with a hypothetical midship bend of which he could be confident, and then of a complete set of lines for the hull.[52] He

48 Adams, *A Maritime Archaeology of Ships*, 111–51.

49 Adams, *A Maritime Archaeology of Ships*, 140.

50 Adams, *A Maritime Archaeology of Ships*, 140.

51 Adams, *A Maritime Archaeology of Ships*, 141.

52 Adams, *A Maritime Archaeology of Ships*, 146–47.

generated the latter by putting the data into an application called *Shipshape*.[53] The provisional results he reports from the performance analysis are most important for us here. They prove that the virtual model of *Sea Venture* is "very stable." Adams tested the model using different heights for the center of gravity, to represent different lading scenarios, and found that the *righting moment*—the force trying to bring the hull back upright—got stronger as the angle of heel increased. Properly battened-down, he says, the hull "would recover from an angle of heel well past 45 degrees. This indicates a considerable sea-keeping ability for *Sea Venture* and ships of her general form." Adams reminds us that what sank *Sea Venture* was leaking from grounding damage, not instability. She had ridden out a hurricane for four days prior.[54]

Adams used Admiralty Library MSS 9 (c. 1620–1625), anonymous but attributed to John Wells, for *Sea Venture*'s midship bend and thus the basis for her hull lines, because the dimensions it specified worked most closely with *Sea Venture*'s recorded tonnage of 300. So, this model was not based on Matthew Baker and thus cannot help us evaluate the stability of Baker's design specifications. Adams does not fault William Baker's use of Matthew Baker's model for *Mayflower II*. He seems to consider William Baker's reasons sound—for one thing, *Mayflower* was considerably smaller than *Sea Venture*, so it is reasonable that her builder would have used a different set of proportions for her. Eric Speth, when working on the designs for the current *Godspeed* and *Discovery*, found that different proportions were necessary for a smaller ship like those than for a larger one like *Susan Constant*. The smaller vessels would not work as scale models of a larger one.[55]

So, an important task of doing a hypothetical reconstruction like this is being able to decide what manuscript source or sources to use as a basis. That requires being able to compare real data—archaeological, and/or recorded data on the specific vessel—to what is in those manuscripts. Using one source

53 ShipShape "is a program for creating, fairing, and interpolating a set of ship's lines. It has been configured to interface with the existing Wolfson Unit Hydrostatics and Stability suite of programs." The program was developed by the Wolfson Unit for Marine Technology and Industrial Aerodynamics at the University of Southampton, UK, where Adams is the director of the maritime archaeology program. See http://www.wumtia.soton.ac.uk/software/shipshape, accessed 24 Jan. 2019.

54 Adams, *A Maritime Archaeology of Ships*, 145.

55 Eric Speth, telephone interview, 14 August 2015—hereafter Speth interview. Capt. Speth has been Maritime Program Manager, Jamestown-Yorktown Foundation, Virginia, for 27 years. Before that, he was in charge of the replica *Maryland Dove* at St. Mary's City. He has extensive experience on at least nine early seventeenth century replicas and more on replicas representing later periods, including service as master of *Mayflower II*.

will give much different—perhaps drastically different—results than will using another.[56]

It would seem, then—so far—that we have no contradictions here, and that it is reasonable to say, as Burningham did, that ships based on Matthew Baker's 1570s model would have less inherent initial stability than alternatives. Adams is telling us that early seventeenth-century English ship design was moving away from what Burningham criticized about Baker's model—indeed, that it had been moving away from that already when Baker came along. We are discussing significant changes taking place in time spans as short as 20 years; this technology was not static. Further experiments with archaeological remains, like Adams', and with virtual models derived from extant plans, promise to yield more data just as valuable.

The design criteria and problems we have considered so far in this chapter have to do with imperatives other than purely commercial ones, although we cannot truly separate commercial considerations from such issues as stability, as those issues bear heavily on cost and safety at sea. Owners, however, were keenly interested in maximizing their vessels' potential to make profitable voyages, and the design criterion they were thus most interested in was capacity-for-size. Merchants generally wanted the greatest capacity they could get for the size ship they were building or buying. Capacity is generally expressed in tons. The evolution of ship measurement in tons, as it turns out, is directly tied to an evolution in ships themselves, as John McCusker has shown.[57]

Three parties, with some conflicts of interest, were interested in the tonnage of a ship—the owner, the builder, and the tax collector. The owners wanted the most capacity they could get—the most cargo they could fit into the ship—but the owners were paying the builder by the ton, so when it was time to pay the shipwright's bill, they could wish for the vessel's tonnage to be less. The tax collector charged the owners duties based on tonnage, so again, the owners could wish the tonnage to be less when duties were due. There is an analogy here to the valuation of a house. The homeowner wants the tax appraised value on the low side, to minimize property taxes. When it is time to sell, however, the seller wants the private appraisal on the high side, to increase the selling price. Tax appraisal values are lower than private appraisal values and selling prices.

---

56 Adams, *A Maritime Archaeology of Ships*, 143–45.

57 McCusker, "The Tonnage of Ships." Also see McCusker, "Colonial Tonnage Measurement: Five Philadelphia Merchant Ships as a Sample," *Journal of Economic History* 27:1 (March 1967): 82–91.

Cargo tonnage means how many tons of cargo the ship could actually carry.[58] *Measured tonnage*—also called "shipwright's" or "carpenter's" tonnage—was an actual measurement taken during construction which the owner had to agree to, and could verify if he wanted. The builder's charges were based on this figure.[59] *Registered tonnage* was legal tonnage—the figure written on all official documents pertaining to the vessel. Taxes and duties were based on that figure.[60]

In the seventeenth century, the formula for calculating cargo tonnage used length, breadth, and depth, but did not take into account *block coefficient*. It was noted earlier that the shape providing the greatest cargo capacity for a given length, breadth, and depth is a rectangular box—a block—but a block will not sail. A vessel designed for speed will have a lower block coefficient than one designed more for capacity. Everyone knew, of course, that a ship did not actually carry what she would carry if she were a box defined by her length, breadth, and depth. Everyone knew that *cargo tonnage* was less than *measured tonnage*, as calculated by the old formula. So, the long-established custom was to discount the measured tonnage for taxes and duties—in other words, basing *registered tonnage* on *cargo tonnage* rather than *measured tonnage*. This reminds us that it is just as important to investigate the extent to which a regulation was adhered to as it is to know what the regulation was and when it was ostensibly in force.

As the years went by, however, the general trend was for block coefficients to increase—for cargo ships to carry more cargo for their dimensions, as recognized by Parliament before 1700, when, realizing that this was costing the state revenue, it disallowed the discount. In practice, this was ignored by owners and builders and overlooked by officials, as so often happened with regulation of British maritime commercial affairs. If it had not been ignored, according to McCusker, taxes and duties would have increased by up to 50%, a crippling burden. By 1700, *cargo tonnage* was equal to *measured tonnage*, and by the mid-eighteenth century, *cargo tonnage* was greater than *measured tonnage*—which meant that it was also greater than registered tonnage, and that was no secret at the time.[61] This means that hull design had changed in a way that increased capacity-for-size. It strongly suggests an increasing trend toward bluffer and fuller hulls. This was doubtless a significant contributing factor

58 McCusker, *Essays*, 26. It does not equate to "tons burthen/burden" as that term was "often applied in situations where the terms were only a legal fiction".

59 McCusker, *Essays*, 26–27.

60 McCusker, *Essays*, 26–27.

61 McCusker, *Essays*, 28–29.

to the productivity of the merchant shipping fleet in the British Atlantic. If merchants could ship more goods in the same size ship, without drastic compromises in seaworthiness or performance, they did not necessarily need a bigger ship with its increased costs in construction, manning, and maintenance. They could respond to population and market growth more efficiently, and merchant ship technology could thus keep up with the changing demands of the world it served at the least possible cost.

The evidence strongly suggests that a major contribution to increasing capacity-for-size in British Atlantic—and French Atlantic—ships in the eighteenth century was the increasing influence of Dutch design. In 1989, archaeologist David Moore performed a conjectural reconstruction of the slaver *Henrietta Marie*, wrecked on a reef in the Tortugas in 1699. Moore used the treatises, descriptions and—in one case—extant plans of similar vessels, contemporary prints, and what we know about common differences between English-built and French-built ships at the time to determine that the vessel was frigate-built in France and was relatively fast. He notes that all three of the main treatises he used give very close measurements for *sternpost rake* and length, *stem rake*, and *transom* width. He concludes that French design was incorporating Dutch characteristics by the time *Henrietta Marie* was built, and speculates that such a vessel had become "a 'frigate'-built' ship with Dutch 'fluyt'-design influences."[62] We get a clear sense from Moore's study of the interplay between technological inputs from rival powers in the Atlantic World. Those powers did have some distinctive tendencies and traditions, but they inevitably blended as ships were captured and studied, foreign agents visited their rivals' dockyards, and shipwrights emigrated.

Moore only had the usual Atlantic wreckage to analyze. As noted, hulls that were deliberately buried stand a much better chance of survival, and one of those was discovered during excavations for a new Manhattan skyscraper in 1982. Warren Riess was eventually able to determine that the remains almost certainly belonged to *Princess Carolina*, built at Charles Town (Charleston), South Carolina, in 1717 for local owners.[63] Riess was able to match the remains to a design technique involving simple circles and proportions that do not match those specified in the treatises for naval vessels and other ships deemed important at the time. He was able to determine that the builder had used the

62 David Moore, "Anatomy of a 17th Century Slave Ship: Historical and Archaeological Investigations of the *Henrietta Marie*" (Master's thesis, East Carolina University, 1989), 140–42. "This is only an observation based entirely on preliminary investigation, however," he notes, "and will require additional research to verify."

63 See Riess with Smith, *The Ship That Held Up Wall Street*.

simplest of fractions, based on the ship's beam of 24 feet (7.32 m). He could match the ship's dimensions to simple straight lines and arcs with radii of 4, 8, 12, and 16 feet (1.22, 2.44, 3.66, and 4.88 m).[64] It was clear that this South Carolina shipwright had used much simpler fractions in his design than those set out in any of the contemporary treatises.

That is not to suggest that his ship was crude. On the contrary, Riess found distinguishing details of construction that struck him and others as highly skilled craftsmanship—and certainly labor-intensive. Most of the hull was re-buried on-site, and a skyscraper built over it, but the bow section, at least, was preserved for study, and may be appreciated as an example of successful construction outside "the published rules."[65] Her bow did not use *cant frames*, as would have been expected, but *square frames*—an older form of construction, so far as we know.[66] To fit the planking properly over these frames, the shipwright had to bevel and shape them very carefully, with great skill. Riess calls them "extraordinary pieces."[67] Cant frames absorb wave shock better than square frames. A bow built with the latter requires more frames, *breast hooks*, *chocks*, and *hawse pieces* between the frames, such that the bow becomes almost solid. This is heavier and uses more wood for the same strength. It is a late example of the use of square frames in the bow.[68] This varies from eighteenth-century warship construction, but supports Lavery's idea that merchant ship techniques tended to be more conservative.[69] "It is not clear whether this was because of strictly traditional forces or because the availability of greater amounts of timber in America made square frames more practical for this

---

64 Riess with Smith, 46. This technique has become known as "reverse naval architecture"—using adequately-extant remains, the investigator determines the proportions and measurements used by the shipwright in the design. Kelby Rose also employed computer-aided "reverse naval architecture" in his analysis of the completely-intact seventeenth-century Swedish warship *Vasa*'s design, overcoming the unusual problem of how to virtually "take apart" an intact vessel for the kind of analysis commonly performed on disarticulated remains. See Kelby James Rose, "The Naval Architecture of *Vasa*, A Seventeenth-Century Swedish Warship" (PhD dissertation, Texas A&M University, 2014). Rose hopes to employ the application he used on *Vasa* to understand other partially intact wrecks.

65 Riess with Smith, 38.

66 Cant frames radiate toward the stem at decreasing angles, as opposed to square frames, which remain perpendicular to the keel and stem, getting progressively smaller to form the shape of the bow. Their outside edges are beveled to allow for the curve of the bow planking.

67 Riess with Smith, 34–35.

68 Riess with Smith, 42.

69 Lavery, *The Arming and Fitting of English Ships of War 1600–1815* (Annapolis, MD: Naval Institute Press, 1989), 7.

ship," Riess writes.[70] As discussed in the last chapter, both the availability of materials and the predilections of individual shipwrights had much to do with the variety of finished vessels we actually find in the mud. *Princess Carolina* exhibits great care in her construction and suggests no cost-cutting shortcuts, although not every piece used for a structural timber was perfect, nor was the frame spacing as uniform as one might expect from studying naval vessels and plans. There is strong evidence, however, that the ship made more than one transoceanic voyage, and that the builder had been "resourceful in using what curved timbers he could obtain and shifting their position a bit when necessary to build a properly shaped, strong ship."[71]

Considering as a whole the extant archaeological literature on British American shipwrecks, it is clear that construction was accomplished by different builders using a variety of techniques and materials to achieve roughly the same end.[72] This lends credence to Burningham's notion that "[s]omeone could decide that ships ought to be designed in some theoretical way. It would

70 Riess with Smith, 46.

71 Riess with Smith, 37.

72 In addition to VanHorn, and Morris et al., see: Adams, "*Sea Venture*: A second interim report—part 1;" Gregory D. Cook, "The Readers Point Vessel: Hull Analysis of an Eighteenth-Century Merchant Sloop Excavated in St. Ann's Bay, Jamaica" (Master's thesis, Texas A&M University, 1997); Amanda M. Evans, "Defining Jamaica Sloops: A Preliminary Model for Identifying an Abstract Concept," *Journal of Maritime Archaeology* 2:2 (December 2007): 83–92; Fred Hocker, "The Clydesdale Plantation Vessel: Savannah River, South Carolina," in *Beneath the Seven Seas*, edited by George F. Bass (New York: Thames & Hudson, 2005), 192–93; Michael Krivor, "Archaeological Investigation of an Eighteenth-Century British Merchant Vessel in Chub Heads Cut, Bermuda" (Master's thesis, East Carolina University, 1998); Brad Loewen, "Ships and Shipbuilding / *Navires et Construction Navale:* Recent Advances in Ship History and Archaeology, 1450–1650: Hull Design, Regional Typologies and Wood Studies," *Material Culture Review/Revue de la culture matérielle* 48 (Fall/Automne 1998), available at : http://journals.hil.unb.ca/index.php/MCR/article/view/17791/22170, accessed 10 February 2015; David Moore and Corey Malcom, "Seventeenth-Century Vehicle of the Middle Passage: Archaeological and Historical Investigations on the *Henrietta Marie* Shipwreck Site," *International Journal of Historical Archaeology* 12:1 (January 2008): 20–38; Moore, "Anatomy of a 17th Century Slave Ship;" Harry Pecorelli, "Archaeological Investigation of the B & B Wreck (38B1672): An Eighteenth Century Plantation-Built Vessel, Charleston, South Carolina" (Master's thesis, East Carolina University, 2003); Christopher R. Sabick, "His Majesty's Hired Transport Schooner *Nancy*" (Master's thesis, Texas A&M University, 2004); James C.W. Southerly, "Cedar on the Reef: Archaeological and Historical Assessments of the Eighteenth Century Bermuda Sloop, Exemplified by the Hunter Galley" (Master's thesis, East Carolina University, 2003); Allan J. Wingood, "*Sea Venture*: An interim report on an early seventeenth century shipwreck lost in 1609," *International Journal of Nautical Archaeology* 11:4 (November 1982): 333–47; and "*Sea Venture* Second interim report—part 2: the artefacts," *International Journal of Nautical Archaeology* 15:2 (May 1986): 149–59.

make no difference to the many shipyards scattered around the Netherlands on muddy shores and river banks."[73] It may be that, with no top-down control over the builders of such vessels, they were freer to improvise than those working in larger, more hierarchical yards or, certainly, in naval yards. Burningham posits an important plausible connection to a wider seventeenth-century Dutch cultural and socioeconomic context when he opines that "the Dutch, because of their less hierarchical society and artisanal ship design could adapt most quickly and successfully."[74] When asked if we might apply his comment to the North American Eastern Seaboard, Burningham said "Yes, in truth we have little idea how the great majority of ships were designed."[75] A group of archaeologists proposed a tentative theory of framing evolution in 1995, based on available evidence, but VanHorn studied the framing patterns of six period British American wrecks, and was unable to discern any clear uniformity therein.[76]

H.H. Holly, writing in 1969 on the remains of the *Sparrow-Hawk*, wrote of the treatises that, while they give us valuable information about the construction of naval vessels, we have no reason to think that the shipwrights building the craft we are considering "adhered closely, if at all, to these ideas. Then, as now, small vessels were built in a large number of small construction yards scattered all along the seaboard. The various builders naturally developed their own ideas ..."[77] The archaeological evidence uncovered and interpreted so far tells us that variation, not uniformity, was the rule for ordinary merchant ships. At first glance, that may not be apparent, but examining how different shipwrights put together their vessels makes it clear. VanHorn compared the keels of the six wrecks in her study to the prescriptions for keels in the Mungo Murray treatise, one of the best-known of the eighteenth-century English treatises, to see if any of them matched his prescriptions. None did.[78]

---

73 Burningham, personal communication, 6 April 2015—hereafter Burningham comm. 4/15.

74 Burningham, personal e-mail, 5 May 2015—hereafter Burningham corr. 5/15.

75 Burningham comm. 4/15.

76 John W. Morris, Gordon P. Watts Jr., and Marianne Franklin, "The Comparative Analysis of 18th-Century Vessel Remains in the Archaeological Record: A Synthesized Theory of Framing Evolution," *Underwater Archaeology Proceedings from the Society for Historical Archaeology Conference 1995*, Society for Historical Archaeology; and VanHorn. VanHorn considered all of the extant North American (and Caribbean) sites and concluded that the evidence could not support a theory of framing evolution such as Morris et al. were claiming, at least for colonial American vessels. More data are needed, and the wrecks to provide that data are still being discovered.

77 H.H. Holly, "Sparrow-Hawk: A Seventeenth-Century Vessel in Twentieth-Century America," *Pilgrim Society Booklet Series* (Boston: Nimrod Press, 1969), 14–16.

78 VanHorn, 178.

Idiosyncrasies aside, however, designs do fit within identifiable general parameters, such as the proportional relationships we have considered. Riess compared the basic shape of *Princess Carolina* with those presented in period works—particularly Chapman's *Architectura*, whose plates are famous for their unique and precise detail. By superimposing the profile of the general cargo area, analogous to the midship bend, of the *Carolina*, Chapman's profile of an English merchant frigate, and Chapman's profile of a Dutch flyboat (*vlieboot*), a common contemporaneous merchantman, Riess clearly shows that the *Princess Carolina* is a combination—whether deliberate or not—of the two forms—just as Moore concluded the *Henrietta Marie* probably was. Considering Riess' comments on the ship's design, we can appreciate how significant it is to have this depth of hard evidence on an ordinary Atlantic merchantman of 1717.[79]

The stem profile, or side-on view of the curve of the bow, was a "true arc of 16-foot [(4.88 m)] radius" at the notch of the main structural member, the stem, where the plank ends were joined to it. The bluntness of this traditional bluff bow "did not allow for a fast ship …" "The stern … was a familiar square tucked stern of an English merchant frigate …" providing "a long keel structure for speed and sailing ability." Riess then makes his most important claim—that the design of this ship "may answer Ralph Davis's question, raised in *The Rise of the English Shipping Industry*, about how British ships became more efficient in the eighteenth century," which I discussed in the Introduction. Davis surmised that British shipwrights studied Dutch and French ships captured in the War of Spanish Succession (1702–13). He also suggested—though he cautioned that this was conjectural—that "the main technical development in English shipbuilding of the early eighteenth century was the adoption … of the hull forms used earlier by the Dutch which made possible a high carrying capacity in relation to the ship's main measurements."[80]

Riess suggests that British merchants and shipowners were aware of the general advantages and disadvantages of frigate-built English ships and Dutch flyboats, and that a successful hybrid would, like the frigate, deliver satisfactory

79 As was typical of such terms, "flyboat" was used loosely.

80 Davis (2012 ed.), 62–63. Davis is writing here of the increase in shipbuilding in north-east England at this time, and the frequent mention of "pinks" in connection with that output. As noted, the word "pink" is related to the word "pinch"; it suggests a stern that narrows as it rises. Although it was a common term, it is vague, like most nautical terms of the time, and could apply to vessels with varying hull forms. So, while flyboats typically had pink sterns, so did other vessels, and Davis freely acknowledges he is on speculative ground here. This is a perfect example of why he was so keen to have nautical archaeology pursue these questions.

speed, agility, and self-defense; while delivering more cargo into shallower areas, like the Dutch craft. By fitting a longer keel, the builder could somewhat counter the poorer windward ability of the flat bottom, as the keel provides both tracking ability (the ability to go straight) and resists leeway (recall the knife-blade analogy). That flat bottom, with its lesser tendency to heel, also required less ballast.[81] Recall that *Bristol Merchant* was ordered with rather flat floors, and that the owners wanted her "to ship in wild roads and with as little ballast as possible…."

The Carolina coast is a perennially shifting maze of sandbars and shallow inlets. Although Charleston is the best and most important regional harbor, and thus the site of the most important port city of the region at the time, the bar at its mouth restricted the draft of ships that could use it. For any other port in the area, such depth restrictions were more severe. So, there certainly would have been good reason to adopt design and construction techniques that allowed for shallower draft while retaining capacity and seakeeping ability—trademarks of Dutch merchant ships, as the Dutch coast is similar.

Can the other noted advantage of Dutch cargo ships Riess notes here—capacity for size—support McCusker's argument that block coefficients increased in British Atlantic merchant ships into the eighteenth century—and support Ralph Davis' conclusion that merchant ships became markedly more efficient as the eighteenth century wore on? Data derived from future archaeological analysis and digital modeling should be compared to the conclusions of Moore and Riess. There are cold-water wrecks around Newfoundland that have not been fully-studied, and four eighteenth-century wrecks have been found buried in the mud on the waterfront in Alexandria, Virginia since 2017. The prospects for gleaning more significant data in the near future are promising.

It is already clear, however, that the merchant ship hull tended to grow fuller in the eighteenth century, to carry more cargo for its size. We need to examine the implications of this trend on the hydrodynamics of those hulls, in order to understand the shifting of compromises accompanying this important development.

We are already familiar with factors selecting for the bluff bow—buoyancy, capacity, the fight against hogging—to which we can add that room in the bow provides for the stowage of long, heavy anchor cable and a number of anchors.[82] Does it knife through the water like a yacht's? No, but these were not

---

81 Riess with Smith, 39–41. They cite Davis, 1972 edition, 286–88. We should note, however, that, while flatter bottoms can provide greater *initial stability*, they generally provide somewhat poorer *ultimate stability*.

82 A point Reynolds made (Reynolds interview).

yachts.[83] Laying all other considerations aside, we are always left with speed versus capacity.

We know that mariners and shipwrights of the time paid proper attention to the flow of water by and under the hull. Most contemporary sources mention the importance of getting a smooth flow of water to the rudder, and not creating strong eddies—what they called "dead water"—behind the stern.[84] Some English ships from the beginning of our period—such as *Mayflower* and *Susan Constant*—seem to give the lie to that, however, as they carried their flat transoms to the waterline, presenting sharp angles to the water and creating just such turbulence at the stern. Much more common later in the period was an upswept run aft, intended to provide fair flow and clean water for the rudder, as seen in Figure 12.

Opinions varied on specific aspects of hull design vis-à-vis resistance. One chief aspect of design, however, was consistent throughout the period, and any study of ship design needs to ask why. Recall that Winne & Hawksworth specified that their *Bristol Merchant* "carry her breadth well forward." They are specifying part of the fish-form hull—the "cod's head and mackerel's tail," as the old expression put it. Fish-form hulls have disappeared from commercial shipping, but we still see them in submarines, torpedoes, and modern touring kayaks. The idea that some basic principle of hydrodynamics renders the fish-form hull a nautical blunder must be rejected. We must assume, as we do with any other aspect of ship design, that it offers pros and cons relative to alternatives, and then set about trying to determine what those are.

Summing up the concept of the form reads something like this. The bluff bow provides buoyancy. It rides up on the waves rather than burying itself, and it pushes the water aside. The rest of the underwater hull then allows the water to flow cleanly back to the rudder. Tapering the hull aft encourages the water's swift travel.

It is not necessary to leave the period to find critics of the feature—it had its detractors at the time, especially late, when a restless spirit swirled around elite British maritime circles. Naval constructor Marmaduke Stalkartt wrote a treatise in 1781, dedicating it (with permission) to His Majesty, with the hope that it could advance the interests of the Empire (and his own, of course).

83 Jim Graczyk, Drew McMullen, and Joakim Severinson all made comments indicating that a modern sailor on a period replica will notice how differently the bluff bow behaves in the water than a modern yacht's bow (Jim Graczyk, personal communication, 6 November 2014—hereafter Graczyk corr.; Drew McMullen, telephone interview, 8 January 2015; Joakim Severinson, written interview, 19 March 2015).

84 For example, see Stalkartt, 2: "Then fix the lower part of the transom clear of the load draught of water, that the boat may have no dead water to drag after her."

FIGURE 12 Upswept after section, showing the run of the hull to the stern
PHOTOGRAPH BY THE AUTHOR

Stalkartt set out from the beginning to defend the bluff bow, the placing of the maximum beam forward of amidships—the traditional shape.[85] Ten years later, however, a contributor to the Society for the Improvement of Naval

85 Stalkartt, i–ii, 30–32.

Architecture's proceedings cited French experiments that, he claimed, proved that bow form was immaterial to resistance.[86] In a letter of 26 June 1791 to the Society for the Improvement of Naval Architecture, Charles Gore claimed to have done experiments proving that the fish-shape theory was wrong, and advocating further experiments to corroborate that. If Gore is the source of the "Observations on the resistance of fluids ..." that follows the letter, which is unclear, although the content is consistent, he described model experiments showing that the solid of least resistance had its maximum breadth amidships, and a sharp entry, and the fish shape did not do as well.[87] Earlier in this collection of letters and essays, we find an even stronger dismissal:

> The idea that ships ought immediately to taper or become narrower from the midship bend or frame, that the closing of the water behind them may push them forward, is a vulgar error; that the shape of fishes ought to be copied in ships, is another; as the analogy does not hold good. Yet those two ideas have occasioned great blunders in naval architecture.[88]

It is worth pointing out, however, that one need not subscribe to the idea of a "push forward" to support the feature of tapering-aft, and whoever wrote this does not explain why copying the fish form is another "vulgar error."

Fluid resistance theory was still an elusive quarry in this period. What is worth investigating is whether well-designed ships made some sacrifice in resistance inherent to the fish-form hull. If so, how much? We have noted—and will examine further evidence for—how unlikely it was for small gains in a ship's speed through the water to make a practical difference in her passage-making. Were the advantages ascribed to the tapering-aft from amidships real? If we find through experiments that there is little or no resistance penalty for this form, that could support the notion that long experience had worked. If

86 Society for the Improvement of Naval Architecture, *Collection of Papers on Naval Architecture* (London: The European Magazine, 1791–1793), Part I, Paper V, 25, in the collection of the Library of the American Philosophical Society. The author of this paper also claimed that the same experiments showed no difference in performance for straight versus curved waterlines—something else Stalkartt had a definite opinion on (see Stalkartt, 9).

87 "Some account of the institution, plan, and present state of the society, of the Society for the Improvement of Naval Architecture: together with the premiums offered by the Society, list of subscribers, rules and orders, of the Society: and 3 original letters, addressed to the committee, from Sir Charles Middleton, Bart. and Messrs. Gore and Patton, on subjects of naval architecture" (London: Society for the Improvement of Naval Architecture, 1792), 17–18, in the collection of the Library of the American Philosophical Society.

88 "Some account ...", 14.

we find that there was a resistance penalty, we will be required to weigh that against other pros and cons of the fish-form hull.

Earlier in this chapter, I tied the importance of proportion to the ontology of the seventeenth century. As the age of Kepler gave way to the age of Newton, however, did the centrality of proportion begin to give way as well? If so, to what—and did that have something to do with Newton's ideas? We do know, for what it is worth, that Newton himself was interested in maritime technology. His invention of a *sextant* is fairly well known. Much less so is the discovery of a shipbuilding treatise he copied in his own hand—that, we did not find until 1994.[89] That was an important find, aside from the association with probably the most important scientific mind of the age, for it contained tables of proportions specifically labeled for merchantmen. We learned from it that the design interpretations of William Baker and Stan Potter, Brian Lavery and Eric Speth, executed in the replicas of seventeenth-century ships operating on the U.S. east coast, fit within the parameters of proportion provided by those tables, even though none of those experts had access to this information when they did their work.[90] So, we have more reason now to trust their insights, and whatever we can learn from studying the designs they produced.

Newton, however, did not write that treatise—he just copied it, around 1700. His profound innovations in scientific thinking only made themselves widely felt in the eighteenth century. Did they make themselves felt in ship design?

Naval architecture as we practice it was born during this period, but that is about the extent of it. The great European scientific minds identified the principal mathematical problems of the discipline, and worked hard to solve them. Probably the most important success in that endeavor in our period was the calculation of the *metacenter* by Pierre Bouguer.[91] The natural philosophers pushed their understanding of fluid dynamics well beyond where it had been, but the application of their insights to the design and construction of ships in their own time was minimal: "Ultimately," concluded Ferreiro, "the invention

---

89 Barker, "A Manuscript on Shipbuilding, circa 1600, Copied by Newton."

90 Potter was the naval architect for the current *Susan Constant*, among other period replicas. See Wrike in Bennett, *Sailing Into the Past*, 120–33.

91 Ferreiro, *Ships and Science*. Bouguer is the central character in that book. The metacenter, "in fluid mechanics, [is] the theoretical point at which an imaginary vertical line passing through the centre of buoyancy and centre of gravity intersects the imaginary vertical line through a new centre of buoyancy created when the body is displaced, or tipped, in the water, however little" (http://www.britannica.com/science/metacentre, accessed 24 Jan. 2019). Now, the concept is central to predicting stability.

of the metacenter did not have any practical effect on ship design during the 1700s."[92]

So, again, what of Newton? As it happened, Newton came up with what he calculated to be a form of least resistance for ships, a problem on which the treatises focused much attention.[93] Newton thought he had come up with a bow shape of least resistance, which the French navy partially adopted.[94] This experiment gave rise to the assumption, both within France and amongst her rivals, that France was the leader in using theoretical hydrodynamics to improve ship design.[95] The Royal Navy, too, took Newton's concept seriously enough to experiment with it, but in 1778, Frederick II, King of Prussia, wrote a letter to Voltaire remarking that the "English have built ships with the most advantageous section in Newton's opinion, but their admirals have assured me that the ships did not sail nearly so well as those built according to the rules of experience." A similar account was repeated by John Charnock in his history of marine architecture.[96]

Using today's naval architecture, Ferreiro demonstrates how Newton got it wrong, corroborating mathematically the empirical observations of the English admirals. He also points out that hull design was less important in determining a ship's performance in this period than the set-up and condition of the rig, the cleanliness or foulness of the bottom, and—most important—the skill of master and crew.[97]

It may be surprising to learn that such a singular genius as Newton could get something wrong that probably seems more pedestrian a matter to us than, say, correctly figuring out why the moon rotates around the earth, what light is made of, and the basic laws of bodies in motion. No other example could more plainly point out something of central importance to all studies of this subject: it is devilishly difficult stuff. Not even Newton could do what would take an unknowable number of mostly unknown people a couple of centuries to do, and only a serious shift in technological need—a need that did not exist in this period—would provide the impetus for the incorporation of the new naval architecture into shipbuilding. They would be building new kinds of

---

92 Ferreiro, 249.

93 See for example Euler, *A compleat theory of the construction and properties of vessels*, 184–85; Henri-Louis Duhamel Du Monceau, a translation of whose 1765 treatise is included in Mungo Murray's, *Treatise on Shipbuilding*, 57; Stalkartt, 9.

94 Their prominent constructor Blaise Ollivier, though, thought little of the idea (Ollivier, 183–84).

95 Ferreiro, 143.

96 Ferreiro, 139. Charnock's work was published at the end of the period.

97 Ferreiro, 143.

ships then, and without a long record of successful precedent to base those on, they would need something else—something that would help them predict what would work and what would not. A major advantage—perhaps *the* major advantage—of today's technical-analysis capability over a period shipwright's is predictive power. Naval architects can predict performance characteristics without building anything. The greater the financial and human risk undertaken in the building of that something—the bigger the ship, the more expensive she is to fit out, the more people she carries—the more compelling such predictive ability becomes. So, it should not surprise us that the hardest push for "naval architecture" was from naval quarters.

There is a sociocultural dimension to the struggle of naval architecture to make itself relevant in this period. As Mokyr notes, "the Enlightenment was an elite movement, and ... much of its thinking served class interests."[98] Ferreiro titled his first chapter "'Mere Carpenters,'" introducing the deep prejudice held by some of the educated elite against artisanal shipwrights. The title comes from a translation of Spanish constructor Jorge Juan y Santacilia's 1783 *Examen maritimo*, in which he claims that "[t]oward the end of the last century ... [t]he Construction of Vessels was abandoned to mere Carpenters; and it was not considered that NAVAL ARCHITECTURE was based on a constant application of Mechanics and Geometry ..."[99] Ferreiro notes that the French constructor Pierre Lévêque, who loosely translated Santacilia's treatise, "added the notion that constructors were incapable of applying theory to design" and, more broadly, that "[m]any contemporary observers of the field also held the opinion that constructors who built ships without having a theoretical basis for their designs were somehow inferior to other naval professionals."[100]

In England, this later eighteenth-century frustration with traditional shipwrightry was also deeply rooted in fear of the French. Victory in the Seven Years War gave the British psyche no respite from anxiety about its chief imperial-naval rival. Those who worried most about the French maritime threat surely, if grimly, felt vindicated by the disastrous War of American Independence, and of course after the French Revolution, the British Empire poured all the resources it could into the Royal Navy, which would find itself fully deployed for most of a 25-year period, with the very survival of the Empire on its shoulders.[101]

98 Mokyr, *Enlightened Economy*, 39.

99 Ferreiro, 23.

100 Ferreiro, 23.

101 The naval establishment wanted control of construction as well as design, and there was a deep mistrust of merchant builders contracting for the Royal Navy. See Roger Knight, "Devil Bolts and deception? Wartime naval shipbuilding in private shipyards, 1739–1815," *Journal for Maritime Research* 5:1 (April 2003): 34–51.

Yes, but why would the British naval establishment fret so much about a French maritime threat when British forces were usually victorious at sea? Ferreiro suggests that British shipwrights would have shared that puzzlement, given the overwhelming success of their warships against their enemies—Royal Navy vessels captured "five times as many enemy ships as its nearest rival, France—all without the benefit of the metacenter or the calculus." So, he speculates that a British ship constructor's response to "the new theoretical developments coming from France, or on the system of rigorous technical schooling for French and Danish constructors" might well have been "So what?"[102] Here we have that cultural divide between those attracted to the promises, however vague, of the Enlightenment and those who did not feel the need to be Enlightened to do a proper job. The "Baconian program" of the Enlightenment assumed that, without "natural philosophers," an economy dominated by artisans such as shipwrights would "revert to a technologically stationary state."[103] Most shipwrights of the British Empire, even those like David Lowell who clearly had no aversion to learning from books, would likely have taken exception to that assumption, and the evidence we have, indicative as it is of adaptability, would support that exception.

There is an instructive irony here: The French—and the Danes, and whoever else was open enough to the tenuous promises of theory-based naval innovation to employ it in their yards—were looking for any leg up that might narrow the gap between them and the British.[104] None of them could hope to match British numbers and sheer force. Perhaps technical advantage would help compensate for that. This is the military-technological thinking of the underdog—of the Confederates with their ironclads and submarine, of Hitler with his wonder-weapons. The irony turns on British paranoia. The supreme maritime power of the day, Britain, feared for that supremacy, paying close and anxious attention to any threats, real or imagined. The British did not need the bow of least resistance or the metacenter or natural philosophers in their dockyards. They saw the French using these things, however, and it only exacerbated their fear of being overtaken technologically by their archenemy. So, they took comfort in embracing the promise of scientific shipbuilding as a matter of what we would call national security—a promise that would, as Ferreiro points out, prove chimerical.

---

102 Ferreiro, 296.

103 Mokyr, *Enlightened Economy*, 61.

104 It is also ironic that the early seventeenth century Dutch merchant ships were built using the most thoroughgoing artisanal methods—the opposite of what the naval architects were about—and yet those vessels were an integral part of a commercial competition with England so successful that it helped set off the Anglo-Dutch Wars. We have discussed Dutch vs. English merchant ship design in that early period.

That the fear was real, regardless of what else was, is made amply clear by the formation and publications of the Society for the Improvement of Naval Architecture in 1791, as the threat of Revolutionary France loomed dark and low. Contributors wasted no time in reminding subscribers of the long-held British fear that French ships were faster than their own. In the Preface to the Society's proceedings for that year, reprinted in *European Magazine*, the author recalls a discussion in a British shop during the Seven Years War, when it was going badly for Britain, lamenting the lack of "Science" in British shipbuilding and contrasting it with the French who had schools and institutes to study it.[105] Four pages later, he writes:

> The candid Ship-builder will readily confess there is not one improvement in our Navy, that did not originate with the French … every officer in the service coveted the command of [captured and refitted French ships] in preference to those built in England.[106]

Ferreiro might suggest that this author would have had some trouble finding a "candid Ship-Builder" who would express such an opinion. In the concluding essay, however, we have it again: "… the French, actually surpassing us in this most important art, have derived many advantages from this superiority in time of war."[107] Ferreiro states flatly that there is no evidence that French ships were inherently faster than British ships.[108] The perception was real, however, and the perception was strong. The rest of the papers and letters in this collection, ranging from thoughtful commentary on technical matters by experienced captains, to an essay by no less a person than Benjamin Franklin, to short rants by disgruntled amateurs, take British ship design and construction to task, questioning, and in some cases railing against, the most basic elements of period ship design—the bluff bow and tapering stern, or "cod's head and mackerel's tail", Hutchinson's dicta on the prevention of hogging, and revisiting bow form and resistance. This was the "restless spirit" in the British naval establishment mentioned earlier—raised by the doings across the Channel, to which all of Britain reacted with growing alarm—and this was, to be sure, the naval establishment—the head of the Society was the Duke of Clarence, naval officer and the future William IV, and the roster of members was title-heavy.[109]

105 "A Collection of Papers on Naval Architecture," Part I, iv.
106 "A Collection of Papers," Part I, viii.
107 "A Collection of Papers," Part I, 64.
108 Ferreiro, 177.
109 This is also the "restless spirit" that would contribute to Britain's "Industrial Revolution," which was beginning during this period. A willingness to question old ways and

At this point, the reader may be wondering why we have ventured so far into Whitehall and the naval dockyards and away from the New England shipwright who might have built Winne & Hawksworth's brig—and his counterparts in, say, the northeast of England. This "naval architecture" subject, though—and the written record of debate and discourse on ship design in this period—is so heavily naval in its orientation that such a side trip was unavoidable if we were to familiarize ourselves with it. To understand the cultural context of ship design and construction at the time, and to place the published primary sources in proper context, we needed to have a look at what the literate were thinking and writing about the subject. This material is the fount of the view that artisanal shipwrightry was a reactionary, hidebound traditionalist's petty fiefdom—that this technology, so vital to the survival and prosperity of the Empire, was dangerously constipated, and that leaving it in the hands of 'mere Carpenters' would not do. The previous chapter claimed that late eighteenth-century British propaganda would have us believe that here was a technology in dire need of rescue by modern, enlightened, educated men with the wider interests of the state at heart. This is the propaganda referred to—these articles were published—the Society saw to that. Ordinary shipwrights were not publishing articles in *European Magazine*.

However, as N.A.M. Rodger points out, once the Society's suggested "improvements" proved ineffectual, it quickly and quietly faded away.[110] Ordinary merchant ships were still designed and built by "mere carpenters," by the ordinary shipwrights of the British Atlantic, despite the nascence of scientific naval architecture and its prominence in the literature. The unknown New England shipwright who may have built and launched *Bristol Merchant* in 1733 or 1734 would have built her according to time-honored and sea-proven principles as laid out in the letter from Winne & Hawksworth to William Jones. He might also have built some of his own idiosyncrasies into her, and if we were to find her hull remains, we might well find, as Warren Riess did, a set of design proportions or framing techniques different from what is common to the published treatises. It would have made no difference to the ability of this modest, ordinary vessel to serve the business interests of her owners well. It is to those interests, and how these vessels served them, that we now turn.

experiment with new ones, combined with Britain's leading economic power, would soon advance the island kingdom's preeminence even further.

110 N.A.M. Rodger, "Navies and the Enlightenment," in Pieter van der Merwe, ed., *Science and the French and British Navies, 1700–1850* (Greenwich, UK: National Maritime Museum, 2003), 5–23.

CHAPTER 5

Merchant Venturers and Merchant Ships

John Winne and Walter Hawksworth, the Bristol merchants who ordered *Bristol Merchant* through fellow Bristol merchant William Jones in January of 1733, were members of a coterie of maritime mercantile businessmen in that port city with commercial ties to British America and to the Africa trade. We do not know much about them, but we can deduce enough from the archival record to identify some typical characteristics.[1] We know that both Winne and Hawksworth were propertied, as we would expect of merchants with enough financial wherewithal to order a ship. They were not great men, but reasonably well-off. Hawksworth owned or co-owned at least two other ships. He was also part-owner of the Bristol Brass and Copper Company.[2] That may seem incidental, but it is actually a useful reminder that most shipowners were moderately to heavily diversified; shipowning was typically not the sole commercial enterprise of the shipowner.

That diversification was part of their "risk-management strategy," which they certainly needed; we have already established how dangerous, both physically and commercially, shipping was. While the Atlantic World and maritime economic history literature already discussed have devoted substantive attention to merchant shipowners, they have not offered an explicit identification of "risk-management" as the best explanation for the bulk of their decisions on all matters financial, commercial, and technological. I will argue that such an explanation best fits all the evidence; that a merchant venture is at heart a series of calculated risks, and that it could only succeed if the management of those risks was also calculated.[3]

1 The archival sources on Winne, Hawksworth, and Jones were located in the Bristol Archives and The National Archives of the UK (TNA). My thanks to Joanna Thomas for retrieving and photographing them for me, and to the Society for Nautical Research for funding that effort.

2 Thanks to Sarah Taylor, Archives Assistant, Bristol Archives for this information. In addition to the two ships mentioned by those authors, Hawksworth also probably owned the *Westbury Galley*, 120 tons, in 1710 (TNA High Court of Admiralty 26/15/4).

3 I first worked on the idea of the ship as a technology of commercial "risk mitigation" for owners in Phillip Reid, "Something Ventured: The Ordinary Merchantman of the British Atlantic as a Technology of Risk Mitigation, 1600–1800," *Journal of Transport History* 38:2 (December 2017): 196–212. Doerflinger, in *Vigorous Spirit of Enterprise*, argues that a ready willingness to take financial risks was a distinctive characteristic of these merchants.

 | DOI:10.1163/9789004426344_007

Another "risk management strategy" was fractional ownership. Winne and Hawksworth intended to co-own *Bristol Merchant*, and they may have had, or intended to recruit, other part-owners. Typically, a vessel had principal owners, who actively managed her career, and passive owners, who trusted the principals to manage the vessel capably and distribute dividends as they were earned. For a vessel the size and cost of *Bristol Merchant*, it is possible that Winne and Hawksworth were to be sole owners. That may have been because they did not need other subscribers to afford the relatively modest cost of the vessel they wanted. It may also have been the other way around: they ordered a modest vessel because they knew they would be the sole owners and *Bristol Merchant* was what they could afford.

Winne is, at least once in the records, identified as "Captain John Winne."[4] This was also typical—for a ship-owning merchant to have served at sea himself. Enterprising seamen with suitable connections ashore and at least a rudimentary education could work their way into part-ownership of a vessel, and if fortune smiled, perhaps eventually move ashore and work as a merchant full-time. A merchant who owned ships might send his sons to sea to learn the business from the half-deck before they inherited the firm. This was true of some of the most prominent New England family firms, such as the Browns of Providence and the Crowninshields of Salem. Practical maritime knowledge, based on experience, would have been useful to those directing the enterprise from their offices ashore.

The records indicate that Winne took on at least three apprentices around the time of the *Bristol Merchant* order—in 1724, 1730, and 1735.[5] These may have served in the counting-house, they may have spent time at sea, or both. In both the naval and merchant services, it was common for boys to go to sea.

As for William Jones, he appears in the Bristol archives as well. His connection to Boston, where Winne and Hawksworth intended to have *Bristol Merchant* built, with Jones as the go-between, is reinforced by legal documents listing him as part-owner of a vessel called *Boston Merchant*.[6] Jones found himself in some trouble in 1736, when he and a co-owner, Joseph Farrell, received a summons from the Mayor of Bristol for bringing two of their ships into the

4 Ancient lease: Corporation of Bristol to Captain John Winne, 24 December 1734, BCC/F/E/6432 2059/7, Bristol Archives.

5 As per Sarah Taylor, Bristol Archives, which holds records of Bristol apprentices.

6 Robert Smith, William Jones, … owners of the 'Boston Merchant', v. Thomas Whitchurch … in a plea of trespass. Court papers: Summonses; writ for arrest; deposition re theft of anchor; pleas and demurrer; counsels' opinion; argument; bill of costs, AC/JS/53/4a-k, Bristol Archives.

Key (quay) without a license.[7] The management of Bristol's trade was heavily-regulated, overseen since the sixteenth century by the Society of Merchant Venturers, a royally-chartered organization descended from a medieval guild of merchants.[8] The records lead us to believe that none of the three merchants associated with *Bristol Merchant* were members of the Society. If correct, that means they were not members of Bristol's merchant elite, but of a more middling sort—like the vessel they were ordering from Boston.[9]

The Society initially had a partial monopoly on Bristol's trade, at the time when such arrangements were believed useful for bringing order and regulation to commerce. It became a philanthropic organization before Jones or any of his fellows were born.[10] The Society was educating sailors' children and caring for aged sailors in the sixteenth century. In the seventeenth century, it was running a school of navigation to train poor sailors. This of course stood to benefit the entire maritime economy of the port. The ventures, then, of the merchant venturers of Bristol extended to group efforts to build up a maritime community with a competent local labor force, well-organized and regulated use of the port, and some provision of care for the most vulnerable members of the maritime community, including sailors too old to go to sea any more. Perhaps the most commercially-important group venture of ship-owning merchants, however, besides the owning of ships themselves, was the underwriting of marine insurance.[11] While insurance may seem abstract and removed from the tangible contrivance of the ship, it was part of a set of defensive measures available to merchant-owners; and, as such, may have influenced some of the tangible properties of the ship itself. In fact, the modern concepts of "risk," and of strategies for its mitigation and management, as we understand

---

7 Summonses from the Mayor of Bristol against William Jones and Joseph Farrell, owners of the 'Stockton' and the 'New Kent', for bringing their ships into the Key without license, 28 August 1736, AC/JS/53/3a-b, Bristol Archives.

8 This connection between an organization of merchants and the state is an example of a cooperation that made colonial trade possible beyond tightly-knit family, religious, and ethnic networks; and thus, extended not just commercial enterprise, but organized state authority, limited though it may have been, overseas. See Christopher Ebert, "Early Modern Atlantic Trade and the Development of Maritime Insurance to 1630," *Past & Present* 213 (November 2011): 87–114.

9 Taylor found no mention of either man as a member of the SMV, and she also confirmed that they are not recorded as having been burgesses.

10 The Society still exists, and it is still the principal philanthropic organization of Bristol.

11 We are most familiar with premium insurance, which became common in the early modern period. An earlier form, called "bottomry loans," persisted into this period. This was a combination of financing and insurance, in which the collateral was the cargo or even the ship, and the loan for the voyage was repaid with interest if and when the voyage was completed and the cargo or vessel sold. See Ebert, 102–3.

and use them, developed hand-in-hand with the development of the insurance industry.

Marine insurance began as a group endeavor by British maritime merchants, such as the Society of Merchant Venturers. Specialized insurance companies would develop in the eighteenth century, but their antecedents were groups of merchants insuring each other's ventures. With both physical and commercial risk environments as daunting as theirs were, it is no surprise that such efforts would arise early and grow along with the shipping industry itself. For these were "ventures" indeed; while the regular trade routes that the Society's members were financing in the eighteenth century were not as hazardous or uncertain as one of the group's earlier investments, the 1631 attempt by Captain Thomas James to find the North-West Passage, the risks were still real and present enough that merchant shipowners needed the "vigorous spirit of enterprise" to take on those risks in hopes of success.

The "insurance industry" in this period, was, in fact, inseparable from the merchant community and thus its pooled financial resources.[12] Ship-owning merchants, then, were the ones underwriting the insurance policies for themselves and fellow merchants, even later in the period. These policies, written by merchants well-versed in the typical risks—and "risk mitigation strategies"—of the ships they insured, placed strict limits on freedom of action by owners and masters. Claims were denied if the underwriters determined that the vessel left port in poor condition, if she failed to sail in convoy as stipulated in the policy, or if she deviated from the voyage itinerary presented to and approved by the underwriters for any reason other than an act of nature or the enemy. In 1823, the ship *Aquila* left London for Van Diemen's Land (Australia) and other Pacific destinations. After long delays, during which the master unlawfully attended to his own personal business interests, the ship sailed for home from Singapore, but was lost. The insurance company refused to pay the claim, owing to deterioration of the vessel during her long sojourn in tropical waters and the master's having sent some of the crew off on his own ventures,

---

12 On marine insurance throughout the period, see Patrick Crowhurst, *The Defence of British Trade 1689–1815* (Folkestone, UK: Dawson, 1977), 81–103. On the development of the British insurance industry and the concomitant development of the concept of risk in business, see Lorraine J. Daston, "The Domestication of Risk: Mathematical Probability and Insurance, 1650–1830," in Lorenz Krüger, et al., eds., *The Probabilistic Revolution, Vol. 1: Ideas in History* (Cambridge, MA: MIT Press, 1987), 237–60; and Julian Hoppit, *Risk and Failure in English Business, 1700–1800* (Cambridge: Cambridge University Press, 1987); for a general history of the concept of risk, see Peter L. Bernstein, *Against the Gods: The Remarkable Story of Risk* (New York: Wiley, 1996).

leaving *Aquila* undermanned.[13] Not only is this an example of how underwriters protected themselves; it reminds us that owners were always vulnerable to masters once the ship left port—yet another risk, and one we shall explore in some detail.

Adrian Leonard argues that London's well-developed marine insurance industry allowed her merchant fleet to sail during wartime, contributing materially to the establishment of British maritime supremacy.[14] He notes that policies issued during wartime routinely required sailing in convoy, and encouraged the paying of ransom for capture, which was less costly than seizure, although this was not necessarily the best outcome for the insurer, as

> [t]he cost of a ransom was less than that of a total loss of vessel and cargo, but the possibility of recapture negating the loss was eliminated, and efforts at evasion and resistance by crews were reduced. Nonetheless, insurance cover of ransoms clearly allowed trade to continue by preventing the permanent loss of vessels from the merchant fleet.[15]

He also notes that Sir Robert Cotton, "a Court Tory," suggested in 1693 that insurance was responsible for vessel losses, as merchants were willing to risk their vessels so long as they were insured, even eschewing convoy protection. Cotton also acknowledged, however, that convoy escorts were insufficient—a point Leonard reinforces, and a problem that continued to impede convoy efforts through the mid-eighteenth century.

A convoy is a simple idea, but difficult to execute.[16] In theory, a naval vessel or group of vessels escorts a group of merchant vessels from a pre-arranged rendezvous point near their respective ports of departure to their destination or to safe waters. Should the convoy be attacked, the naval vessels can fight off the attackers, but with luck, the mere presence of the escort will discourage predators from even attempting an attack. In practice, the following difficulties consistently presented themselves. First were the difficulties intrinsic to ocean sailing—the vagaries of weather, including storms, fog, and contrary winds. These could and did delay or even prevent the formation of a convoy, or cause stragglers to be left behind. They also made signaling difficult, and

13 Mark Howard, "Richard Mount, London Shipowner," *Mariner's Mirror* 103:1 (February 2017), 94.

14 A.B. Leonard, "Underwriting Marine Warfare: Insurance and Conflict in the Eighteenth Century," *International Journal of Maritime History* 25:2 (December 2013): 173–85.

15 Leonard, 176.

16 On convoys and their evolving role in Atlantic maritime security, see Crowhurst, *Defence of British Trade*, 43–80; and French, "Productivity in the Atlantic shipping industry."

visual signaling was the only means of communication between ships beyond shouting distance. Second, it is a truism that a convoy is only as fast as its slowest ship. Even once it got underway, a convoy was a much slower way to make a passage for the faster vessels in it. That meant increased exposure to both natural and human threats, larger wage bills for owners, and more risk to the integrity and thus salability of cargoes. Speaking of business risks, arriving convoys could easily and instantly glut a market, driving down prices for goods and profits for owners.

If a convoy were attacked, there was no guarantee that the naval escort would be able to protect all of the merchantmen, especially if one or more of the escort commanders chose to chase the predators, in hopes of prize money, rather than remain close to their charges. It was not uncommon for enemy cruisers to take prizes from convoys.

In sum, convoys were far from a perfect security solution for merchant ships in dangerous seas. This was particularly true before the mid-1700s, as there were not enough suitable escort vessels, nor adequate naval infrastructure abroad, to support a convoy system adequate to protect the bulk of Atlantic trade.[17] When weighing the pros and cons of sailing in convoy, owners might have considered the relative speed and maneuverability of their vessel, as well as its value and the value of the cargo. This was certainly true of later periods, but further examination of merchants' correspondence and convoy records will be needed to establish whether there was a tendency to eschew convoy participation for speedier vessels as opposed to their slower, more cumbersome counterparts. If so, that would tie the defense strategy of the convoy directly to ship technology.

After the 1750s, as Great Britain solidified its imperial power in the Americas, the effectiveness of the convoy system improved; and, along with a more consistently effective projection of British naval power, we see both an increase in the availability and affordability of marine insurance, and a sharper general decline in the arming of merchant ships.

One of the most compelling arguments developed by economic historians is that the need to defend merchant ships consistently limited their efficiency

17 On the distribution of British warships throughout the Empire over the course of the period, quantifying the increase in naval resources devoted to convoys and Atlantic operations in general, see Sarah Kinkel, *Disciplining the Empire: Politics, Governance, and the Rise of the British Navy* (Cambridge, MA: Harvard University Press, 2018), Kindle edition, locations 610–30. Kinkel's totals are derived from tables in Rodger, *Command of the Ocean*, 612–15.

as commercial transportation throughout the period.[18] One might rejoin that war is of course bad for commerce, so a maritime region plagued with chronic warfare—especially warfare that emphasized maritime predation—would be bad for the shipping business. The truth is more complex. War was profitable for some merchants some of the time and disastrous for others at other times. Great Britain's "fiscal-military"—or perhaps "fiscal-naval"—state thrived in the eighteenth century not just in spite of war, but in important respects, because of it.[19] We know that the power of the Navy to protect commerce and the availability of affordable insurance both increased during the eighteenth century, and that unarmed Atlantic merchantmen were more common after 1750 than they had ever been. At the same time, privateering and naval cruising against enemy commerce remained important features of Atlantic imperial conflict. If war had not been so endemic to the Atlantic in this period, the practice of arming merchant ships would have been nowhere near as prevalent; the nineteenth century makes that clear. Such was not the case, however. This does not mean that all ships were armed, even in wartime. Certificates of clearance and entry kept by port officials specify the armament of the ship. The records do show a positive correlation between armament and wartime, supporting the belief that protracted maritime peace would have permitted owners the luxury of dispensing with guns—which indeed they frequently did in peacetime.[20]

What concerns us here is how all of this might have shaped what a British Atlantic merchant ship actually was, at any given time during this period. How did intangible external threats and defenses—privateers and cruisers, guns and convoys and insurance policies—affect the tangible properties of the ship itself, as well as her owners' accounts?

In Edward Barlow's account of his maritime career, he recounts sailing on six vessels in the late seventeenth century, all of which were armed, with

18 Most prominently, Shepherd and Walton, whose work was discussed in the Introduction. They did not, however, pay any attention to any ways in which technological change might have affected efficiency, even modestly, and they claim no appreciable increase in average size, which others have strongly disputed—see Menard in Tracy, 258–62; and McCusker, *Rum and the American Revolution*, 836–38, note 27. We are operating under the hypothesis that a combination of factors—including modest average size increases—combined to increase the cost efficiency of the merchant ship in the mid- and late-eighteenth century.

19 See N.A.M. Rodger, "War as an Economic Activity in the 'Long' Eighteenth Century," *International Journal of Maritime History* 22:2 (December 2010): 1–18.

20 The data compiled by Shepherd and Walton clearly indicate that merchantmen were, as a rule, more heavily armed during wartime—see 76 and 196–97. French's research adds that, even so, the general trend was for reduced armament after the Seven Years War in both peacetime and wartime. See French, "Productivity," 631.

ton-per-gun ratios ranging from a low of 10.71 to a high of 21.25 (see Table 2). From official documents found in merchants' papers in Philadelphia and New England, I can present the same information for 10 vessels, from 1727 to 1809 (see Table 3). Of those 10, only three were armed. The largest ship sailed by Barlow was *Queen Cathrane*, 400 tons, mounting 34 guns, for a ton-per-gun ratio of 11.76. The largest ship in my list is the Brown East Indiaman *Ann & Hope*, 550 tons, mounting only 12, for a ton-per-gun ratio of 45.83. As Shepherd and Walton demonstrated using data from the Colonial Office records, ton-per-gun ratios rose dramatically from 1696 to 1773, with far more unarmed vessels sailing in the late 1700s than before. This was especially true of colonial-built vessels vis-à-vis British-built vessels, but the latter, too, demonstrate the same overall tendency. Shepherd and Walton do not delve into possible explanations for the difference. Perhaps it was a carryover of the seventeenth-century tradition of Thames-built ships being well-armed for the Mediterranean trade—a trait evident on Barlow's *Cadiz Merchant*, on which he made two voyages to and from Jamaica and England between 1680 and 1682 (see Figure 3). It may also have to do with the higher cost of construction and fitting-out of British vessels, motivating owners to spend extra money on defensive armament. Too, British-owned vessels generally made more voyages to American ports from the British Isles than did colonial-owned vessels, and the retention of stronger armament may be related to real or perceived dangers from privateers and enemy cruisers prowling the Channel and the Western Approaches to the home islands. Of course, none of these possible explanations necessarily excludes either of the others. An investigation into each of them would be worth pursuing.

TABLE 2 Ships, tons, and guns from Edward Barlow's journal

Name	Date	Tons	Guns	Ton-per-gun ratio
Queen Cathrane	<1662	400	34	11.76
Reall ffrenshippe	1668	140	10	14
Mayflower	1676	150	14	10.71
Rone (Rouen) Facter	1677	85	4	21.25
Guannaboe	1678–79	350	20	17.5
Cadiz Merchant	1680–2	280	24	11.67

TABLE 3 Ships, tons, and guns from Philadelphia and New England merchants' papers, 1727–1809

Name	Date	Tons	Guns	Ton-per-gun ratio
Rainbow[a]	1727	40	none	NA
Active[b]	1783	95	none	NA
Eagle[c]	1784	100	none	NA
Astrea[d]	1785	300	none	NA
Ann & Hope[e]	1798	550	12	45.83
Cruger[f]	1798	?	8	?
Gadsden[g]	1799	?	16	?
John Langdon[h]	1806	24	none	NA
Anna[i]	1806	62	none	NA
Jane[j]	1809	4	none	NA

a Customs clearance for schooner *Rainbow*, 1727, Shipping Papers, box 1, folder 2, New Hampshire Historical Society (hereafter NHHS).

b Massachusetts certificate of clearance for brigantine *Active*, 23 April 1783, box 1, folder 1, Derby Family Papers, PEM-Phillips.

c Virginia certificate of clearance, brig *Eagle*, 11 June 1784, box 1, folder 2, Jones and Clark Papers, 1784–1816, HSP.

d Dominica certificate of clearance, ship *Astrea*, 1785 or 1786, box 1, folder 6, Derby Family Papers, PEM-Phillips.

e Charter agreement, ship *Ann & Hope*, 27 July 1799, box 475, folder 9; Estimated building costs, ship *Ann & Hope*, box 475, folder 1, Brown & Ives, JCB.

f Privateer's commission from President John Adams, brig *Cruger*, 1798, box 24 (OS), folder 1, Derby Family Papers, PEM-Phillips.

g J.F. Fleming to Jones and Clark, 19 February 1799, box 2, folder 8, Jones and Clark Papers, HSP.

h New Hampshire certificate of clearance, brig *John Langdon*, 1806, box 1, folder 12, Tredick Family Papers, 1753–1835, NHHS.

i New Hampshire certificate of clearance, brig *Anna*, 5 July 1806, box 1, folder 12, Tredick Family Papers, NHHS.

j New Hampshire certificate of clearance, schooner *Jane*, 28 June 1806, box 1, folder 7, Shipping Papers, NHHS.

The central technological question, however, is what effect reduced armament may have had on the design and construction of the ordinary Atlantic merchant ship—and thus on her cost, her capacity, and her operation. With all the evidence pointing to a general increase in capacity-for-size, and connecting that trend to general changes in rig, did a reduction in armament allow for lighter, cheaper scantlings, and perhaps even more capacity-for-size, as well as foregoing the cost of the weapons themselves? Rather than speculate, we should attempt to quantify the differences between a vessel armed in a typically seventeenth-century fashion, and an analogous craft either lightly armed or unarmed. Such a comparison could take more than one form. One would be a comparison of a typical Thames-built English merchantman of the mid- to late seventeenth century with a Dutch-built *fluit* or other unarmed or lightly-armed vessel of comparable size. It is accepted that the Dutch vessels were cheaper to build, but it would be instructive to know the details of why that was. Another possibility would be to compare two British Atlantic vessels diachronically—say, a 400-ton Thames-built vessel of 1680 with a 400-ton vessel of 1780. It is highly likely that, as we saw in comparing Barlow's vessels with later American-built ones, the later vessel would be more lightly armed, or unarmed. What differences in design and construction could be attributed, with confidence, to the difference in armament?

Until we can answer this question, we are assuming that a decrease in armament equaled an increase in efficiency—but what does that mean? Does it mean higher cargo capacity, cheaper construction and outfitting, lighter overall weight and thus less sail power—and less labor—to drive the ship satisfactorily? Shepherd and Walton do not tell us this—they cannot, because they were not asking the technical questions necessary to do so. Nor do they make an explicit claim about the relationship between guns aboard and crew requirements, although they consider the two issues sequentially. To understand that relationship, we must know how many sailors it took to operate a vessel's complement of guns, versus how many sailors it took to safely operate the vessel herself. With enough data on that, we could mount a convincing argument as to the nature of that relationship, and whether decreases in armament contributed to the increase in ton-per-man ratios that we know took place in the eighteenth century—or, perhaps, it was the other way around: the desire to increase ton-per-man ratios to save labor costs, made possible by adaptations to rigs and by the intangibles of increased security and insurance, pressured for the reduction in armament—which in turn was considered an acceptable risk, given those same intangibles. The seventeenth century shows us that, in the absence of acceptable security at sea and an accessible, affordable insurance market, or perhaps even in the absence of one of those, owners were not willing to forego armament. That lends much credence to Shepherd's and Walton's argument that security at sea played a key role in the reduction of armament on board.[21] It should now be clear, however, that without deriving quantitative data from asking and answering technical questions, such as specific manning

21 Recall from the Introduction, however, Menard's argument that better ships were more important than security at sea—60% to 40%, to be precise. As we shall see, though, a

requirements for specific operations—handling a two-masted square rig, or operating sixteen cannon—we cannot move beyond assumption and educated guessing as to how these changes related to each other, and we cannot point to any causality with any confidence.

As a first move forward on that, consider the following. Jim Graczyk, speaking from his experience on the replica Armed Sloop *Welcome*, gives us a clear idea of one way to gather the necessary information on manning guns. The replica *Welcome* (now retired) needed a crew of ten or eleven to "man her under sail." She carried seven guns—three deck-mounted swivel guns and four deck guns. "For us a full compl[e]ment in the gun crew was four gunners, two powder monkeys, one Linstock holder and a master gunner. Under good conditions some of the sail crew would man several guns." Research in the records of the original ship revealed that, when operating as an ordinary merchantman, she carried a crew of five or six. When taken into the Provincial Marine, her minimum crew was eight. On military sortie, "she had twelve sailors and twelve soldiers of the King's 8th regiment of foot."[22] The exact numbers from vessel to vessel will of course be found to vary, and indeed would have varied some even from voyage to voyage, given the vagaries of manning in general. It is clear, however, that even disregarding the extra personnel carried aboard *Welcome* on military sortie, when she was taken into the Provincial Marine, at which point we can assume she was expected to be able to defend herself adequately, she carried a crew at least 25% larger than she did when operating as an ordinary merchantman. Using operational period replicas based on sound research, we can compile data on the number of crew it takes to operate a vessel on a voyage from port to port with and without defensive capability. We know how to work those ships and we know how to work those weapons. We could then do the same for a number of vessels. The data thus derived could then be compared with that derived by Shepherd and Walton, and by Ralph Davis, from the official shipping records. If we find that it does indeed require a crew larger by a certain range of percentage to operate a vessel armed to a given extent than it does to operate one more lightly armed or unarmed, then we can claim with confidence that reduction in armament and reduction in crew went hand-in-hand, regardless of other factors influencing the vagaries in the manning of ships, even if we cannot also prove the exact causal relationship—whether it was the reduction in armament permitted by increased security that drove the reduction in crew sizes for given vessels (or,

---

linchpin of Menard's concept of "better ships" is not something we can accept at face value.

22 Graczyk corr.

put another way, in the increased ton-per-man ratios), or whether it was the feasibility of reduced crew sizes—increased ton-per-man ratios—that drove the reduction in armament. Indeed, those two changes may have simultaneously pressured for each other.

The guns themselves were heavy. *Ann & Hope*, built by Benjamin Tallman in 1798 for Brown & Ives of Providence, Rhode Island, had 12 9-pounders.[23] Each of those was seven to nine feet (2.13 to 2.74 m) long and weighed between 2,828 and 3,472 pounds (1,282.76 and 1,574.87 kg), excluding their stout oak carriages.[24] If we split the difference and assume each gun weighed 3,000 pounds (1,360.78 kg), that adds up to 36,000 pounds (16,329.36 kg)—18 modern (U.S. "short") tons—of guns. Each carriage would have weighed a few hundred pounds. In 1800, 18 tons of rice would have cost £434 ½ in one of the five principal U.S. port cities—less than pork or tobacco.[25] So, for the weight of those guns, it is fair to say that *Ann & Hope* could have carried—on repeated voyages—a cargo worth at least £434 instead of guns that cost, with their carriages and "implements," £730. (Her full suit of sails—and this was a large, three-masted ship with a relatively tall rig—cost £650.)[26] As noted, *Ann & Hope* was lightly-armed for a ship her size compared to Barlow's *Queen Cathrane* of a century earlier. So, even a fairly light complement of armament amounted to a significant expense. It is not difficult to understand, then, why owners might be tempted to avoid that expense if they found the risk of doing so to be acceptable. In September 1798, a painter named Muscomb submitted a bill to the exceptionally-successful shipowner Elias Hasket Derby of Salem, in which he itemized, among other charges, ten shillings for painting six wooden guns and carriages. Wooden guns were fake—dummy guns. The vessel carrying them

23 Estimated building costs, ship *Ann & Hope*, box 475, folder 1, Brown & Ives, JCB.

24 David McConnell, *British Smooth-bore Artillery: A Technological Study to Support Identification, Acquisition, Restoration, Reproduction, and Interpretation of Artillery at National Historic Parks in Canada* (Ottawa: Minister of Supply and Services Canada, 1988), 86., accessed at https://sha.org/assets/documents/British%20Smooth-Bore%20Artillery%20-%20English.pdf, 3 October 2018. McConnell lists the weights in hundredweights (cwt). I am assuming the imperial hundredweight of 112 pounds (50.80 kg).

25 A cwt of rice cost $4.50 in each of the five principal U.S. port cities in 1800. Based on the total building and outfitting cost of *Ann & Hope* listed in her owners' records in both pounds and dollars, we can equate one pound in 1798 to $3.33. One modern U.S. "short" ton is 17.86 cwt, assuming the 112-pound imperial cwt. So, the cost of a modern ton of rice in 1800 in a U.S. port city would have been $80.37, or £24.14. 18 tons of rice, then, would have cost £434.52. United States, Bureau of Labor Statistics, *History of Wages In the United States From Colonial Times to 1928* (Washington: United States Govt. Print. Off., 1934), 21, accessed at https://babel.hathitrust.org/cgi/pt?id=uc1.32106007458745;view=1up;seq=35 3 October 2018.

26 Estimated building costs, ship *Ann & Hope*.

would be displaying a visual bluff, and her crew and owners would hope that a predator did not choose to call it.[27] Dummy guns, and fake gunports, which were just painted squares, were cheap camouflage, but of course they would be of no use against an attack.

The cannon—real or fake—carried aboard a merchant ship represented the owners' and crews' dependence on their own devices for defense against maritime predation. It stands to reason that, should the threat of such predation diminish, or external means become more dependable for meeting that threat, owners and crews could abrogate some or all of that on-board defense, and save the cost and the risk of it.

The unanswered question, however, is, what was the relationship between insurance and armament? Can we compile data that suggest a correlation? As noted, insurance policies did not stipulate that vessels be armed. Lloyd's Register does not specify whether a vessel was insured, although it does list the vessel's armament. Adrian Leonard has compiled a database of insurance policies written for specific vessels for specific voyages. Cross-referencing the vessels in his database with those vessels' listings in Lloyd's Register might well allow us to hypothesize whether or not owners were eschewing armament in favor of insurance. Armament decreased—that we know—so we need not be concerned with the inverse. If we were to find no discernible correlation between whether a vessel was armed and whether it was also insured, that would, however, weaken the hypothesis that the increased availability and purchase of insurance was a significant causal factor in the decrease of armament late in the period.

In any case, such an inquiry will not answer the central technological question posed earlier: did a reduction in armament affect other aspects of the ship's design, construction, and performance both as a sailer and a business asset, and if so, how? As always, we must consider that question keeping in mind that all such aspects are interrelated, as are the factors influencing whether, and how, they changed. An important example would be the issue of capacity. Round, full ends, as we know, increased buoyancy in those ends, which helped support the weight of armament and resist hogging and sagging (longitudinal hull deformation). They also, however, increased cargo capacity vis-à-vis sharper ends, and helped the ship ride up over waves and stay drier. If a ship were built to be unarmed, then, that does not mean she would necessarily be built with finer ends, because those bluff bows and full sterns were advantageous for other reasons. As for scantlings and the structures of hull

27 Painter's bill, Muscomb to Derby, September 1798, box 9, folder 5, Derby Family Papers, PEM-Phillips.

and decks, it has been assumed that substantial armament required substantial increases in both, leading to increased cost of construction and additional weight above the waterline. That assumption is sound. While merchant ships were not armored—they did not have vertical timbers between the frames, forming almost a solid wall within the planking, as warships did—they still needed significant structural reinforcement to house guns. Specifically, the gun ports had to be framed with timbers above and below, and if a frame did not happen to fall at the edge of the port, timbers to fill out the space between the port and the nearest frames. These framing timbers, in addition to reinforcing the hull where an opening had been cut into it, served as anchors for the big iron ringbolts to which the gun carriage tackles were secured. Those had to take the shock of recoil. Additional ironmongery was needed in the form of hinges through the hull for the port lids, and rings for the ropes that lifted and lowered them. Those frames, too, were not adequate to reinforce the port openings; this also required thicker outside planking above and below the port, as well as thicker inside planking below the cut-out. The deck planking under the gun had to be reinforced to support the gun's weight, and the deck above the guns needed large openings to vent the smoke, or it would quickly blind and choke the sailors working the guns. Those openings, too, had to be reinforced with structural timbers.[28] So, yes, the shipwright's bill was going to be higher for a ship built to carry a complement of carriage guns, even if all other specifications for the hull were the same. If we add that increased cost to the cost of the guns and carriages, and consider the space and weight taken up by barrels of gunpowder and racks of iron shot, it is clear that owners stood to gain by operating in a risk environment that allowed them to forego all of this.

Does that mean, however, that, once ships began to be built without such armament, they were built lighter and cheaper? Traditional ship archaeology and shipbuilding agreements are our twin best hopes of answering that question, and they should be able to answer it. We know that building costs, in general, went up, so there certainly would have been incentive to cut those costs where possible without compromising seaworthiness or longevity.

While this discussion, like so many in this book, raises more questions than it answers, it is worth remembering what the evidence does make clear: that those creating and using this technology adapted it as they saw fit, in response to changes in the marketplace, the need to adjust one aspect of ship technology to accommodate a change made in another, and changes in taste—in what

28 I am indebted to Joakim Severinson, master shipwright of the Swedish Ship *Götheborg*, for the information on structural reinforcement. Joakim Severinson, personal communication, 1 November 2018.

looked proper. Thus, we are safe in discounting the possibility that, if ships were not built lighter, and thus cheaper for that reason, without armament than they were with it, such a decision (or non-decision) was due to inertia or stagnation. While it is certainly true, as modern economists finally acknowledged, that humans do not always make such decisions rationally, it would go against all that we have learned about these ships, and those who used them, to ascribe stasis in this area to an irrational decision not to innovate that would have cost all shipowners money unnecessarily. Perhaps the bounty on ships over 100 tons offset the cost of building those ships so that they could be armed for defense.[29] Perhaps accepted standards of seaworthy construction dictated the scantlings and construction we can observe, even if lighter and cheaper timbers and methods might have sufficed. Too, there is no reason to assume a uniform approach to the issue. There is evidence, as we have seen, for some propensity of American builders to build cheaper ships that did not last as long, and for some British yards to turn out more expensive ships that rewarded their owners with a longer service life. It is possible that merchants' records, especially their correspondence, might yield clues as to why merchants and shipwrights made the decisions reflected in the archaeological record, but it may require much searching, as merchants spilt precious little ink on technological matters.

What they *did* spill ink on were their ledgers and accounts, which consistently reveal meticulous attention to detail, accounting for every cent paid out and taken in, leaving us in no doubt of what they traded, and for what price, nor of what sundries went into everyday business and the maintenance and repair of ships, and what all of that cost. Also fortunate for us is that ship owners took pains to keep expenses separate by vessel. So, we can see, even for an ordinary little sloop, what material and labor went into her construction, her fitting-out, provisioning, and maintenance. As with their letters, however, the overwhelming bulk of merchants' journals and account books are much more concerned with cargoes for sale than ships in service, reinforcing the already-strong impression that the vessel was a rather stable quantity while salable cargoes were anything but. Prices and qualities fluctuated, demand fluctuated, credit with which to buy goods fluctuated, and merchants were constantly challenged to enter a profit rather than a loss in their ledgers.

While journals were kept in the office, *bills of lading* and *cargo manifests* were ships' papers. Assuming they were not fraudulent, they tell us exactly what was on board for a voyage, who owned it, how much it cost, how much the owners were charging to ship freight for others, how much of each item

29 The bounty was offered for the construction of vessels large enough that they could be conscripted by the Royal Navy in an emergency.

was carried, and what units of measurement were used for each. Cargoes were almost always mixed. That helped to spread risk, given the unpredictability of markets. There were exceptions of course—tobacco ships from Virginia would carry little else to Britain after the harvest, but they would certainly return with mixed cargoes. Sack ships would carry dried cod from the Grand Banks fisheries in large volumes, every year, to Europe and the West Indies, but then typically load wine and other Mediterranean products or tropical produce.

Bills and *invoices* make up much of the bulk of ship owners' papers. They bring the everyday details of maritime business almost to life. Frequently just chits of paper with a few creatively-spelled words and a few numbers scrawled on them, they were written by everyone from a glazier submitting his bill to an owner for repairing stern-gallery windows, to a Brown, Crowninshield, or Derby issuing a "pay-to-the-order-of …" From stacks of these, we learn what tradesmen charged for what work at a certain place at a certain time—and how many of those trades a ship owner relied upon—sailmakers, glaziers, masons to build and repair the ship's brick stove, painters, rope makers, coopers, blacksmiths, spar makers, block makers, riggers, caulkers, joiners.

On 15 August 1759, sailmaker Eleazer Moses billed Timothy Orne for making a complete mainsail, foresail, a set of jibs—flying jib, middle jib, and jib (innermost)—main topsail, gaff topsail, and tarpaulin. The total was £63:16:10 old tenor; £08:10:2 3/4 current lawful money.[30] "Old tenor" refers to Massachusetts paper money withdrawn from circulation in 1749 to combat inflation; note the drastic difference between the two sums. Such tradesmen's bills remind us that currency was a perennial problem. Everyone wanted pounds sterling, but most had to settle for something else much of the time. There were colonial currencies whose value fluctuated by location and people's assessments of their worth in the near future. As important as any "colonial currency" were sugar and its sister products, molasses and rum—especially rum.[31] People along the waterfront in this world paid each other in the produce of the sugar cane all the time. Of course, no one in the shipping centers of the northeastern seaboard produced the raw products themselves. They imported as much of them as they could from the West Indies, and relied on those imports in large part to keep their maritime economy going and growing. It is no wonder that these ship owners were willing to go to great—including supralegal—lengths to procure the sugar products they wanted, and to offer their own goods in exchange for them wherever the best markets were—and the best markets were so frequently the off-limits French islands, with no eastern

30 Bill, Eleazer Moses to Timothy Orne, 15 August 1759, Orne Family Papers MSS 41 Box 1, Folder 6, PEM-Philips.

31 See McCusker, *Rum and the American Revolution.*

seaboard colonies of their own to supply them with provisions and building materials, and with booming sugar production and no home market for rum in a country where brandy was so well-protected by law and taste.[32] Rum distilling was the number-one industry in New England in this period. The success of the British sugar plantations, and the lobbying power of the planters, lie behind the famous rum ration issued in the Royal Navy from the eighteenth century until 1970.

Sugar products could help owners buy ships to carry sugar products. Of particular interest in the stacks of merchants' bills and invoices are bills of sale for ships. These reflect the cost savings owners might realize by buying used versus new. On 2 September 1799, Brown & Ives bought a six-year-old brigantine called *Maria* for $3,200. On 14 February 1801, a George Robinson offered to sell them a new one, when she was finished, for $6,500.[33] No size difference between brigantines eight years apart in age would account for that disparity. Most of it was depreciation. Obviously Brown & Ives thought *Maria* was worth buying, even though six years was plenty of time for a ship of this period to deteriorate badly. The importance of evaluating the condition of a used ship before making an offer on her is made quite evident in merchants' correspondence on the matter. In 1777, Nicholas Brown and Christopher Starbuck were considering buying an 80-ton sloop called *Success*. The memorandum of sale, which we would call a listing, claimed she was "a good Sailing Vessel," seven years old, standing rigging one year old, which might mean it was already due for replacement, depending on usage and maintenance, running rigging "pretty good," with sails fit for a voyage to the West Indies—her one-year-old standing rigging and "pretty good" running rigging could reasonably be expected to make at least the one-way trip, if her owners were being honest—as noted, owners frequently sent these smaller vessels down-island in hopes of selling them there rather than bringing them back.[34] Note also here that typically vague description of the vessel's performance—"a good Sailing Vessel."

Ships, especially larger ones with full rigs, were expensive—to own and to buy. Owners wanting to sell might have to wait some time for a buyer, and

---

32 See Boucher, *France and the American Tropics*, and Kenneth Banks, *Chasing Empire Across the Sea: Communications and the State in the French Atlantic, 1713–1763* (Montreal: McGill-Queen's University Press, 2002), which explores similar themes as Steele's *English Atlantic* for the French empire. The French maritime historian Paul Butel wrote a survey of the French Caribbean carrying the history into our own time in *Histoire des Antilles françaises, XVII*^{e}*–XX*^{e} *siècle* (Paris: Perrin, 2002).

33 As this is after U.S. independence, the currency has shifted from the pound to the dollar.

34 Memorandum of sale, sloop *Success*, 1777, box 682, folder 4, Nicholas Brown and Company, JCB. Standing rigging is the rigging that holds up the masts and the bowsprit. Running rigging is the moving rigging that controls the yards, booms, and sails. All of it has to be adjusted for stretch and kept tarred to prevent rot.

meanwhile, the ship continued to cost money in upkeep. For over a year, in 1807–1808, Jacob Smith, working for Brown, Benson, & Ives of Providence, tried to sell the ship *George & Mary*. Just as he thought he had a deal, the ship was damaged in a gale by another ship riding alongside her. His letters to his employers concerning the drawn-out affair span three folders in the collection.[35] Seven years later, Smith was working on another deal—this time, to buy a ship called *Two Catharines*. Smith had been writing to the firm about other ships, but not with the detail or enthusiasm he did about this one. *Two Catharines'* owners' representative, Captain Dennis, reported that she "has always had a ship keeper to keep her And untill last fall and this spring and Summer the Bow lumber port has allway been kept Oapen and Cabbin Windows ...".[36] A ship keeper stayed aboard a moored ship to take care of her, and by keeping her lumber port and windows open, she remained ventilated. Alas, however, it seems all Smith's efforts to acquire this ship for his employers came to naught—his final letter on the subject notes that he made an offer according to their wishes but it was declined, and that they were so far apart he was sure they would not accept the counteroffer he made. He then moved on to other business.[37]

Note that the priorities of all parties in this exchange were the condition of the ship, her price, and what conveyed with her for that price. The vague "Very Good Sailer" is the only performance-related comment in the exchange. Her capacity received more attention, with Smith seeming to imply that the seller exaggerated it but that it was generous for her size nonetheless. Merchants wanted the greatest cargo capacity for size they could get, and it was those cargoes, their protection and sale, that preoccupied them, as their correspondence makes clear.

Merchant venturers depended on a regular stream of correspondence—with their ships' masters, with each other, with agents and factors in distant ports, with their attorneys—to keep abreast of markets as best they could, and to keep up with their own business affairs, spread as they were across transoceanic networks.[38] That correspondence makes up the bulk of the records they left.

---

35 Jacob Smith to Brown & Ives correspondence, 1796–1834, box 306, folders 3–5, Brown & Ives, JCB.

36 Thomas Dennis to Jacob Smith, 16 October 1815, box 306, folder 7, Brown & Ives, JCB. Bow lumber ports were like gunports cut in the bows, which could be sealed for voyages, through which long timbers such as mast timbers—basically entire tree trunks or sections of them—could be loaded into the hold.

37 Jacob Smith to Brown & Ives, 17 October 1815, box 306, folder 7, Brown & Ives, JCB.

38 *Factors* were partners or employees of an owner or firm who was entrusted to conduct business in destination ports on behalf of the owners. Thus, the master himself was a

One of the most instructive types of outgoing correspondence was to the ship's master.[39] It was a loosely-standardized set of instructions for a voyage, called Sailing Instructions. An extant Sailing Instructions letter yields a lot of information for a short, matter-of-fact document. They are so common and important that an example, with comment, is called-for:

> Capt John Lovitt Salem New England June 20 1758
>
> You being appointed Master of our Sloop Andrago, Our Orders are that you proceed the first good wind with our s.[ai]d Sloop & Cargo for Barmuda upon your Arrivall there make report & entry according to Law and Dispose of our Cargo to our best advantage & make us Returns in a load of Salt if you can purchase it so as that you are well assured it will pay a good freight here if not proceed with our Sloop & the Effects of our Cargo in Cash or good Bills from Barmuda to Charlestown in South-Carolina & there lay out our Cargo in good Rice some pitch & Tarr & Staves or any other goods that you are well assured will be most profitable to us, If we should not have Effects sufficient to fully load our Sloop, you may fill her up with freight goods. Or if you think it best for us you may draw upon us for as much as will fully Load our Sloop & we will punctually pay your Bills, Write us by all Opportunities make all [the?] Dispatch possible & in all Cases & Circumstances do that which you shall think best for our Interest. We wish you a good Voyage & are your Friends & Employers.
>
> P.S. If you should be Taken going or Coming
>
> Bezaliel Toppan
>
> we would have you Ransome at the best
>
> Tim^{o} Orne
>
> rate you can ---------
>
> BT
>
> TO
>
> The above a True Copy of Orders Rec.^{d} from
> my Employers
>
> John Lovett[40]

---

common factor. *Agents* were contractors who performed the same function; they were not employees or partners. The terms "master" and "captain" are somewhat, but not entirely, interchangeable in the merchant service, but not at all in naval service. The master of a merchant vessel was directly addressed as "Captain," but it is more correct for us to refer to them indirectly as masters.

39 On the role of the master in shipping management, see Davis, 153–67 (Ch. 8).

40 Toppan and Orne to John Lovitt (or Lovett), 20 June 1758, Orne Family Papers, MSS 41, Box 1, Folder 3, PEM-Phillips.

Unpacking this short letter will prove it a treasure chest of insight into how all of this worked.

As sailing instructions go, this was a short one, certainly run-of-the-mill, for a typical island run in a typical sloop with typical cargo. There were thousands of letters just like this one. That makes it more valuable. The instruction to sail with the "first good wind" reminds us that these voyages did not take place on a timetable as ours do. Ships might sit in port for weeks waiting for a wind and tide combination suitable for making way. As much as owners may have wanted to control such things, they could not. Nor could they control the ship's business in destination ports as much as they might have liked—that is clear in agreements like this. Note the leeway given to the master to conduct trade. Merchants tried to stay on top of markets in destination ports by devouring the latest business newspapers, correspondence, and hearsay, all of which were in constant circulation around the Atlantic ports. Even armed with the latest prices and market conditions, however, the speed of communication was too slow to allow owners the luxury of dictating all their ships' business from home.[41] The market conditions that made the voyage attractive in the first place had all too frequently shifted or evaporated by the time the vessel made port—a pervasive theme of incoming correspondence. So, owners gave the vessel's master a degree of latitude that would be wholly unnecessary now, and that meant the master had to act as the owners' factor, whether or not he was actually one of the owners, which he sometimes was, and whether or not the owners had another factor, a business agent, or "correspondent" in the destination port.[42] Sailing Instructions that bound the master too tightly—that constrained his choices too much—placed greater risk of a financial loss on the voyage (provided the master was a trustworthy factor). As it was, such loss occurred all too often. Masters' letters to owners routinely complain of poor market conditions and offer justification for the masters' decisions, in an attempt to shield themselves from culpability for a voyage's failure.

The first choice stipulated in the instructions to Lovett is to Bermuda for salt. Salt was always in demand, so that part was a sensible risk, but if the price were high enough at Bermuda, that cargo would not provide the owners a profit. The owners were placing the onus on Lovett not only to judge the price of salt in Bermuda, but to judge the market conditions for selling it back home. Besides,

41 For an overview of communication, see Steele's *English Atlantic*. On the maritime business press, see McCusker, "The Early History of 'Lloyd's List'," *Historical Research* 64:155 (October 1991): 427–31.

42 Sometimes—more commonly early in the period than later, as a general trend—a supercargo served as the owners' agent on the voyage. That person, too, had to be someone the owners trusted.

the salt in Bermuda had already come from elsewhere—most likely Turks & Caicos—so who brought it there, and on what terms, affected that market.[43] Regardless of whether Lovett chose to load Bermudan salt, however, he was to comply with Customs laws and procedures. Owners did not want their vessels fined or seized for failure to declare themselves and their cargoes according to the law.[44] On 22 April 1789, William Tredick wrote sailing instructions for his son Henry, in command of the schooner *Dove*, destination Martinique or St. Lucia—both French. William of course wanted cash for the cargo if he could get it, and if not preferred that Henry go to the Turks Islands (Turks & Caicos) for 10 or 12 hundred bushels of salt. He could sell the schooner if offered 420 pounds "lawful money." "By all means Be Careful of Contraband goods Not to Run any Risk …", warned William.[45]

Should Captain Lovett decide not to buy salt, Toppan and Orne suggest that he is still to sell the outward-bound cargo in Bermuda for "Cash or good Bills."[46] That was not always the case. Owners frequently gave masters the option of *not* selling the cargo if prices were not to their liking, and ships might sail to multiple ports before selling the cargo, or they might sell part of the cargo in one port, where the price for that commodity was attractive, and sail somewhere else to sell the rest, looking for a better deal. This of course drove up cost and risk of damage considerably for the voyage. The master had to weigh the potential gain for doing so against that. In this case, a nearby alternative to Bermuda was Charlestown (today's Charleston), and the commodities specified for that option were the chief products of the Carolinas—rice,

---

43 On the salt industry connecting Bermuda, Turks & Caicos, and the wider Atlantic, see Cynthia M. Kennedy, "The Other White Gold: Salt, Slaves, the Turks and Caicos Islands, and British Colonialism," *The Historian* 69:2 (Summer 2007): 215–30; and Neil Kennedy, "Impermanence and Empire: Salt Raking in the Turks and Caicos Islands," in Peter E. Pope with Shannon Lewis-Simpson, eds., *Exploring Atlantic Transitions: Archaeologies of Transience and Permanence in New Found Lands* (Woodbridge, UK: The Boydell Press, 2013), 80–92.

44 If they intended to smuggle, owners would typically abide by the usual practice of putting such language as we see here in the official sailing instructions, but then they might deliver illegal instructions orally. Regardless, owners used these agreements to shift the onus of any illegal activity onto the master. This might help mitigate legal repercussions in the event that the vessel ran afoul of the customs laws, whether inadvertently or intentionally.

45 William Tredick to Henry Tredick, 22 April 1789, box 1, folder 1, Tredick Family Papers, NHHS.

46 Bills of exchange were basically I.O.U.s asking the recipient to pay the bearer the amount presented and charge the issuer. These, though, were complicated by exchange rates and the perceived, or actual, creditworthiness of the issuer, and were also traded as commodities themselves, like stock options and securities. Owners also sometimes gave masters the option of selling the vessel if good terms were to be had.

naval stores, and barrel staves. The other options offered by the owners were to take freight—goods shipped by others for a fee—or to borrow money ("draw upon us") on the owners' credit to buy a return cargo. Next, the owners instruct Lovett to keep them as well-informed as possible by writing to them frequently. Masters typically did this, although there was no guarantee that their letters would reach the owners in a timely manner. Both outgoing and incoming correspondence are full of complaints about letters not received and decisions delayed or made without requested advice. Owners had no idea what had happened to their ships and cargoes without letters. The desire for mail was urgent and insatiable.

Toppan and Orne then urge Lovett to conduct the voyage as quickly as he can. Time cost money—a longer voyage meant more time to suffer crew desertion for better wages or personal reasons, more time to succumb to disease, more time to suffer accident. More time underway meant more wear and tear on the vessel and cargo, and the tying-up of capital in her. The owners close the letter by reminding Lovett of how much trust they are placing in him.

In a postscript, they acknowledge the possibility of interception by privateers or enemy cruisers—this was during the Seven Years War. Maritime predators frequently demanded ransom to let the ship go rather than take her as a prize—it was less profitable than winning a successful judgment in a prize court, but much easier, much faster, and less risky. It allowed the predator to continue the cruise and intercept many more vessels, it took away the risk that the prize would be re-captured by the enemy, and insurance policies would reimburse at least some of the ransom payment.[47] In such case, there would be nothing for it but for Lovett to get the best terms he could to preserve his own freedom and that of his ship and most of his crew—though it was common practice to take one or two crew members hostage, to be held until the ransom bill was paid (along with a fee for the hostages' sustenance).[48]

What comes through most clearly from one of these letters is the sense of risk. So many things could go so wrong—market conditions, bad business decisions, disease, accident, weather, war—that it is no wonder owners fretted constantly about their ships at sea and cargoes for sale. The risks these people took seem outrageous to us. Yet they had no choice but to accept them and try to mitigate them if they were going to engage in maritime commerce. The ship itself was just one of those risks. The risks we associate with these ships, that seem so stark to us—their plank-on-frame wooden construction, their wind-dependent power source manipulated by human strength with little

47 Leonard, "Underwriting Marine Warfare," 176.

48 Chet, *The Ocean is a Wilderness*, 57. Chet in turn cites multiple sources.

mechanical advantage, their relatively small size by our standards, the exposure of their crews to the elements and to physical injury or death during normal operations—seem less aberrant when considered in the broader context of risk laid out so clearly by a simple letter like this.

We could hope that letters from masters back to owners would contain explicit mention of vessel performance and the master's technical management of the voyage. Alas, such is rarely the case. The closest they usually come is to comment on the duration of the passage and whether the master considers that duration "good" or "indifferent"—not very good. The overwhelming concern of these letters is business—the business of the master acting as the owners' factor. The following is as good an example as any I found. It is from Capt. George F. Blunt from Cape Francois in Saint Domingue (now Haiti) to owner William Hale, 29 August 1799, describing a poor market at Trinidad and his decision to sail from there to Grenada, after selling only the cargo of lumber on deck—not in the hold—for cash:

> … The wind being far Northerly, a strong Lee Current and the Brig still Very Crank, I could not gain sight of Grenada, and the first Land I could fetch was the West End of St. Croix, the only offer I had there was 40 Doll.s for my Boards only [which is what he got in cash at Trinidad for his deck lumber], and that payable in Rum, there I gained Intelligence of the Cape being open to us, and thinking it not prudent to run to a French Port with Naval Stores on board I sold my Pitch @$4. Tar & Turpentine @$3 & Varnish @55 Cents per Gall. and took Rum in paym.t I then proceeded on for this Infernal port, from whence we had at St. Croix the most flattering prospect of making a grate Voiage on Lumber I arrived here yesterday, and the Current prices of this Day. are, Boards $15, Shingles $1 1/2, Staves & Headings all together $16, Beef $7 1/2 Fish 2@3 if good …[49]

Mainwaring's *Sea-mans Dictionary* says that a vessel is "crank-sided when she will bear but small sail, and will lie down very much with little wind …"[50] and "a crank-sided ship can never sail well by the wind."[51] That is because a sailing

49 George Blunt to William Hale, 29 August 1799, box 2, folder 3, William Hale Correspondence, 1791–1839 (1984–025), NHHS.

50 Mainwaring, *The Sea-mans Dictionary*, location 1997.

51 Mainwaring, location 3188. For a better sense of this: In a letter from Capt. John Crowninshield to owner Elias Hasket Derby from the ship *Henry* on 4 July 1791, he writes that, although otherwise a good sailer, the ship "is however very crank & tender & has in fact come so far almost upon her b[eam] ends, & her Spars are so [stighty?] (The top masts) and she has so little spread to her rigging, that we have sometimes been obliged to

vessel must be able to stand up to the wind to sail to weather, as the wind has its greatest heeling force from forward of the beam (amidships). So, Blunt is complaining that his brig will not go to weather well, and thus his options for his next landing are further restricted by that.[52] Blunt, as his other correspondence in this collection reveals, is a perennially unfortunate master, a perennially self-pitying one, or both. He bewails his misfortunes and takes pains to ensure that his owners know that nothing is his fault. Is the brig crank? We cannot know. Perhaps her rig was out of tune. Perhaps she was loaded out of trim or balance. Perhaps she was just a crank design. Perhaps she was not crank at all and the conditions forbade making Grenada. These masters' letters, however, represent a consistent effort at self-justification very much in the writers' self-interest, so we should consider them with that in mind.

Most masters' comments to owners on vessel performance—if any—were offhand and vague. After a passage from Massachusetts to Canton in 1800, Capt. Bently of the Brown East Indiaman *Ann & Hope* wrote home that "… our ship Behaves Very Well and tite.."—which is all he had to say about her performance on a voyage of five months and ten days.[53] In 1805, John Odiorne wrote from Norfolk to James Locke in New Hampshire that "We have at last arrived here after a long passage of nine days, we are all well and the Ship behaves extremely Well, the Ship Answers your expectations as it respects Sailing to the uttermost, you will without Doubt have received Capt. Adams' letter on this …"[54] The following year, from Alicante, Spain, he writes: "The Ship continues good and strong and sails exceedingly well upon an even keel we out sail Capt Robt Follinsbe [?] and Capt Rob.t [?] Bagley by one half which you must thnk is very gratifying to me.---------------------------"[55] These are typical.

This one is not typical, and perhaps here the exception proves the rule:

> The Rangers sailing does not answer the general expectation oweing in a great measure, to her being too deep, very foul, & overmasted. her ballast Laid too high, on Account of its improper Quality, for a Ship of this

double reef when we might have carried top Gall[t] sails—we make just enough water to keep her sweet…." (Crowninshield to Derby, 4 July 1791, box 1, folder 9, John Crowninshield Papers, PEM-Phillips). To double reef is to drastically shorten sail area. To "make water" is to leak; to "keep her sweet" is to keep enough seawater cycling through the bilges to keep that water from becoming too stagnant and fetid.

52 Meaning she will not sail as close to the direction from which the wind is coming as she might otherwise, or as another vessel might.

53 Bently to Brown & Ives, 18 January 1800, box 475, folder 9, Brown & Ives, JCB.

54 John Odiorne to James Locke, 30 July 1805, box 1, folder 6, Shipping Papers, NHHS.

55 Odiorne to Locke, 7 July 1806.

> construction; this, with the extraordinary weight of her lower Masts, occasioned her being very crank. I am paying my whole attention to remedy these inconveniences as much as possible; I am shortening the lower Masts, shifting the Main Mast further aft, and mean to Ballast with Lead as that Article will stow under the lower tier of Water, the less quantity will be sufficient, of course the Ship will be so much the lighter, and Sail so much the faster; and we shall then, I hope, be able to Stow the Cables under the Platform.[56]

That was written by John Paul Jones to John Wendell, 11 December 1777, at the conclusion of USS *Ranger*'s first passage as a new U.S. warship. Jones was expected to take prizes and engage the Royal Navy in combat at public expense, and the performance of his vessel had to be a top priority to both him and his paymasters. Merchant masters did not bother their owners with details like this. The rest of the letters from merchant masters from which the brief quotes above were excerpted are preoccupied almost exclusively with details on market conditions, prices offered, declined, or accepted for goods, and any legal issues the master is handling or seeking advice on from the owners. That material consistently supports the impression we get from the Sailing Instructions from owners—that they are preoccupied with the ship's business, not the ship. It is rare to read praise or complaint about the vessel itself. Where it exists, it is more likely to comment on the ship's condition than on her performance qualities. Those seem more or less taken for granted, as a known quantity. She does what her owners and masters expect her to do. If only the markets could!

Once the vessel made port, the ship's log shifted to the business of selling and unloading cargo, procuring and loading a new cargo, trips into town—or farther—on ship's business and for mail, repairs and reconditioning of the ship, watering and provisioning, and concerns related to the retention, desertion, and recruitment of crew, as well as their health, or lack thereof. These sections remind us that, while voyages were long by our standards, time in port was even longer. While maritime economic historians have shown that efficiencies in this area improved over time, contributing materially to the growth in shipping productivity, some of the basic realities of the entire period militated against, and limited, such improvement.[57] While populations grew,

56 John Paul Jones to John Wendell, 11 December 1777, box 1, folder 7, Langdon Family Papers, 1762–1972 (M 1981–111), NHHS.

57 For the inherent inefficiency of trying to load cargoes from scattered plantations, see James Hedges, *The Browns of Providence Plantations: Colonial Years* (Cambridge, MA: Harvard University Press, 1952), 28. Hedges describes trying to load molasses at Surinam. The best-known example of a particular trade that improved its cargo-handling and

demanding more goods and services, and the density of population centers and plantations grew concomitantly, allowing ships to dock or anchor closer to more sources of product, those changes only went so far. Density was still low enough in most places—and land transport difficult enough—that ships typically spent weeks or months waiting to finish loading and unloading even late in the period. Financial, technological, logistical, and political obstacles stood in the way of artificial port and harbor improvements in the period, although major projects were attempted—and to some extent successfully concluded—later in the period. These, however, were most important in major English port cities—London, Liverpool, and Bristol.[58] Business networks grew and gained experience, but communication was still no faster than ships could sail, although there were more ships to deliver mail and news. Market volatility meant that masters might drop anchor fully intending to fulfill their sailing instructions, only to have that hope dashed by prices and product availability. Owner William Hale's papers contain a printed letter soliciting his business from the House of Jacques & Theodore Rocheteau & Company of Surinam, agents:

> ... There is no market, more exposed to sudden changes, than that of Surinam, and it has often happened, that in the short space of a week, sometimes of one day, a voyage, that would have been the most advantageous, had the vessel arrived, but a little while before, proves ruinous, from her arriving two or three days later; it would there, be essential, to keep up a correspondence, with some good character there, in order, to be advised, what may best suit the market, at any particular time.[59]

Like most advertisements, this one claims more than was likely to be delivered. While it was always advantageous to have a reliable correspondent at a

---

loading efficiency dramatically over the period is the Chesapeake tobacco trade. See Jacob M. Price, *Tobacco in Atlantic Trade: The Chesapeake, London, and Glasgow, 1675–1775* (Brookfield, VT: Variorum, 1995); and John F. Wing, "Shipping Productivity in Maryland's Tobacco Trade, 1689–1759," *International Journal of Maritime History* 20:2 (December 2008): 223–39.

58 See Francis Hyde, *Liverpool and the Mersey: An Economic History of a Port, 1700–1970* (Newton Abbot, UK: David and Charles, 1971); Gordon Jackson, *The History and Archaeology of Ports* (Tadworth, UK: World's Work, 1983); Adrian Jarvis, ed., *Port and Harbour Engineering* (Aldershot, UK: Ashgate, 1998); and Alan F. Williams, "Bristol Port Plans and Improvements of the 18th Century," *Transactions of the Bristol and Gloucestershire Archaeological Society* 81, 1962.

59 Rocheteau & Co. to William Hale, 1799, box 2, folder 1, William Hale Correspondence, NHHS.

destination market, if the market at Surinam really were that volatile—and it may well have been—it is doubtful that a correspondent would be able to shield owners and masters from that, given the speed of communication versus the speed of the market's ups and downs.

Perhaps demand for lumber was high on a certain island when you shipped your load of lumber, but if you knew that, your competition probably did, too. If yours was the tenth schooner to drop anchor there with a load of lumber within the month, chances were that your master would either find your cargo unsalable there, or would be selling at a loss. Logs and masters' correspondence are full of just that misfortune. If the master decided to sell at a loss and procure a cargo to make up for it, that would likely take weeks or even months. He, or your factor or agent in that port if you had one, would have to find a buyer, assess that buyer's liquidity or credit, come to terms, do the paperwork, have the cargo transported to the vessel, have the crew and perhaps the longshoremen, in a larger port, load the cargo—which could be accomplished only after the original cargo had been unloaded, either to begin incurring warehouse fees ashore or, with better luck, delivered to the buyer—and that unloading had to be done with care to avoid possible capsize from instability. If the cargo needed to be fully unloaded before a return cargo was available, the crew was frequently put to work loading ballast, which was exhausting and time-consuming, and they might have to unload that ballast again once a new cargo was procured. Indeed, it was worth finding cargo that could serve as ballast. Recall poor Captain Blunt, sailing all over the islands trying and failing to find good prices for his cargo. In 1799, he wrote from Antigua, reporting that he had loaded 1,150 bushels of salt "@ 3/3. it will save buying Ballast at 1 Dollar per ton. & tho high will Answer as well as Rum …"[60] If a cargo could serve as ballast, that was worth considering as part of its value.

If all this were finally accomplished, you might run into trouble with the officials and have to waste even more time trying to sort out your clearance and port duties. The sad saga of Captain Blunt has gone from bad to worse. From New York, he writes of his latest misfortunes in Jamaica, from whence he has just landed after a 37-day passage:

> … to Crown the Voiage and Compleat my unhappyness, was refused a Clearance by the Mean and Dirty Colle[cto]r for having 6 Puncheons Saint Croix Rum on board … and had been knowingly and fairly reported for Exportation, and Bonded in the penalty of £2000, at Port *Antonia* where I first Entered the Vessell, in Jamaica, I Immediately made my

60 Blunt to Hale, 7 February 1799, box 2, folder 1, William Hale Correspondence, NHHS.

> *will*, and set off for Kingston by land to wate on the Govonor , (thinking never to return from such a Journey in such a Country), after daneing attendance on the Govonor 6 or 7 Days I obtained his Special order for my Immediate Clearance, and returned to Montego Bay on 3d Nov.r and saild on the 4th, But previous to all this, Mr. Bolt of Saint Anna, who had promised me the Freight for New york as far as the Word of a Gentleman was binding Sold his Sugar and Coffee and Disapointed me there, -------------- after getting on the Coast of America finding the weather Cold and Windy, my riging & sails much shattered from beating through the Gulf. and being short handed, (John Brown having deserted at Montego Bay) 15 Chances out of 20 of Being Blown off to the West Indies again the Rivers being froze up and many Dificulties attending the procuring another Cargo at Portsmouth the low prices it bairs (*Our Country Lumber*) in the West Indies, the Wretched payments we meet with here, *Rum* the sacrifices we must have Made to fitt her away again Out of *Rum*, putting all those together and suming up the Whole, I judged it Most prudent for the Intrest of all to make this port, hopeing you will Not Censure the step, ---- here we can realize the Cash, and obtain a higher price than at any port in the United States, ---------- I Could have Chartered the Brig at Montego Bay. to bring a load of staves and heading from Wilmington or Savannah at 50 Dollars Freight payable in Rum, I would not accept it.[61]

Blunt then proposes a venture to Wilmington, North Carolina. The owners confer amongst themselves and reject it firmly. It is apparent that by that point whatever confidence they had in Blunt's business judgment was limited. Some of that may have been that all his vacillations came through in print, in his copious outpourings to the owners. Most of it was surely that the owners had sat home through a series of miscalculations and consequent losses.

On 19 June 1794, the Brown brig *Friendship* left Providence for Bayonne, France. Her voyage log ends 18 July, for a 30-day crossing—unremarkable for a transatlantic passage in that direction. Master Henry Olney writes in the log that he has unloaded her full cargo of flour, rice and fish by 8 August. Three days later, they took on three boat loads of ballast—not doing so would have been a serious and avoidable risk. By the 13th, they were working on repairs and maintenance. Those lasted the rest of that month and, apparently, all of September. On the last day of that month, Olney writes that they are just sitting there waiting to get paid for the cargo they unloaded two months ago, and for official permission to load brandy—a protected commodity in France. That

61 Blunt to Hale, 12 December 1799, box 2, folder 5.

loading did not begin until just before Christmas. On 13 February, they were waiting to load more. Same for 21 March. Same for 2 May. They finally sailed for St. Thomas on 26 May. They had been in Bayonne for over ten months.[62]

These, then, were some of the endeavors, risks, and obstacles faced by ship owners as they attempted to make their ships pay, as revealed in the archives. Masters and crews shared in the challenges, risks, and reward of the enterprise, too, although they were living them aboard the vessel and in destination ports, rather than in their offices at home, waiting for news by mail and hoping that news was good. It is time to explore how those aboard experienced the ship.

---

62 Logbook of brigantine *Friendship*, 10 June 1794–1795, box 733, folder 1, Brown, Benson, & Ives, JCB.

CHAPTER 6

Sailing and Surviving: People and Labor Aboard

The previous chapter made clear that the capabilities and fortunes of the ship's master had direct bearing on the immediate fortunes of the merchant owners. Masters were in a middle position between owners and crew—and between owners and their ship. While acting as the factor of the owners—and sometimes owners themselves—they also had to supervise the running and care of the ship and the management of her crew. Here, we will consider how masters and crews experienced the ship, how mariners acted aboard and upon the vessel, and how passengers—whether voluntary or captive—experienced existence aboard one of these ships—an experience profoundly alien to the uninitiated.

From the responsibilities of the master, including the navigation of the ship, to the work demanded of the crew, to the conditions aboard ship for passengers in starkly varying circumstances, the experience of humans aboard early modern merchant ships has been addressed by historians studying labor, migration, slavery, and applied science. Primary sources, written by mariners and travelers alike, provide rich material from which to write such histories, and the data compiled for quantitative analysis of the slave trade and migration have revealed the true scale of human movement in the Atlantic in this period.[1] This chapter will tie the experiences of those humans aboard ship as closely as possible to the ship itself as a large, complex material object operating in a hazardous and inhospitable fluid medium. It is important to point out how thoroughly skilled labor was integrated into every aspect of vessel operation, and the experiences of passengers will help to highlight that by contrasting the "risk-management" capabilities, and thus expectations, of those who possessed the knowledge and skills of maritime technology and those who did not. The concept of "risk management," discussed in the previous chapter, is perhaps the most comprehensively helpful one for a modern student to

1 On the master's duties and navigation, see Hicks, *Voyage to Jamestown*. For important literature on the labor and identity of sailors, the experiences of slaves and slaver crews, and first-person accounts by period mariners, see the Introduction, 2, n. 4. For recent work on the experiences of transatlantic passengers in this period, see Stephen R. Berry, *A Path in the Mighty Waters: Shipboard Life & Atlantic Crossings to the New World* (New Haven, CT: Yale University Press, 2015). The monumental on-going quantification of the transatlantic slave trade is the *Trans-Atlantic Slave Trade Database*, which may be accessed at http://www.slavevoyages.org.

 | DOI:10.1163/9789004426344_008

understand this technology and its relationship to those who used it. In this chapter, I will discuss "risk" and "risk management" on board the ship, rather than in the merchants' books, although there was ample overlap between the two. This human-centered discussion of risk, centered on danger and safety aboard, is a necessary prologue to the detailed discussion of operational technology in the next chapter.

The master of any of these ships, no matter how small and humble, had his quarters in a stern cabin. In a multi-decked vessel, that cabin would be on the main deck, or even above it. The location gave the master quick and easy access to the quarterdeck, poop deck, or aft section of the weather deck in the simplest vessels. The traditional location for directing the vessel's operation allowed the master (or mate—second-in-command—if the master was off-watch) to view the entire deck forward, as well as the entire rig. It was also the location of the helm—the steering station. The ship's hierarchy, too, helped dictate the location of the master's quarters. The ship's motion—especially pitching—is generally least in the stern, and in most conditions, the stern would be the driest location for accommodations. The upper stern could be safely fitted with windows for light, something those in the forward section had to do without.

Masters' cabins tended to be large enough for the master to take his meals there, and to spread out his charts and logbook for work. On smaller vessels, he was probably the only member of the crew to have private sleeping quarters. On the larger ships, the mate might also have a cabin, and passengers of some wealth and social standing might enjoy that privilege too.

The rest of the crew, however, inhabited the forecastle or fo'c'sle, whose only sources of light and air were deck hatches and *companionways* (short staircases from openings in the deck above), which had to be secured in foul weather, such that the fo'c'sle would soon become foul as well. When conditions permitted, wind scoops made of canvas were suspended from above, catching the breeze and funneling it below. In hot, clear conditions, sailors routinely slept on deck when off-watch to stay cool and breathe more easily. Unlike the master's cabin on larger vessels, which was frequently equipped with its own private *head* (toilet—really just a small closet with a seat over a hole open to the water below), the head for the crew was in the head of the ship, a wet and boisterous place to be in any sort of sea but the calmest.

The area between the officers' quarters aft and the crew's quarters forward was known as *amidships* (also called the waist). This was a space for deck work. Most of the sail-handling that did not require going aloft was accomplished here. Novice seamen were generally assigned their work stations here, as tasks required less skill than those aloft. The rig of a square-rigger was the work space of able seamen—those who had acquired at least intermediate skills.

They constituted the middle ranks of those aboard; their work space would rarely see a novice, on the one hand, or a master, on the other.

So, while not nearly as rigid (or rigidly-enforced) as on a warship, the hierarchy aboard divided the ship into distinct spaces inhabited by members of the ship's company whose status was determined largely by their skill and experience, although masters and mates also owed their positions to connections ashore—connections that, as we have seen, they maintained as best they could through correspondence. Indeed, the ability to write—and read, and compute—was a prerequisite for command.[2] Literacy and numeracy formed a threshold between the ablest seaman and the officer. No amount of sea experience and skill would make a man eligible for command unless he acquired those skills. For, not only was he expected to correspond with the owners and act as their business factor in far-flung ports, he was responsible for navigation, and that required an understanding of arithmetic, geometry, and trigonometry. The navigational skills of masters varied widely, and the aggregate of that skill in the fleet increased in the eighteenth century along with the increase in accuracy of the tools and techniques of navigation. In any ship at any time in the period, however, the master had to be able to practice, with at least rudimentary competence, the reckoning and recording of the ship's position by landmarks, *soundings*, compass course and speed, and celestial bodies.[3] He had to keep the ship's log, both a practical and legal record of the ship's progress, the sea and weather conditions, and the goings-on aboard.

We are fortunate that entire ships' logs do survive in some number, especially (as usual) for the eighteenth century, and most especially for the late eighteenth century. It is worth taking note of what a ship's log contained, as they were standardized in format and content. In the late eighteenth century, printers even produced blank ones for sale. They are on heavy paper and stoutly bound, contributing to their longevity in a hostile environment of moisture, heat, and rough conditions.

2 On the division and use of lodging space aboard an early modern ship, see Niklas Eriksson, "Lodging in a *Fluitship*: The Material Setting of Everyday Life on Board *Anna Maria* of 1694," *Journal of Maritime Archaeology* 10.1007/s11457-015-9134-3. http://dx.doi.org/10.1007/s11457-015-9134-3 (March 2015): 1–19 (print journal 10:1 (April 2015), 45–63.

3 *Soundings* are measurements of the depth and samplings of the composition of the sea bottom, taken simultaneously by lowering a lead weight attached to a long, knotted line—called "heaving the lead." The bottom of the weight was concave, and filled with tallow or wax, so that it picked up a sample of the bottom material—sand, mud, gravel, etc. Not only did the depth of water tell the master much about where the vessel was, but bottom compositions were also part of a navigator's body of knowledge and recorded on charts. A vessel was said to be *on soundings* when the water was shallow enough for the depth to be measured with the lead line, and at that point the master would order a sharp lookout for land.

TABLE 4 Transcription of pages from log of snow *George*, 1805–6

H	K	HK	Courses	Winds	Remarks on Monday November eighteenth 1805
2	...	...	...	...	At 4 AM Weighed Anchor from
4	...	...	...	...	Ready Island in Company with
6	6	–	E B S	NW	several Sail, fresh Wind NW
8	5	1	–	WNW	At 4 Made Cape Henlopen bore
10	6		–	–	West. Dist 4 Leagues off—All Sails
12	6	1	...	NW	Set—At 11 in Stearing Sails—
2	6	1	...	–	At 7 AM set top G Sail, at 11 took
4	6	1	–	–	in top G Sail—& took two Reafs
6	6	1	–	–	in each top Sail—
8	6	1	–	–	Three Ships in sight—
10	7	–	–	NNW	Ends with fresh gales—
12	7	–	–	–	NOTD—

Course	Dist	Dif lat	Depar	Lat by DR	Lat by Ob	Mer Dist	Lon made	Lon in	Bearing & Distance
E ½ S	141	14	140	38.32	11–1	140	2 – 58	72 – 07	– –

next page

H	K	HK	Courses	Winds	Remarks on Tuesday November 19th 1805
2	7	–	E B S	NNW	The first part of these 24 hours a hard gale
					at 2 PM Dbl reef[d] the Main Sail. At 3 ship[ped]
4	7	–	...	...	a heavy sea which caused a Barrell of
6	6	1	...	...	Beef in the Cabin, that was placed at the
					Weather side to give way—it came down
8	6	1	E B S ½ S	...	with much Velocity & swept every thing before
10	5	1	ESE	...	it—Our Stove, tho small, was fixed firm nearly
					nearly in the center of the Cabin with large
12	5	1	...	...	16 Nails drove thro each leg into the floor, also
2	5	1	...	...	an Iron band which came over the tops
4	5	1	...	...	of & each end secured to the floor in like manner,
					seemd not in the least to obstruct its
6	5	1	...	...	passage—However the Stove is no way
8	5	1	...	...	injured & will be fix in its former station.

Table 4 is a transcription of an excerpt from the log of the snow *George*, underway from Wilmington, Delaware to Cork, Ireland in November 1805.[4] On this page, we see the end of one day's entries at the top, followed by the next day's entries, for Tuesday, November 19th. The columns are consistent in logs throughout the period. To the far left is the time, in two-hour increments—ship's time was kept noon-to-noon, since that was when the master took his sun sight, weather permitting, to establish the vessel's latitude as best he could.[5] Sand glasses and bells kept time and told the officers when to change the *watch*—the portion of the crew on duty. Running twenty-four hours a day while underway, ships needed a time system that, like so many aspects of life on board, was distinct from that used on land.

The next columns are for measured speed—Knots and Half-Knots. So, a "5" in the K column and a "1" in the HK column means 5 ½ knots. This was measured using a *chip log*—a flat piece of wood tied to a string with knots tied in it at regular intervals—hence the measurement of speed in "knots"—wound around a reel. The chip was thrown overboard from the stern, and the line let out, while an assistant watched the sand glass. However many knots ran out in a measured interval of time was the speed. The concept is similar to taking a person's pulse.

---

4 For what *George* would have looked like, see Figure 2, No. 2 (top row, second from left). I have transcribed this excerpt exactly, except where noted in brackets. *George* (ship) logbook, 1805–1806, Am. 6823, HSP.

5 While navigators (and surveyors) employed various instruments to accomplish the task, the principle of a "sun sight"—or a sight of any celestial body—was the same. The instrument, if properly used and without too much jostling, measured the angle between the celestial body and the horizon. From that angle, the navigator could calculate, with the help of astronomical tables carried aboard, a line of position—a line upon which his vessel was located at the instant of the sight. A sight of the sun taken exactly at noon local time—which means when the sun was directly overhead—would yield the vessel's latitude. Longitude was much more difficult to establish. Because the Earth is rotating so fast—about 1,000 mph (1,600 kph) at the equator—a navigator using a sight on a celestial body must enter the astronomical tables with an exactly-correct time, or introduce an error measurable in miles (km), rather than yards (m). A chronometer suitable for sea duty was not available until the end of the period. The alternative, a set of complex calculations called "lunars," since they were based on the position of the moon, only became available with the publication of the Nautical Almanac of 1767. See George Huxtable and Ian Jackson, "Journey to Work: James Cook's Transatlantic Voyages in the *Grenville* 1764–1767," *Journal of Navigation* 63 (2010): 204–14. Longitude by lunar distance, though, was beyond the abilities of all but mathematicians, skilled astronomers, and the best-trained navigators—like Cook. It was not a typical procedure aboard ship. Because latitude was so attainable, and longitude was not, ocean navigators from the sixteenth century onward practiced "running down the latitude," in which the master sought the latitude of his destination and then tried to remain on that line until he reached it. See Derek Howse, *Greenwich Time and the Longitude* (Oxford: Oxford University Press, 2003).

This is speed through the water, not speed over the ground or absolute speed. Since absolute speed is distance covered divided by the time it took to cover that distance, absolute speed is better. Calculating absolute speed, however, required fixing the vessel's position, and at sea, that required knowing both latitude and longitude. By keeping up with speed through the water and time underway since leaving a known fixed point, masters could estimate their position using deduced, or "dead," reckoning—abbreviated in the log as DR—but this was vulnerable to significant error caused by currents from ahead, astern, or abeam, slowing or speeding the vessel, or pushing it off course at an angle, or a combination of both—and by leeway, the sideways-pushing effect of the wind, and by the inherent inaccuracy of chip logs.

This is a good place to discuss vessel speed, as it is central to ship technology in a broader sense than at first may be apparent. While the log kept underway could only provide speeds over the ground or absolute speeds once fixed points of land were observed, those fixed points were noted in the log, marking the official beginning and end of a voyage. Finding the distance between those, and dividing that by the time the vessel spent underway from one point to the next—passage time—gives the average absolute speed of the voyage. Because sailing vessels almost never sail the most direct route from one point to the next, owing to their need to take advantage of wind direction, or fight it, sailors—and experimental archaeologists—find it useful to employ the concept of *velocity made good* (VMG).[6] Let us suppose that the destination is upwind—not unreachably upwind, but far enough upwind that the vessel must sail hard on the wind (close-hauled) to reach it. This is the most uncomfortable and usually least-efficient way to sail. Unless forced to, by the presence of a lee shore or other hazard, masters would rarely elect to sail close-hauled. They made better speed and subjected the vessel and her crew to less punishment if they bore away some. In our example, however, doing so means no longer sailing directly toward the destination. In general, that is preferable, as vessel speed will increase, perhaps making up for the less-direct route, or coming close.[7] So, if the straight-line distance between beginning and end point is, say, 2,000 nautical miles, the vessel might actually cover, say, 2,300 nautical miles to make the passage.[8] Her average speed through the water is 2,300 miles

6 See Whitewright, "Technological Continuity and Change"; and Anton Englert, "Trial voyages as a method of experimental archaeology: The aspect of speed," in Blue et al., *Connected by the Sea*, 35–42.

7 The answer to that question can be computed trigonometrically, and some masters would have known how to do that.

8 A nautical mile in this period was defined as one minute of latitude along any meridian of longitude. So, 60 nautical miles was one degree of latitude. Because the Earth is an oblate

divided by the time it took her to do it. VMG, however, is not the speed the vessel is making through the water—it is the speed she actually makes toward her destination over the course of the voyage, no matter her course changes and the changes in wind speed and direction. So, in our example it is the straight-line distance to her destination—2,000 miles—divided by the time it took her to get there—the time it took her to actually sail 2,300 miles through the water. Since, by that calculation, it took her more time to cover less distance, VMG is lower than average speed made good over the voyage. VMG is a true indicator of vessel performance, taking into account the vessel's sailing characteristics, the judgment of the master in route planning and in taking advantage of wind shifts and changes in sea conditions. We can calculate VMGs from the information presented in most log books, and we can use VMG calculations from replica trials to make comparisons. Passage times are generally unsuitable for evaluating relative vessel performance, subject as they are to so many external and variable forces.

Passage times have something important to tell us nevertheless. If we compare log speeds and average speeds over the course of a voyage, we see a clear picture of the speed ceiling of ocean sailers. Any good ship, properly laden and equipped and well-sailed, could make a good turn of speed. True tubs were rare in the Atlantic, as they would have pushed the capacity-performance continuum too far in the former direction to be satisfactory. It was common, in steady breezes and reasonably well-behaved seas, for vessels to make *hull speed* and even to sustain it for periods of time ranging from hours to days.

Hull speed is worth a brief explanation. A displacement hull—any hull that does not plane, so any ship hull other than that of a hydrofoil, cannot climb out of the trough of its own bow wave—the wave it makes as it pushes through the water.[9] That wave's length, then, limits the speed of the hull through the water. The hull speed is easy to calculate—1.34 multiplied by the square root of the vessel's length at the waterline. So, the hull speed of a vessel 80 feet (24.38 m) at the waterline—typical of a middling commercial ship in the eighteenth century—is just under 12 knots. Even respectable average speeds on ocean voyages, however, were but a fraction of the vessel's hull speed—2 to 3 knots was typical. A host of impediments lined up against a vessel's making anything close to hull speed over the course of a passage, including surface friction from

spheroid, latitude varies between the Equator and the poles. In the 20th century, an exact definition of the nautical mile as 1,852 meters was eventually adopted.

9 A vessel can, however, certainly exceed hull speed by surfing, and frequently will do so at sea. Surfing is accelerating down the front of a wave. The wave does not have to be breaking. The hull speed of the author's own 28-foot sailboat was 6.6 knots, but she would surf under full sail at 9 knots.

bottom fouling, periods of light winds or calms, storms which kicked up rough seas through which the vessel had to labor, foul winds—winds from unfavorable directions—contrary or side-setting currents, and gear failures.

The snow *George* dropped anchor at 5 pm on Thursday 19 December in Cork harbor. Calculating her average speed for the passage gives us 3.64 knots.[10] Note the speeds her master recorded in the log excerpt. 6 and 7 knots were common. Elsewhere in the log, he records speeds up to 10. This is the norm for these logs. If conditions had allowed these ships to maintain their best speeds over most of a passage, passages would have been at least twice as fast if not more. In reality, they almost never did, as the vagaries of weather and sea conditions retarded a ship's progress even when the crew did what they could to maximize her speed at all times and in all conditions.

So, minor differences in the performance speed of a vessel were likely to make very little difference in the passage times she could accomplish. The 3.64-knot average speed calculated for *George*'s transatlantic passage is perfectly respectable for the period. It is difficult, in that light, to imagine any incentive to exert effort, expense, risk, and loss of relative capacity in an attempt to increase performance speed in an ocean-going merchantman. What a vessel like *George* really needed, beyond a fair hull and stout construction, were good sails, sound rigging, proper maintenance, proper lading, and enough hands to accomplish the never-ending hard work of putting up and taking in sails to keep her moving at her best speed in the Atlantic's exhausting swings between dead calm and howling gale.

Returning again to *George*'s log: The Course column is for entering the compass course. Courses were read and recorded using a 32-point system. Given the limits of precision of every other means of measurement available, there was no need to bother with the precision offered by the 360-degree system on a modern compass. On a sailing vessel in motion in open water, it took skill to read an average course correctly at all, as the compass card swung one way and the next with the motion of the ship in the waves and wind.[11] The courses we see here are East by South, East by South ½ South, and East South-East. Just as with speeds, these are estimates, averaged from a period of time underway,

---

10 Distance traveled: 5,193 kilometers. 1 kilometer = 0.539 nautical mile, so 2,799 nm in 33 days, minus 1/2 day on the front end and 7 hours on the back end, added together = minus one day total, leaving 32 24-hour days. That's 768 hours. 2,799 nm / 768 hours = 3.64 knots average speed, for a ship capable of measuring 10 knots through the water.

11 On a modern small yacht, it is futile to expect to steer a compass course with more precision than five degrees in any sea—if the water is at all rough. That is slightly less than half a point by the older compass.

whose trustworthiness depended much on sea and weather conditions and the skill and alertness of the man at the helm. Winds were measured using the same point system, but described in terms of the direction they were coming from, rather than direction heading toward. So, a wind blowing out of the north, toward the south, is a north wind. A ship sailing toward the north is sailing a course of North.

At the end of a day's entries, we find the master's best estimates of how far the ship has sailed that day, in what direction(s), and where she is. In the Remarks column we can read his observations about the weather, sail changes, gear breaking and being repaired, which was almost constant—"hands occupied in mending rigging" or something similar is one of the most common remarks. Sometimes Remarks include crew behavior and morale, and the master's own assessment of the voyage's progress. Here, as so often happens in these, we are reminded of how violent this experience was. A barrel of beef broke loose when the *George* was hit by a particularly vicious sea, and it did much damage rolling and crashing about the cabin, the master notes. It could have killed someone.

The Remarks for Monday the 18th tell us that the *George* began this voyage from Reedy Island in Delaware Bay, making for the Atlantic in company with other outward-bound ships. Twelve hours later, the snow exited the bay and bore west with "all sails set." This was a sailing maneuver; Ireland lay far to the northeast. Seven hours after reaching the open sea, they took in their "Stearing Sails"—an older term not for "steering sails," but for "studding sails"—extensions to the main square driving sails set on booms at the ends of the yards to maximize driving power in light to moderate winds. So, the wind was building to some strength. The next morning, they set their topgallant sail (probably the main topgallant), and flew it for four hours before climbing up and taking it in again. They then took two reefs in each topsail—the next sails down the mast from the topgallants. Soon, they were sailing in "fresh gales," typical of the east coast of North America in mid- to late November. These gales would usually spin most or all the way across the North Atlantic on the prevailing westerly winds, so any vessels headed from North America to Britain or Ireland would likely be dogged by them periodically throughout the journey.

The next day's Remarks open with the *George* making way through a "hard gale;" the master ordered a double reef in the mainsail, which means at that point the vessel was flying minimal sail—storm canvas. It was under these conditions that the barrel of beef broke loose. For the remainder of the voyage, sailing conditions would range from storms to ideal winds to calms and back to storms, and the crew would make and take in sail in response, climbing up and down the rig, out onto the yards and back in again, hauling on lines and tackles on deck, as the snow plunged, heeled, raced along nicely, or drifted with her sails slack.

In the original log, carefully-drawn straight lines, neat entries, and the elegant sweep of penmanship offer an impression of tidiness and control that utterly belies the reality of what was really happening on a ship like *George* on the Atlantic. Here is, literally, an attempt to impose order on chaos, a strong and widespread human impulse, particularly obvious here, and especially urgent at sea—the primordial chaos. The neat system of habit of the logbook, whose entries were commonly made as the ship lurched and rolled and pitched through the roiling seas, also belies the reality that the navigational information recorded there could only be trusted to a strictly limited extent. A diligent master usually had a good idea of where he was, but clouds and fogs and driving rain could blind him long enough that his estimation of the ship's position was in serious error, and that could put him and the lives and property in his care on a reef or a lee shore. In the event of such a disaster, or a mishap less dire, the efforts of all on board would be directed toward damage control and, if it came to that, survival.

Among official documents and merchants' papers, we sometimes find *crew protests*. These are not protests in the common modern sense of the term, but statements made before a magistrate or notary by a delegation of a ship's crew after an accident or loss. They describe in detail what the crew had to say about how the accident or loss occurred and—most importantly—what attempts they made to avoid or mitigate it. Crews were legally obligated to do their utmost to protect the ship and the cargo up to the point where their own lives were in immediate danger. Protests were a crew's first official attempt at self-exoneration—with a letter to the same effect frequently sent to the owners concurrently—so they should be read with that agenda in mind. They are still valuable primary sources of information on vessels and crews in crisis situations, and thus on the risks inherent to the enterprise. A protest made at Liverpool on 20 August 1810 describes an accident whereby the *Anna* was fouled by a brig that dragged anchor while anchored with both bowers (largest main anchors) in the River Mersey, and because of the strength of the tide, her crew could not free her cable until the tide changed, and much of the cable was damaged. The docks were full and there were about two hundred ships at anchor. She was allowed to ground, which was normal in England, when they could not get her into the docks, but because the mud and sand washed away under part of her, she began to leak and later they found that she had hogged. She had about four feet (1.22 m) of water in the hold. They hired a surveyor and she was declared irreparable. They were under a pilot for all the maneuvers.

By making this protest, the crew was claiming they had done all in their power to prevent the damage, which was the fault of the other ship, the lack of dockage in Liverpool harbor, and nature. The ship was under a local pilot (expert guide who took temporary command to bring the ship into a harbor or river mouth).[12] Should the protest hold up, the owner would be obligated to pay the crew their wages, and they would be free from any prosecution before the law.

Whether from a cracked spar falling from aloft, an uncharted shoal in a gale, a tropical fever, or a splinter from an enemy shot, the risks to a mariner aboard ship were real and many. In fact, the culture of these sailors—the group identity they fashioned for themselves over time—was in large part a response to the risks posed by sailing these ships on these seas.

As a rule of thumb, roughly four to five percent of voyages in this period ended in the loss of the vessel.[13] That is a significant risk of disaster, and a much higher percentage of voyages encountered significant hardship and risk to life and limb. In order to parse the specific risks that we should consider, it is helpful to start by distinguishing between unavoidable and avoidable risks. Unavoidable risks were the risks imposed by nature—storms, calms, and fire, either from lightning or from open flame aboard. There was no alternative to fire for on-board lighting and cooking.

It is when we consider the actual work of the sailors that we make the transition from completely unavoidable to partially-avoidable risk, and thus begin to gain insight into their risk culture. The easiest, and safest, way to get to the top of the rig was to use what was called the "lubber's hole," an opening cut in the mast top platform that allowed the sailor to go further up the rig without climbing the *futtock shrouds*, which jut out from the mast to go around and above the top platform, requiring the sailor to climb them while angled backward from the vertical. This was much more dangerous, but could be quicker for a skilled seaman. More to our point, "lubber" was a derogatory term for someone—especially a novice seaman—who did not possess the skills of a more experienced hand, and the acquisition of those skills required the acceptance of a higher level of personal risk.

Once the sailor reached the yard, he had to climb out onto it to work the sail. This is much easier and safer to do with *footropes*, which are attached below the yard, allowing the sailors to put their feet in bights of rope similar in concept to stirrups, while resting against the yard itself, for more support as they

---

12 Crew protest, brig *Anna*, 20 August 1810, box 1, folder 16, Tredick Family Papers, NHHS.

13 Amy Mitchell-Cook, *A Sea of Misadventures: Shipwreck and Survival in Early America* (Columbia: University of South Carolina Press, 2013), 30.

handled the sails. Footropes, it would seem, came into common use sometime in the mid-seventeenth-century—perhaps as late as 1680.[14] Our authorities on period rigging and sail-handling think that footropes were introduced and spread, not as a safety measure as we would conceive of it, but to make it less awkward to furl and douse sails once the standard technique for doing so had changed.[15] In other words, if they are correct, footropes were adopted to make work more efficient, not safer.[16] If the new technique was also safer, that was incidental. Before footropes, working sails from the yards meant standing on, sitting on, or straddling the yards while handling big, heavy sails. This was quite dangerous, especially in foul weather, when everything was wet or icy and the ship was rolling and pitching. On smaller square-rigged vessels, upper yards with their sails attached were lowered to the deck so that the crew did not have to climb the rigging, but this was impracticable on larger ships. It would be easy to assume that, once fixed yards became the norm on most vessels, requiring trips aloft to accomplish most adjustments to sail area, the need for footropes became immediately apparent for crew safety. Again, the evidence does not lead in that direction.

A task similar to going out on the yard was going out on the *bowsprit* to handle forward headsails. Another long spar like a yard or a boom, the bowsprit jutted out over the water from the bow of the ship. As the ship plunged, the bowsprit would be doused with spray or even briefly hit the water. On today's sail-training ships, knotted rope netting is fitted under the bowsprit to help catch a lowered sail or a dislodged sailor, but this was a late nineteenth-century innovation.[17] This was a dangerous job, especially on eighteenth-century sloops and fast schooners with extremely long bowsprits and multiple tall forward headsails.

On sailing vessels today, crew wear harnesses and clip on to *jacklines*—stout ropes or cables secured to strong points on deck—to move around the decks, in bad weather or at night, or to perform tasks in precarious, exposed positions. This precaution prevents falling overboard should the sailor lose grip or footing, and has been employed routinely since the last days of ocean-going commercial sail in the early twentieth century. Should a sailor go overboard on a modern vessel, crewmates can deploy a life-buoy on a line as an immediate rescue attempt; a complement of these devices is stored all around the deck

---

14 Harland, *Seamanship*, 25.

15 Harland, *Seamanship*, 104; Harland also cites R.C. Anderson.

16 Speth also believes from his own experience that footropes were introduced for labor efficiency (Speth interview).

17 Harland, *Seamanship*, 81.

for such purpose. Sailors who find themselves in the water can deploy a useful skill to help save themselves—swimming. In the early modern period, most sailors could not swim.

The surviving documents on the original schooner *Sultana*, including her draught, show no indication of railings at the waist—the exposed central work area of the main deck. The U.S. Coast Guard would never have allowed the replica to be built without such railings.[18]

We employ a number of other safety features today, from fire extinguishers to automatic personal strobe lights, but those mentioned above were certainly within easy technological reach of early modern mariners. They were not absent because the technology to deploy them was not available or economical. They were absent because the dominant seafaring culture discouraged such things. The willing acceptance of personal risk—a certain sense of fatalism—was a key ingredient of that culture.

Sailors could have refused to work under the typical conditions found aboard ship. Indeed, sailors were known for desertion, mutinies, and even strikes. The reasons behind these revolts, when they involved risk and safety, let us know what these sailors considered acceptable risk, what they considered appropriate "risk management," and what they did not. In general, their standards seem quite low by modern standards; early modern people's general standards of comfort, safety, and health were vastly different from ours. They were closer to those of ancient peoples.

Sailors deserted or mutinied because they found the on-board discipline too harsh and arbitrary, the food too bad or rations too meagre, the labor unreasonably demanding—perhaps because the vessel was undermanned—the original terms of their employment altered, such as by a master's decision to prolong the voyage or make for a destination other than the one originally agreed-upon, or because they found more appealing employment elsewhere. Sailors struck over wages—either because they considered them unfairly low or because they had gone long-unpaid, a chronic problem in the Royal Navy. Strikes were not organized as they would be later, under labor unions, and were not common, but they did occasionally occur rather spontaneously, frequently accompanied by violence.[19] A more common strategy for avoiding service was

18 Drew McMullen, personal conversation, 16 September 2019.

19 Perhaps the most noteworthy example was the 1775 Liverpool Seamen's revolt, which climaxed with a bombardment of the Liverpool Town Hall by sailors employing guns removed from ships.

the withholding of labor—the refusal to sign on for a voyage. Sometimes, informal groups of sailors would make such a decision collectively.[20]

On the other hand, sailors routinely served on vessels whose structural integrity was severely compromised by rot and lack of proper repair—by the neglect of distracted, miserly, or financially-strapped owners. Sailors agreed to serve on voyages sure to expose them to severe risk, such as winter passages in the North Atlantic or slaving voyages to the lethal disease environment of the West African coast. They did grumble on occasion about a poorly-kept or ill-equipped ship, but their chief gripe in such cases was likely the extra labor it required of them, for which they would not be compensated any better than they would have been otherwise.

Sailors also agreed to serve on ships they might find themselves defending by force of arms, knowing there was a better-than-slim chance they would be attacked. They might be killed, maimed, or captured. They routinely accepted the meting-out of what we would consider brutal corporal discipline, so long as they felt that that discipline was fairly administered.

It is clear, then, that the physical-risk tolerance of early-modern Atlantic mariners was high by our standards. What about their strategies for "risk management"? In general, there seem to have been two. The first was a combination of skill and courage—the high value placed on the acquisition of the complex set of skills needed to serve a ship as an able seaman, and the courage to use those skills without hesitation for the proper operation of the ship and the safety of the crew and cargo. The second was a strong emphasis on a balance between the work performed and the risks accepted versus the compensation offered. Sailors who had not acquired the skills of an able seaman, and who thus were not as invested in the labor culture of the ship, were not paid as well as those who had, and who were thus invested. They might malinger if they could get away with it—and they routinely did. They might desert if they decided the work and risk were too great for the wages due—or if they were resigned to the assumption that they would never see those wages. It was up to the most skilled and experienced of the able seamen to enforce discipline on the ordinary seamen and encourage them to do their duty and thus improve their skills. Waisters—those novice or unskilled seamen who did not climb aloft, but were restricted to working sheets and halyards and braces on deck in the waist—mid-section—of the ship—did not take on the same level of

---

20 This nearly happened at Bristol in 1750 when ship's carpenter Robert Barker and other mariners who had signed on for a slaving voyage to Africa seriously discussed backing out because they believed the ship to be too heavily-laden to be safe. See Robert Barker, *The Unfortunate Shipwright*, 6.

personal risk as the able seamen did, but neither did they receive the prestige within sailor society, or the financial compensation, to which able seamen were entitled. Sailors who made some commitment to the trade willingly accepted personal danger. They accepted that the sea was a harsh mistress; they did not expect the sea to treat them fairly. They did, however, expect fellow mariners to treat them fairly, and could be quite particular about what that entailed. They knew what was under the control of human beings who took to the sea, and what was not. They made little attempt to mitigate the risks posed by nature, but they would go to great lengths—including the courts of law—to mitigate the risks posed to them by other human beings.[21]

Crises always provide instructive examples of human behavior, and we can certainly make some useful observations by considering the behavior of crews on ships in dire distress—especially as compared to that of any passengers who may have been aboard. The skills and courage required of a competent seaman were of course fully-employed in a crisis situation. Sailors, who were not known for their piety, had other things to do besides pray fervently for deliverance as their vessel labored through a hurricane or attempted to beat off a lee shore. They were working the pumps, repairing rigging, attempting to steer.[22] Some would turn to piety *in extremis*, once they felt there was nothing more they could do, but others never did, and instead would turn to a last-ditch attempt to save themselves, or at least numb the terror of imminent destruction.

The "risk-assessment" and "risk-management" of passengers must be considered separately. Passengers were not members of the mariners' subculture, and were often bourgeois or even aristocratic, so class differences must be taken into account as well. Those unaccustomed to shipboard life, and especially those accustomed to comforts ashore, found the ship a trying place, especially if the weather were less than perfect. Even in a calm sea, a ship under sail leaned one way or the other. Life on the tilt meant nothing was exactly horizontal or vertical. If the ship were sailing dead downwind, she would not

21 The records of the High Court of Admiralty, a large collection in the National Archives (UK) at Kew, contain a strong representation of actions brought by mariners against owners and masters. For examples, see Earle, *Sailors*, and Rediker, *Between the Devil and the Deep Blue Sea*, especially 205–53.

22 A notable contemporary account of dire peril and survival may be found in Charles May, "An Account of the Wonderful Preservation of the Ship Terra Nova of London," in Awnsham Churchill and John Churchill, *Collection of Voyages and Travels, Some Now First Printed from Original Manuscripts, Others Now first Published in English, In Six Volumes, With a General Preface, giving an Account of the Progress of Navigation, from its first Beginnings*, six volumes, third ed. (London, 1747), Google Books facsimile. For recent scholarship, see Mitchell-Cook, *A Sea of Misadventures*.

lean over to one side as she did when sailing closer to the wind, but in a following sea, she could take on a corkscrew motion that is notorious for inducing seasickness. Seasickness, resulting from a disturbance of the inner ear's perception of equilibrium, could, and frequently did, incapacitate passengers, especially in the first few days of the passage. If those afflicted did not recover soon enough, they risked becoming dangerously dehydrated and malnourished. Those capable of moving about had to watch for low, heavy deck beams that did not give at all when struck by human heads. Passengers had to hold on and crouch to move about, and would never be able to do so with the ease that seasoned sailors developed—a physiological adaptation that gave them their distinctive rolling gait ashore. Accommodations were cramped. Everything was damp. Passengers were in the way on deck if the crew were trying to make major adjustments to the rig, or handling cargo. In a serious storm or other emergency, passengers were sent below to the wet, lurching, swaying, crashing darkness.

Surviving shipwreck narratives were generally written by those well-removed from sailors in both culture and social class , so their accounts of how mariners behaved, and why, vis-à-vis their own attitudes and behaviors, must be approached skeptically.[23] Maritime law required passengers to assist as they were able in the event of impending disaster, but they were obviously limited to the most unskilled tasks aboard, such as working pumps, shifting or jettisoning cargo. It is also striking to a modern reader that, according to the shipwreck-survival accounts we have, female passengers almost never participated in any efforts to save a distressed ship or even to save themselves, but most often remained physically passive until they were rescued or died. Their survival rate in cases of shipwreck was much lower than that of men, despite the adage of "women and children first," which became a priority only in the mid-nineteenth century. The evidence is overwhelming that gender roles emphasizing male dominance and capability versus female dependence and physical weakness were so thoroughly internalized—by both sexes—that they trumped the preservation of life itself. That is powerful acculturation.

A sense of helplessness combined with abject terror surely encouraged an outpouring of desperate piety, and passengers—especially female passengers—usually resorted to that in crisis. Barred by unyielding adherence

23 See Mitchell-Cook, *A Sea of Misadventures*, which examines and analyzes one hundred such narratives from the seventeenth to the early nineteenth centuries. It should be noted that class distinctions existed not only between passengers and crews, but between different sorts of passengers and between officers and those "before the mast"—the working sailors. A preponderance of passengers were of course poor immigrants crammed into the holds of ships in that trade, but they did not tend to be literate.

to gender roles from taking a physically-active role in "crisis-management," female passengers' "risk-management" involved comforting children and attending to the "spiritual" role of pleading for divine intervention and engaging in religious ritualistic behavior, such as prayer and Bible-reading, intended to prepare the souls of those aboard for death and the avoidance of subsequent eternal damnation.

It is worth noting that overt piety was an important means of establishing and preserving what we would think of as bourgeois respectability, so it is no surprise that ships' masters also routinely engaged in it along with the passengers.[24] Sailors did so much less consistently. Some were committed Christians and behaved accordingly, but mariners in this period were a "notoriously irreligious and profane group."[25] As Marcus Rediker observed:

> Seamen were always perched on the curling lip of disaster; they were ever at the mercy of nature. Yet the very potency of the elements and the consequent need to battle them were key features in promoting a less religious view of the world among seamen. Self-help and solidarity, so utterly essential to survival, eclipsed religion in cultural importance and value. Irreligion became a basis for community, a triumph of sorts over a menacing natural situation.[26]

It is worth keeping in mind, too, that the few published mariners' memoirs we have were written by those who eventually attained command, and thus bourgeois status. How they characterize and judge the piety or lack thereof among more ordinary seamen—including their former selves—must be interpreted accordingly.[27]

Few passengers had the experience of the sea that mariners did, so their frame of reference on maritime peril was not finely-honed, nor were they capable, in most cases, of operating the ship's rig or steering system effectively. Without the recourse of the crew to their default "risk-management strategy" of skill and the courage to use it, passengers tended to resort to appeals for supernatural intervention. Passenger accounts of surviving crises at sea sometimes feature somewhat antagonistic attitudes on the part of passengers, who were not part of the ship, toward the crew, who were, and who were sometimes seen by the passengers as unnaturally cavalier about the danger at hand

24 See Berry, *A Path in the Mighty Waters*.
25 Mitchell-Cook, 57.
26 Rediker, *Between the Devil and the Deep Blue Sea*, 179.
27 I am thinking here especially of Samuel Kelly's memoir, already cited.

and frequently dismissive of the passengers' open displays of pietistic fervor in the face of what they were sure was their imminent destruction. If the crisis carried through until the destruction of the ship was deemed inevitable, the crew were just as likely to break into the liquor stores for a final bender and/or abandon the vessel than they were to join the passengers in pleading for divine deliverance.[28] This highlights the difference between the mariners' subculture and the wider culture. It also highlights the difference between an active and a passive "risk-management strategy." A master might exert great effort to retain control of his crew as disaster unfolded. Whether or not he was successful depended in large measure on the esteem in which he was already held by the crew, and the crew's own assessment of their chances of survival given a certain course of action versus another. Although maritime law required crews to do everything they could to save the ship, her cargo, and her passengers, so long as their own lives were not thus put in peril, sailors sailed for pay, and most were not willing to risk death if they thought they had a chance of survival by abandoning their work and their ships in a crisis.

Chances were better of avoiding such a dire eventuality if the ship were sound and well-equipped for the voyage she undertook. Unfortunately, keeping unfit or marginally-fit ships in service was not uncommon, and crews and passengers made decisions about what ships to serve on or book passage in based solely on *caveat emptor*, given the absence of any regulatory apparatus governing safety standards on commercial ships. In the eighteenth century, the marine insurance industry developed a classification system for evaluating the structural condition and seaworthiness of a vessel, both for assessing the vessel's value and for determining how much risk of damage or loss was posed by her condition and fitness for sea. As it first developed, the system placed primary emphasis on the vessel's age, as indicative of the likelihood of rot, structural weakening, corroded or loose fastenings, and cumulative damage. As vessels aged and their insurance classification changed, they were frequently put into less-arduous service, perhaps more than once, before being abandoned or broken up. Such practice attests to the fact that maritime interests accepted the general correlation between the age of a ship and her

28 Benjamin Franklin mused on the abandonment of flooding vessels by their crews, who, exhausted from pumping and convinced the vessels were lost, would take to the boats, only to find that the vessels did not sink after all. Franklin considers the possible explanations for this, including the eventual equalization of air and water pressure inside the hull and the buoyant nature of some cargoes. He does not attribute the crews' actions to cowardice, but to a decision, based on observation and past experience, that their situation was hopeless, when in fact in these cases it was not. See "A Collection of Papers on Naval Architecture," Part I, Paper IV, 23–24.

seaworthiness. The system would also come to take into account how the vessel was constructed and rigged and equipped. Lloyd's Register of Shipping is still in use for the same purpose.

No one suffered more onboard merchant ships than those whose circumstances rendered them the most helpless: poor immigrants and the enslaved. Their "risk-management strategies," like those of more privileged passengers, were much more circumscribed than those of the sailors operating the ship, and, *in extremis*, were too frequently limited to hoping and praying for survival, given that in most cases they had no access to the means of controlling any aspect of the ship's operation. In the case of the enslaved Africans enduring the Middle Passage, violent revolts were not uncommon, and frequent suicides by jumping overboard make more sense as "risk-management strategies" when we learn that some Africans believed their spirits could return home in this way, whereas if they reached a foreign shore, their spirits could not.[29] Normally, we take for granted that the risk of death is the risk people accept any other risk to avoid, but I have pointed out that such does not seem to have been the case when prescribed and internalized gender roles ran up against the threat of imminent death. When it came to accepting slavery, or permanent exile from their homeland, the same was true; a significant number of imprisoned Africans chose to risk death rather than endure slavery—or whatever other unknown fate might await them. If they were chained below during a crisis, however, they might well drown where they lay. In such a case, there was no available "risk-management strategy" beyond recourse to faith in an alternate reality.

As for the crews of slavers, they were taking a substantial risk by signing on for such a voyage to begin with. The disease environment on the coast of Africa was, for a European sailor, the most dangerous he would be likely to encounter anywhere. "Beware, beware the Bight of Benin, where few come out, though many go in." The old rhyme has survived, and it is obviously not about the transported African prisoners, as more than twelve million of them "came out," but about Europeans, who were almost certain to die if they remained on the coast long enough, and might well do so even after brief exposure to the mosquitoes carrying malaria and yellow fever. So, although the ship would be a hell for those chained below, she might well prove a hell for those charged with operating her as well. Adding to the disease risk, and to the certainty of having some of their number incapacitated by disease after reaching the Slave Coast, with workloads on the healthy sailors increased accordingly, crews had

29 Harms, *The* Diligent, 261–62. Of course, some occurrences of this behavior were surely due to despair and/or terror.

to accept the risk that they could be injured or killed in the event that the prisoners revolted. They were also taking profound psychological risks; the horrors they witnessed, and participated in, brought out the worst in many of them, as it did in their masters. Brutality toward the African prisoners by slaver crews was rife—as was brutality toward those crews by ships' masters.[30] Such behavior was far more prolific in the "Africa trade" than in others. Still, once sugar took off in the 1640s, this trade played an important role in the sustenance and growth of the British Atlantic economy. Merchant ships had to feed a growing stream of Africans into the people-grinder of chattel slavery to keep the cane mills rolling, which in turn, as we have seen, kept the shipyards building—and, for that matter, the entire Atlantic economy going. To get a quick perspective on the scale of the Middle Passage, assume that each slaver carried 100 prisoners. That was actually a fairly high number, so our rough estimate will probably be conservative. To transport twelve million prisoners one hundred at a time would have required 120,000 voyages, most of which occurred in the period 1640–1800. That works out to a rough average of 750 slaving voyages a year from Africa to the Americas; it ignores any other legs of those voyages. Since it was unlikely that any vessel would make more than one such voyage a year, a fleet of 750 ships would be required to accomplish that scale of transport. Thus, even if it had not been so in any other respect, or in any other circumstances—which of course it was—the merchant ship in the British Atlantic would still have been a carrier of violence and death on a huge scale—more so than its naval counterpart by orders of magnitude. For millions of people who sailed on such ships, they were a conveyance to the grave, whether that grave was at the bottom of the sea or on a foreign shore.

The transport of emigrants to the Americas also became big business—especially the Scots-Irish migrations after the mid-eighteenth-century.[31] For those packed together in steerage, misery and mortality were high, even when the ships kept the sea out.[32] Desperate people took desperate measures—sailing across the Atlantic in steerage—because they decided the risks inherent in that were less than the risks they faced if they stayed where they were. Some turned out to be right. Others did not.

Human violence of another sort—maritime predation—presented another grave set of risks to ships and crews at sea, as we have seen. It is a testament to the chronic nature of warfare in the Atlantic that the arming of ships decreased only slightly overall during the course of the period, even though the

30 See Christopher, *Slave Ship Sailors and their Captive Cargoes*.

31 Paul Butel, *The Atlantic*, trans. Iain Hamilton Grant (London: Routledge, 1999), 181.

32 See Berry, *A Path in the Mighty Waters*, 231–37.

effectiveness of the Royal Navy at protecting merchant shipping increased over time, especially after 1750, and the marine insurance marketplace grew substantially. The popularity of privateering—the commissioning of private ships of war, at private expense, during wartime—meant that merchant shipping was vulnerable to predation regardless of whether any naval cruisers were about.[33] Privateers and naval cruisers captured a prodigious number of ships during the imperial conflicts of the seventeenth and eighteenth centuries. This was not a peripheral risk; crews and passengers who sailed during wartime stood a good chance of encountering a predator on an Atlantic voyage—even a short inter-island run in the West Indies. How did they manage this risk, both physically and psychologically?

One way, of course, would have been to avoid sailing during times of conflict, but conflict was so frequent, and wars so drawn-out, that this was impracticable. Maritime intercourse in the Atlantic world could not stop during wartime, although it did change character and was made considerably more difficult. One way in which it changed character was in the shifting of resources to privateering, since, while it did not stop, maritime commercial activity was at times severely curtailed by warfare, with owners unwilling to take the commercial risks of sending their vessels to sea in the face of a powerful seaborne enemy. With some trades suspended or constricted for an indefinite period of time, serving on privateers was a way for sailors to remain employed. Privateering, however, was high-risk; most of those who participated did not come close to getting rich from it. Engaging in it meant exposing oneself to potentially-fatal violence as well as the natural risks of the sea.[34] While patriotic zeal and the hope of quick gain spurred a flurry of enthusiastic activity in the early stages of a conflict, that zeal tended to wane as the war dragged on and the big payoff failed to materialize.

At least privateer crews could take some comfort in knowing they were likely better-armed than their intended quarry. For unarmed or lightly-armed merchant ships, it was frequently far better to surrender quickly to a privateer or naval cruiser, once it became clear that she could not be outrun. Merchant crews, as noted, did not typically have any interest in dying for their masters' reputations or the owners' property, and that included fighting to the death to avoid capture. Rational "risk management" meant sizing up the odds and

---

33 For recent work on privateering, see Sheryllynne Haggerty, "Risk, networks and privateering in Liverpool during the Seven Years' War, 1756–1763," *International Journal of Maritime History* 30:1 (February 2018): 30–51.

34 See Haggerty; she compares the risks of privateering versus slaving as perceived by Liverpool merchants, based on their participation in both.

making a decision accordingly. Nevertheless, we know that the arming of merchant ships was ubiquitous early in the period and still common at the end of it. We need to know what labor requirements and risks a ship's weapons—both her own, and an enemy's—presented to her crew, should active defense of the ship be required. We should review the principal types of guns carried aboard merchant vessels and how they were used, in order to understand the physical demands they made on the ship and the crew.

Even the smallest vessels commonly carried small swivel guns on their upper (weather) decks. These were usually mounted on or by the rails. They were used against people, not ships, and could be operated by one person. Filled with scatter shot, they worked like large shotguns to kill, maim, or at least discourage those who might be preparing to board the ship or use their own deck weapons against her crew. Slavers had such weapons ready to use against their own prisoners, should they manage to escape their manacles and mount an insurrection aboard.

Swivel guns did not make significant financial, physical, or labor demands. Larger guns did. Carriage-mounted guns were secured by block-and-tackle arrangements to the ship's structural timbers. They sat on the weather deck, where they could be fired through the railings or ports cut in them, and on lower decks, where they could be fired through ports that were opened for that purpose and kept closed the rest of the time. Sailors had to muzzle-load these guns—load them from the outer end—with a bagged powder charge and a projectile—on these ships, usually a solid iron ball of between 6 and 12 pounds (2.72 and 5.44 kg). These were rammed down the barrel with a ramrod, the gun was "run out" by the crew's hauling on the blocks and tackles, and aimed by adjusting the angle of the barrel with a lever. The powder bag was punctured by plunging a thin spike through a hole at the breech, and the gun was fired by lowering a smoldering wick to that hole. The gun recoiled, jerking back on its tackles, after which the barrel was wet-swabbed to prevent a smoldering ember or the hot metal from prematurely exploding the next powder charge, and the process repeated. Some of the crew had to serve as "powder monkeys," fetching the powder bags from the powder magazine and carrying them as quickly as possible to the gun position. It took at least two sailors to actually operate the gun. This was obviously a labor-intensive process. Should the ship come under fire from an enemy, the crew were at high risk of serious injury or death not only from the projectile itself but from flying splinters of the hull and parts of the rig falling to the deck. Even in the age of cannon broadsides, too, hand-to-hand combat was still normal once the two vessels came into physical contact.

It is clear, then, that while crews might have risked a greater chance of capture by sailing on an unarmed ship, there was another set of risks they avoided by doing so; and, while the crew had no control over whether or not a ship was armed, they did have some control over which ships they sailed on. It would be instructive to know something about merchant sailors' attitudes toward guns aboard—guns they would be expected to use, at their own risk, should the vessel be singled out by a predator. It is not a subject they dwell on in their memoirs, probably because the presence of armament was ubiquitous on ships of the period. The fact that so many mariners served on both merchant and naval vessels in their seagoing careers would have added to the sense of normalcy of serving on armed ships.

A lone merchant ship captured by a privateer was likely in such a vulnerable position because her owners or master had chosen to eschew joining a convoy and elected instead to "run," hoping to make it to their intended destination unmolested. If a ship's luck held, the voyage might well be a success, with high wartime freight rates and prices sweetening the deal. If her luck did not hold, she would encounter a privateer or naval cruiser. At that point, a number of "risk-management" options might be available. The most attractive was flight. Depending on the relative speed of the vessels, the wind direction, and the position of any nearby land, the merchantman might be able to elude the predator. The skill and experience of master and crew were critical here. They might also choose to employ subterfuge at this point, perhaps by hoisting false colors. They would be aware, however, that the predator might well do the same. Seasoned professional knowledge of different vessel types and physical features distinctive to the building practices of different maritime powers might help the crew determine the predator's likely allegiance regardless of what colors she flew.[35]

A strategy of destruction of the enemy's commerce has been a prominent feature of more recent maritime warfare, but in this period, there was more of an emphasis on profiting from the enemy's commerce. This reduced the physical risk to crews and passengers, and in most cases, the interaction between a predator and an ordinary merchantman was less violent than an encounter between two warships. So, the particular culture of maritime predation meant that fewer merchant ships were destroyed and far fewer of their sailors killed than in, for example, the 1940s.

As with the culture of maritime predation, most physical "risk-mitigation" was external to the ship. Some of these attempts were given a great deal of

35 For details on those distinctive features, see R.C. Anderson, "Comparative Naval Architecture, 1670–1720, Part I," *Mariner's Mirror* 7:2 (1921): 38–45; "Comparative Naval Architecture, 1670–1720, Part II," *Mariner's Mirror* 7:6 (1921): 172–81; "Comparative Naval Architecture, 1670–1720, Part III," *Mariner's Mirror* 7:10 (1921): 308–15.

attention, particularly the improvement of navigation, which had direct royal support under Charles II (1660–85) and was the subject of treatises and published admonitions and the establishment of schools and curricula in each of the maritime empires and in the young United States at the close of the period. The art and science of navigation did indeed undergo rather dramatic change over the course of the period, in the development of more accurate and easier-to-use instruments, more accurate and comprehensive astronomical tables, and more accurate and complete charts.[36] By the early eighteenth century, there was a strong call for another "risk-management strategy" external to the ship: more and better lights to mark navigational hazards. That cry grew louder after Sir Cloudesley Shovell's Royal Navy squadron wrecked in the Isles of Scilly in 1707 with staggering loss of life. The resulting push did not call for safer ships or more capable officers and crews; it called for the reduction of risk from hazards external to the ship and completely out of the control of those aboard. More and better lighthouses and fixed lights would help ships avoid the situations discussed earlier, such as embayment or being driven onto a lee shore, from which ships frequently could not save themselves regardless of the soundness of their design and construction, the standard of maintenance to which they were kept, or the skill and courage of their masters and crews.

The early modern maritime enterprise was inherently high-risk in every way, for everyone involved in it, from owners to sailors to passengers. It had been since ancient times, and would remain so into the twentieth century. All participants in maritime travel accepted risks that we would not, because they had to in order to participate. Most shipowners who profited from maritime ventures did so modestly, not exorbitantly, and plenty of them lost whatever they invested. They were not exposing mariners to high physical risk and arduous labor while reaping fat profits themselves, in most cases. Besides, mariners had, and used, power to negotiate the terms under which they would accept the risks they themselves considered inherent to the trade, and they routinely rejected conditions they considered beyond the pale of what was acceptable—generally speaking, work conditions under human control that were unnecessarily harsh, rather than the physical risks imposed by the sea on a sailing vessel.[37] Their response to the latter was a labor culture based on the acceptance of risk, the reliance on skill to mitigate it, and the encouragement of a rather insouciant, whistling-past-the-graveyard fatalism and pride in their

36 For a recent scholarly assessment of research on the history of navigation, see Bruyns, "Research in the History of Navigation: Its Role in Maritime History."

37 Sailors also frequently carried aboard a small amount of trade goods that they were free to sell for their own profit.

distinctive identity as a subculture held together by this shared experience of the dangers of the sea and the fortitude to accept those dangers and work through them—literally. It was an active, not a passive, fatalism, and we see that clearly in the contrast in attitudes and behaviors of sailors and passengers during a crisis, as passengers had no recourse to the active "risk-management strategy" of the able seaman, while it was a point of pride to the latter that he not succumb to the fear of the sea. When all his skill had been rendered moot by the sea, breaking into the liquor stores or taking to the boats as a response to imminent death, rather than praying and obeying, was in a sense a refusal to break faith, even to the bitter end, with this distinctive labor culture from which the committed seaman derived his identity. He might be afraid, but he was more afraid to show it. As Benjamin Franklin observed, he was a coward "only in one sense, that of fearing to be thought afraid."[38]

Sailors, then, accepted the risks of the sea in part because they had in their own hands the ability to mitigate those risks. They were an integral part of their ships in a way that modern merchant sailors are not. Their knowledge, strength, and dexterity provided components of the operational technology of the ship that would later be transferred to mechanical contrivances. The numbers, skill, and strength of the crew was as important a factor in the performance of that ship, broadly considered, as anything else.

When Captain John Crowninshield reached Bordeaux in *America* in early February 1803, he wrote an utterly scathing letter home to the family firm, excoriating the owners with bitter sarcasm for sending the ship out on the cheap, at the expense of himself and everyone else in her.[39] He says he did not have enough hands to "carry sail handsomely" and so his passage was much slower than it should have been, and that although the crew did their best to take in sail for rough weather, "we are half man[n]ed" and it is not fast enough, "every rope breaks" and their sails were torn up. "… [T]his is the real effects of small economy—". He says a proper-sized crew would have made a difference of 20 days underway:

38 Franklin, "A Collection of Papers on Naval Architecture," Part I, Paper IV, 23–24.

39 John Crowninshield to George and Richard Crowninshield, 27 January 1803, Letter Book No. 7, Letters from France, 1803, John Crowninshield Papers, PEM-Phillips. I do not know whether Crowninshield actually sent the letter; it is in his papers, but he could have written it in a fit of pique and thought better of sending it. On the other hand, he might have sent it after making a copy for his own record; that was typical. We do not know exactly when he arrived at Bordeaux, but from other correspondence we can conclude that it was between 2 and 9 February.

> you all ought to know what sailors are after a disagreeable, hard, labourious, tedious long passage----------ask them what kind of a ship is she------why Sir as damned a thing as ever swam-----how so does she leak----ask her pump ports------does she strain---ask her seams--------does she sail fast---see what a passage we had-----------how is her rigging & sails-----we had neither when we left home & if possible less now, ... [but] with a full crew every thing goes smooth & easey & in fact such is all before the wind.... then ask each -- what kind of a ship are you in----by God sire as fine a ship as ever swam-- -- does she leak---not a drop---- ------ does she strain, never------is she well found---certainly ------does she sail fast, very --------this is the very ship we want, & she will sell well-----will the other I fear not----"

He goes on to castigate the owners for not providing adequate spare rigging and sailcloth. He compares the paltry sum it would have cost with the substantial cost in labor of the longer passage:

> but if you ment we should run under last reefed topsails all the way then I grant we were compleatly man[n]ed------- --------- ...
>
> ... I hope to God you will have a poverty auction ------- even the T[op] G[allant] top rope was either taken out or left on the w[h]arf------------- T[op] G[allant] sheets would not hold a 10 knot breeze------"
>
> "puttyed ports (for fear of hurting the paint if paid with pitch) is an excellent thing to keep out water on a cold Europe winter passage-----[no, it is not—he is being facetious]...
>
> ... what will you say---how many excuses have you to make-----------the America not in in 50 days.... what think has become of John --- --- have you heard of the America yet-------------... then an old man picking his teeth they dont say much but they fear she is gone-------
>
> (how much better & to our credit & perhaps her exit would have been the more honourable---for one of the boys coming from the post office allmost out of breath-------sir the America is in Bordeaux in 16 days---by heavens is it possible)

Later he uses superstition as another way of commenting on an ill-fated passage:

> ... & don't you remember after they sailed we remarked that one of the white hens stood up & crowed in the long entry------& that our great black cat sat to the fire backwards & wisked her tail *left handed.* & at the same

> time what was that which hapened to Johns white horse-------all ominous of squalls---then you will begin to realise how astonishingly oversighted you was to send us to sea in so miserable a plight --- ---

If he is not exaggerating, and the passage took 50 days, that is 21 days longer than the same passage he made in 1796, a passage he described as "very long and disagreeable"; perhaps he believed a 16-day passage possible.[40] Here, Captain Crowninshield is pointing out that, because he was undermanned for the conditions in an ill-equipped ship, he had to run the vessel under reduced sail even when she could have carried more, because he did not have the manpower to spare for the constant physical labor of going aloft and changing sail. Apparently, his dismal prediction that the ship would be difficult to sell once he reached France was borne out. He had hoped to sell her for $30–$40,000, but after deducting the costs of work on her, provisions, insurance, and port bills, he worked out her net worth at $21,400, which he calls a small sum for her.[41] He makes a compelling case that penny-wise-and-pound-foolish was a dubious "strategy" for "risk management" in this business, and that the suffering for that strategy would fall heavier on those aboard than those who saved their pennies at home—though, in the end, they, and their accounts, were likely to suffer too. His account underscores the importance of adequate manning, even from the owners' vantage point, and even though more men meant higher wage and victualling bills.

If the ship could be made easier to sail, however, without unduly impairing her performance, then owners might be able to save on those costs without resorting to the unhappy situation aboard *America*. We have discussed evidence pointing to the strong likelihood that changes in sailing rig allowed for modest size increases without increasing sail-handling labor requirements. If nothing else, that means we cannot assume that a higher ton-per-man ratio meant more work for the individual sailor. That is important for labor history as well as technological history. For the sailors aboard, as Crowninshield's complaints point out, there was no "set it and forget it" when it came to working the ship's rig on an Atlantic voyage. Setting too little sail for the wind conditions meant less speed than could be obtained by better-matching the sail plan to the available wind. On the other hand, the available manpower had to be adequate to reduce that sail area quickly once the winds picked up, or

40 John Crowninshield to George and Richard Crowninshield, 25 November 1796, Letter Book No. 11, box 4, folder 8, John Crowninshield Papers, PEM-Phillips.

41 John Crowninshield to George and Richard Crowninshield, 31 March 1803, Letter Book No. 7, Letters from France, 1803, box 4, folder 8, John Crowninshield Papers, PEM-Phillips.

the rig could be damaged, prolonging the voyage for that reason, and adding a high risk of serious injury to the crew. Ships' logs of Atlantic passages make clear just how often the crew might expect to make sail adjustments—most of which, on a square-rigger, required work both aloft and on deck. Logs like that of the *George* record periods of near-constant sail changes in reaction to changing conditions—adding sail as winds grew lighter, taking it off as storms built, with luck before something broke. The goal was to maintain as high an average speed as possible over the course of the voyage. As we have seen, however, the logs make clear how serious the impediments were to translating a high average speed for any given portion of the passage into anything close to a high average speed for the entire voyage. We are safe in assuming that, for most merchant ships most of the time, "the search for speed under sail," to use Howard Chapelle's evocative titular phrase, did not drive technological decisions. It is the problems raised by Captain Crowninshield's experience crossing the Atlantic in *America* that promise our best explanations for why period mariners made decisions to retain or adapt important technological features of their ships. In the next chapter, we will explore how the material, tangible, observable properties of the ship reflected the wants and needs, predilections and limitations, of those who sailed her.

CHAPTER 7

Working the Ship: the Technology of Operation

If there is one modern technological concept that has no application to one of these ships, it is "automation"—a term by which we refer to having machines do work for us, whether that work is computational or mechanical. It is even only partially correct to call these ships "wind-powered," as the sails could only harness the wind's power when skillfully deployed and adjusted by the crew, using their own computational and mechanical power. Block-and-tackle purchases, pumps, windlasses, and capstans offered some mechanical advantage to amplify human power, but they still required significant expenditure of effort—and cognitive function—to operate. Human muscles, guided by human brains, performed all tasks aboard. Understanding those tasks, and how riggers and sailors adopted and adapted somewhat different means of accomplishing them over time, allows us to hypothesize about consistency and change in how these ships, and their crews, worked. Further, such an understanding allows us to test those hypotheses, and to understand the risk environment of, and the culture of risk endemic to, the Atlantic sailing workforce.

Testing hypotheses about technological continuity and innovation in merchant ships—rather than merely noting that a hypothesis "makes sense" and allowing it to become an assumption, as has been done repeatedly, requires moving beyond the use of conventional sources of evidence. This chapter relies heavily on applying information supplied by the operators of period replicas to technological questions already raised, and that information in turn raises more questions about prior assumptions that the literature presents to the uninitiated reader as foregone conclusions.

As we consider skilled labor and the operation of ship technology, it is important that we keep in mind a fundamental difference between warships and merchant ships: that warships could and did carry much larger crews than their merchant counterparts, so they could employ more labor-intensive technological features than merchant ships of similar size. Before taking command of the operational replica U.S. Brig *Niagara*, Captain Walter Rybka had commanded the 1877 iron three-masted bark *Elissa*, a cargo carrier designed and rigged for economy of operation, and when he reported aboard the replica War of 1812 brig, he guessed that, with her tall rig and light purchases intended solely for speed, power, and maneuverability for short-term emergency service, she would be about 30% more work to sail, even though the two ships shared some

 | DOI:10.1163/9789004426344_009

FIGURE 13 U.S. Brig *Niagara* under full sail
PHOTOGRAPH BY LANCE WOODWORTH

basic dimensions. "I was absolutely dead wrong," Rybka says. "It was 200% more work." Warships did not place priority on economizing crew, as cost of labor was not a primary concern. They could have highly labor-intensive rigs that optimized speed and maneuverability with less regard for workload than a merchant ship. The lighter purchases on *Niagara*—with two-to-one mechanical advantage rather than three-to-one, as those on *Elissa* have, are much harder to work, but work much faster, as less rope has to go through the blocks to get the spar or sail to move the same distance. So, the *Niagara* can maneuver faster—but only because she has the human power on board to do so.[1] In the photograph of *Niagara* (Figure 13), note how much canvas she is spreading for her hull, which is low and sleek. Any experienced mariner seeing the original under sail like this would have identified her immediately as a small warship, even at the limits of visibility.

It is to economy of operation that we should look for the optimizing of the merchant ship's rig and the workload required to handle it. As noted, the square-sail rig of the typical ocean-going merchant ship was divided into more

1 Walter Rybka, telephone interview, 21 April 2015.

sails in the eighteenth century. We know that it was easier for the same number of crew to hoist, trim, and furl a smaller sail than a larger one.

Today's Jamestown, Virginia, is the site of—as it turned out—the first permanent English settlement in the Americas, although it was closely followed by one at Cuper's Cove, Newfoundland, established in 1610 by Bristol's Society of Merchant Venturers, and St. George's, Bermuda, in 1612. *Sea Venture*, wrecked at Bermuda in 1609, had been on her way to Jamestown to resupply the beleaguered, pitiful settlement there, and the castaways made their way to Virginia after building new ships on the island. The Commonwealth of Virginia maintains carefully-researched, fully-operational, conjectural reconstructions of the three vessels that originally landed at Jamestown, under the overall charge of Eric Speth. In order of size, the Jamestown ships are *Susan Constant*, the largest; *Godspeed*, in the middle; and *Discovery*, the smallest (see Figure 14, p. 204). All three are representative of small ocean-going English ships from the beginning of the period—about 1605. Eric Speth observes that

> [t]o set *Godspeed*'s mizzen with a full crew we station one crew at the bowlines, two crew (1 [port side], 1 [starboard side]) at the foot brails and martnets, one crew on the topping lift, two crew at the halyard, and one crew at the sheet for a total of seven crew. Due to the number of lines that are handled, we need the same number of crew on the smaller *Discovery*. Because of *Susan Constant*'s size and weight of sail/rigging we add one more crew to the bowlines, two more crew to the halyard for a total of ten crew.[2]

So, it takes the same number of crew to set the mizzen on *Discovery*, even though she is smaller than *Godspeed*, because her rig is set up differently. That means that it is possible to increase the size of the ship without increasing the number of crew necessary to accomplish a basic operation. *Susan Constant* requires three more crew to accomplish the same task, not just because she is larger than either of the others, but also because of the weight of her sails and rigging. Speth has also, in the past, commanded the larger reconstruction *Mayflower II* in Massachusetts, and notes that it requires considerably more manpower to operate her lateen mizzen than it does on the two smaller Jamestown ships.[3] Given that more sails mean a more complex, and expensive, rig, and more tasks to accomplish to adjust the entire rig, our best hypothesis as to the primary pressure for splitting up the sail plan—and probably for the

2 Speth, personal e-mail, 27 July 2018.

3 Speth interview.

FIGURE 14 Jamestown replicas *Godspeed* (left) and *Discovery* at the wharf
PHOTOGRAPH BY THE AUTHOR

abandonment of the lateen mizzen, too—was to increase the size of the vessel without increasing the manpower requirements of her rig. (All four of the ships just discussed are rigged similarly.)

As for the replacement of the lateen mizzen, however, Speth raises another possibility for us to consider—a possibility that does not, it is important to note, depend upon the notion that the lateen mizzen was ineffective. First, we should note that the disappearance of the lateen mizzen was not a single-stage process; at first, the part of the sail that projected forward of the mizzen mast was cut away, leaving the yard still projecting as it had, but bare forward of the mast. Reynolds speculates that the full-length mizzen yard was retained after the sail was cut back because it was easier to trim, as the yard "could be warped over with better leverage (especially in high winds), than by trimming the sheet." That could be done forward of the quarterdeck area reserved for the officers, he says. The hierarchy on a ship was certainly strong enough to serve as a social factor selecting for a technological preference. "The full-length yard also balanced the weight" of the sail, making it easier to attach the yard to the mast with simpler, cheaper fastening methods.[4]

Over sixty years ago, William A. Baker wrote that tacking the mizzen was simpler than expected on the new *Mayflower II*.[5] Burningham wrote that "… on *Duyfken* … the mizzen … would have been helpful in a few specific circumstances had we not been able to use the engines."[6] Speth agrees with Reynolds that the mizzen is "handy," comparing favorably to the later gaff spanker.[7] The *Duyfken* experience at sea with the mizzen makes sense to him, and he notes that the sail is a balancing and maneuvering sail, and would be furled for downwind sailing—which is most of the sailing these ships did, *Duyfken* included—as it would move the center of effort of the entire sail plan too far aft otherwise, making it more difficult to steer the ship on a steady track.[8] Sharon Dounce of the *Kalmar Nyckel*, a larger ship than Reynolds' *Half*

---

4 Reynolds, personal e-mail, 5 February 2015. Parrels and halyards are made of wood and hemp rope, materials and devices already familiar and already aboard.

5 Baker, *The New Mayflower*, 117–18.

6 Burningham corr. 5/15; and see Burningham in Bennett, *Sailing Into the Past*, 116.

7 Speth interview.

8 See following discussion of tiller and whipstaff steering; Speth says the tiller and whipstaff are difficult to use if the sail plan is not correctly balanced. That is true of a modern wheel system on a modern yacht, so it stands to reason that it would be truer of these larger, heavier vessels (Speth interview). Center of effort of the sail plan refers to the focal point of the wind's force on the sail plan as a whole. If it is too far aft, the wind will tend to push the ship's stern to one side or the other, so the ship will slew rather than track straight downwind. This is not only potentially dangerous, as it could cause a *broach* in heavier seas. Broaching is being pushed broadside to the waves by a strong following sea, so that the ship is in immediate danger of

FIGURE 15 *Kalmar Nyckel* replica of Dutch pinnace
PHOTOGRAPH BY JON CASPAR, COURTESY OF THE KALMAR NYCKEL FOUNDATION

Moon or Speth's Jamestown ships, writes that the mizzen on her ship "balances the sail plan forward when sailing close hauled." She too notes that it is not useful off the wind. They use it for tacking, but not wearing (jibing). The balance it provides for maneuvers greatly assists with steering the ship.

As for handling it, she writes that the sail on its yard

> would likely have been shifted from side to side of the mizzen mast.... [W]ith *Vasa* researchers, we practiced and timed it, as it fit into a tack. Got it down to about 90 seconds, and with available personnel of a normal size crew. It's our easiest sail to set, by far.[9]

---

being rolled over. As noted, it is also true that the corkscrewing motion of an unbalanced vessel sailing deep downwind is literally sickening to human beings.

9 Dounce, personal e-mail, 24 April 2015. Hocker points out that, while it may well be possible to execute some maneuvers without the helm, it is clear from contemporary sources (Smith's *Seaman's Grammar* and Mainwaring's *Sea-mans Dictionary*) that other maneuvers required the helm, especially those that involved using the helm in opposition to the sails (Hocker corr. 8/15).

We can say, then, that the lateen mizzen functioned well on seventeenth-century ships as a balancing and maneuvering sail, and that its long yard was manageable given the rest of the sail plan.

To Speth, the most sensible explanation for the retention of the full lateen spar once the sail itself had been cut back to the mast is that the spar lent itself to *counter-bracing*. Counter-bracing means forcing the yard, and thus the sail, over to the windward side, causing the stern to rotate and encouraging the ship to come through the wind—a technique Speth has used often to tack a ship in light winds or to tack a ship that did not otherwise tack well. Once staysails came into use, however, says Speth, that forward-projecting yard would have gotten in their way. Adding sails to a rig usually means making significant adjustments elsewhere so that those sails may be properly handled and so that they—or other sails near them—receive clean wind, unblocked by neighboring canvas.[10] Be that as it may, if Speth's hypothesis was a sufficient explanation for the change, it took some time to take effect, as we see clear evidence that staysails were in use well before the disappearance of the lateen mizzen—see for example Barlow's sketch of the *Cadiz Merchant*, from 1682 (Figure 3). The truth is that none of the three explanations offered here—the increased labor demands on modestly-larger vessels, the need for a newfangled way to join the cut-back yard to the mizzen mast, and the interference of the lateen yard with after staysails—necessarily excludes the others. In fact, the technological history of something as complex as a ship teaches us that multiple simultaneous pressures are usually present. It is entirely possible, perhaps even likely, that all three were in play. However, while we do not have adequate evidence to rank them with any confidence, the factor of size versus manpower is the one possible explanation connecting this particular adaptation to others we are considering, and we should keep that in mind. More experiments on working replicas could quantify the different labor requirements for various sizes of lateen mizzens, yielding data that we could add to a more general compilation of manpower needs to operate the same types of sails given certain differences in size.

The masters' accounts of the lateen mizzen in operation emphasize that it was a balancing and maneuvering sail. As Burningham implies, sailing ships equipped with auxiliary engines use those engines for the sorts of close-quarters maneuvering, such as making way through an anchorage or a restricted channel, for which the original vessels would have relied on their balancing

10 Speth interview. Staysails—fore-and-aft (triangular) sails hoisted on the stays—the standing rigging that holds up the masts—became increasingly common in the eighteenth century until they were ubiquitous on ships, brigs, and snows.

and maneuvering sails to change direction relatively quickly. Sails at the bow and stern could be used to turn one or the other end and thus help to rotate the vessel partly or fully. However, maneuvering also required effective steering gear to turn the rudder. Another significant change that took place over time—specifically, in the early eighteenth century—was a changeover in the steering system of medium-sized and larger ships from *tiller-and-whipstaff* to wheel steering. This change, too, may well have had to do with size increases, although as with the adaptation of the aftermost sail, other pressures may also have been at work.

The tiller is the long horizontal handle attached to the top of the rudder, used for turning it side-to-side. On larger vessels, the tiller would be underneath the weather deck, and the helmsman would actually turn it by using another handle, a vertical *whipstaff*, running through the deck and attaching to the end of the tiller. This is direct and introduces little margin for failure, but it does not provide much mechanical advantage. The wheel uses block-and-tackle arrangements to steer the belowdecks tiller with mechanical advantage, but it is more complex and has more moving parts vulnerable to failure—especially the rope upon which it relies to work.

As noted, ship historian and curator Alan McGowan ties the adoption of the wheel to the adoption of the jib, claiming that the wheel provided precise enough steering to allow maintaining control while sailing as close to the wind as the jib allows.[11] Recall that his claim was accepted by Ian Steele, and then by Russell Menard in the latter's attempt to prove the importance of "better ships" to increased productivity in the mid-eighteenth-century tobacco trade. Applying practical experience to that claim, however, exposes problems with it. Losing control while sailing close-hauled could result in an accidental tack, damaging or destroying the rig, and possibly wounding or killing crew, or it could cause the vessel to come to a sudden stop and lie helpless. Burningham, on the other hand, writes that

> *Duyfken* steers just as precisely as a wheel-steered ship with a good helmsman at the helm and just as badly with a poor helmsman.... The challenge is to prevent square riggers from tacking themselves when sailing on the wind. If you luff up aggressively, or the wind shifts significantly, and the sails on the foremast get aback, the ship will be heading round to the new tack before the watch officer can [splutter expletives].[12]

11 McGowan, *Tiller and Whipstaff*, 43.

12 Burningham corr. 5/15. To "luff up" is to point the ship toward the wind enough that the sails are barely filling, or even beginning to flap some as they lose smooth airflow.

The danger Burningham describes here, of letting the ship come up into the wind unintentionally, is that of being *taken aback*, wherein the sails backfill, but the yards are still braced on the original side. The ship loses forward way quickly, and can even start to move backward. The sudden shock of backfilling can damage the rig. It is important to note that putting the ship aback was done intentionally to accomplish various maneuvers; if done under control, it helped the ship change direction handily, or stop forward motion, to avoid a dangerous situation ahead. The danger was in letting it happen accidentally, especially in strong winds. That danger, writes Burningham, is not made worse by jibs, as McGowan believed, since any square-rigged ship can be caught aback through error or sudden wind shift while close on the wind, regardless of what headsails she carries.[13] This may seem to suggest that a fore-and-aft rig would have the advantage here, but that would be an unfortunate oversimplification, as it would obscure the important fact that square-rigged vessels could tack or wear more reliably than their fore-and-aft-rigged counterparts. Yes, square sails were easier to catch aback accidentally than fore-and-aft sails, but square-riggers did not need to back their jibs to tack, whereas fore-and-aft-rigged vessels "generally had to back the fore-staysail, but they were less reliable in stays."[14] That means they were more likely to be caught *in irons*—unable to come through the wind, losing their forward way as their sails could not fill on either side. Even a modern sailing yacht can be caught in irons if the tacking maneuver is improperly executed, and such a craft is far sleeker, lighter, and sharper than any early-modern merchant vessel. The square sail was more effective at turning the ship's head through the wind when backed—whether accidentally or deliberately. In fact, the schooner with one or more square topsails was "more reliable in stays" than a schooner not so equipped and thus relying entirely on fore-and-aft sails.[15] The danger, then, to the fore-and-aft-rigged vessel was that she would fail to avoid the danger that precipitated the tacking maneuver—or, as no less a mariner than James Cook tells us, fail to *wear*—come through the wind with the wind behind rather than ahead.

In the mid-1760s, after Britain gained Canada in the Seven Years War, Cook was charged with surveying the coasts of Newfoundland and the Straits of Belle Isle between Newfoundland and the mainland.[16] He was provided a

13 Burningham, personal e-mails, 30 August–2 September 2019 (hereafter "Burningham corr. 8/19").

14 Burningham corr. 8/19.

15 Burningham corr. 8/19. Such "topsail schooners" were common. See Figure 19, p. 265.

16 Cook was as accomplished a surveyor as he was a navigator, and the charts produced from his surveys were used until relatively recently. His surveying work was carried on by others until interrupted by the War of American Independence. See Stephen J. Hornsby,

Massachusetts-built schooner for the task, but Cook found the schooner rig poorly-suited to the work, and successfully petitioned the Admiralty to have the vessel re-rigged as a brig. He claimed that the schooner was unreliable in stays—coming through the wind to change tacks—and that, because her fore-and-aft sails were insufficient to lay the vessel properly aback, she would run herself onto a lee shore before she would wear—come through the wind and escape. In fact, that very mishap occurred, he reported, and he felt fortunate that the vessel survived it. In the interest of brevity, he foregoes describing "many more" advantages of a brig over a schooner.[17]

Far from being a matter of only narrow technical interest, Cook's experience and his opinion carry profound implications for the understanding of technological choice and its ramifications. First, they give the lie to the notion that fore-and-aft-rigged vessels were to be preferred for coastal work where maneuverability was of paramount importance. In fact, they call into serious question the notion that fore-and-aft-rigged vessels were more maneuverable, all other things being equal. Cook was not asking for a brig because he needed more downwind speed on long passages. He was asking for one to provide safer maneuvering around rocky lee shores as he did close-in coastal surveying. So, the square brig rig offered him a more, not less, maneuverable vessel. That alone would require us to look for other explanations for the popularity of the fore-and-aft rig for this size vessel than maneuverability for coastal work. There is more, however. George Huxtable and Ian Jackson show that Cook's transatlantic passages in the brig-rigged version were not only faster than the first one he made before the rig conversion, but the brig was able to make fast passages on the direct route west from England to Newfoundland, against the prevailing winds and current—and against prevailing modern wisdom about how to cross the Atlantic under sail.[18] Huxtable and Jackson point out that some of this may be due to differences in current and wind patterns from the eighteenth century until now, and that this bears investigation. They also point out, however, that while Cook was clearly an exceptional navigator and mariner, there is no reason to think that his choice of route was exceptional. By the 1770s, hundreds of ships a year were making the fishing runs to and from the Grand Banks, and it is likely many or perhaps most of them were using this route. Unfortunately, the logs we would need to prove that have not survived.[19]

Surveyors of Empire: Samuel Holland, J.F.W. Des Barres, and the Making of the Atlantic Neptune (Montreal: McGill-Queen's University Press, 2011).

17 James Cook to Navy Board, 22 January 1765, in J.C. Beaglehole, *The Life of Captain James Cook* (Stanford, CA: Stanford University Press, 1974), 82.

18 Huxtable and Jackson, "Journey to Work."

19 Huxtable and Jackson, 213.

Cook's passages in the brig, which Huxtable and Jackson precisely reconstructed from his logs, prove that the square-rigged vessel could make ocean passages comparable or even superior to the same vessel with a fore-and-aft rig, without the advantage of downwind sailing. The fact that such a huge fleet regularly used these routes at the same time means that Cook was certainly not the only one who knew this.

When we remember how important hull form was to the sailing performance of a vessel, it is no surprise to learn that the use of jibs, or other fore-and-aft sails like staysails, did not make a vessel more weatherly. Jibs were useful for steering, they reduced *weather helm* (the tendency for the bow to round up into the wind), and they were relatively easy to set and trim. We have no evidence, however, that they allowed a vessel to sail closer to the wind. Sharp, sleek hulls, with deep keels to resist leeway—like pilot boats or pleasure yachts—could sail effectively close to the wind better than any cargo ship (or warship, for that matter). And, even if the jibs did not luff (spill wind) when the ship drew closer to the wind, the after square sails likely would, and that would mark the limit of the vessel's weatherliness in those particular conditions.[20] As for triangular staysails, they were deployed for more power off the wind, not close to the wind. If carried hard on (close to) the wind, they would backwind the square sails.[21] That reminds us that no sails on the ship operated in isolation from the others; any particular sail plan had to be deployed in such a way that the most important sails had free and clean air.

We are thus presented with two important research tasks. First, we need to look more closely at technological choice between square and fore-and-aft rigs in ordinary small to mid-size merchant ships, and to consider more seriously the likelihood that factors other than the respective sailing attributes traditionally attributed to each were of paramount importance to those making those choices. That means conducting detailed work on the difference in labor demands, and labor cost, between the two. The results of such work should offer a fresh contribution to the ongoing evaluation of ton-per-man ratios and their significance to the economic history of the British Atlantic. Second, we need to set aside our traditional assumptions about square and fore-and-aft rigs, some of which have been passed down in the literature and some of which are based on modern yacht sailing. We need to accept that the evidence is telling us that we still have much to learn about how these rigs performed in the

20 Burningham corr. 8/19.

21 Burningham corr. 8/19. Burningham notes that Basil Lubbock presents clear evidence for this use of staysails on the mid-19th-century tea clippers in his *The China Clippers* (Glasgow: James Brown & Son, 1914), 146.

seventeenth and eighteenth centuries, and start looking both for experiences in the archives, like Cook's, and to experiments aboard accurate period replicas and using virtual modeling.[22]

As we re-consider the fore-and-aft rig in light of its capabilities and disadvantages relative to the square rig, we must also consider continuity and innovation in steering systems, as steering a vessel was accomplished through a close coordination of sails and helm. We read that wheels replaced tillers and whipstaffs for steering after 1700 because wheels were so much easier and more effective to use, and authors muse about why it took so long to make that transition. However, replica masters report that the tiller and whipstaff method works fine. Technical ship history tells us that the technological impediments in the way of developing effective wheel steering were daunting and took a long time and many failed attempts to overcome. With a satisfactory system in use and difficulties in the way of adopting an alternative, the tiller and whipstaff were under strong selection to persist. That point does not seem to have made it into general academic maritime history.[23] Also, while it is doubtless true, as Ian Steele writes in *The English Atlantic*, that wheel steering "dramatically increased rudder control on larger vessels"[24] due to increased mechanical advantage—provided the system was functioning properly—it is important to know *how much* larger those vessels needed to be for that advantage to manifest itself, all other things being equal, if we are to understand the relationship between wheel steering and ship size.[25]

The wheel offered advantages beyond the mechanical. Mounted on deck rather than below, it offered a less-restricted view for the helmsman, especially of the sails. With no deck over his head, the helmsman on a wheel-steered vessel could look up and see what the sails were doing while he steered, while his counterpart on a larger tiller-and-whipstaff-steered vessel enjoyed no such

---

22 It is also worth investigating why schooners, such as the one Cook started his surveying with, were so popular as Grand Banks fishing vessels. It is likely to do with the fact that they sailed out to the Banks from New England or Atlantic Canada, fished, and sailed back, rather than crossing the Atlantic.

23 Ian Steele cites John Harland and Alan McGowan on this subject—strong sources on such matters—but he does not mention the technical difficulties Harland discusses in his *Mariner's Mirror* article See Harland, "The Early History of the Steering Wheel," *Mariner's Mirror* 58:1 (1972): 41–68.

24 Steele, 49–50.

25 Samuel Kelly recounts that he once had to cut the steering rope when it took a riding turn over another on the drum, jamming the steering as a pilot was maneuvering the ship through some rocks, and use the tiller to steer the ship (in these systems the wheel actually drives a tiller belowdecks). He says it is the only such incident he was ever aware of, though. See Kelly, 265–66.

advantage. McGowan noted this, and we have the evidence to back it up.[26] "The limited view problem," writes Fred Hocker, research director of the *Vasa*, "is acute in both *Vasa* and *Kalmar Nyckel*." Hocker points out that, on *Vasa*, the helmsman could not see the horizon, and thus could not judge if he was holding course except by the compass. As everyone who has spent time steering any sort of sailing vessel knows, steering solely by the compass is undesirable, as there is a lag in its movement which tends to encourage unnecessary steering corrections. These in turn produce a wandering, somewhat serpentine course, slowing the ship down both in the water and over the ground. On *Kalmar Nyckel*, the helmsman cannot see the sails, so he cannot be ordered to maintain a course such as will keep the sails full as they are set.[27]

To what extent did that issue select for the adoption of the wheel? Discussion of steering issues continues among replica sailors. One view holds that seventeenth-century vessels did not rely on their rudders as more modern vessels do, and as we assume vessels must. Sail trim, according to Reynolds, is the primary means of directional control, with the rudder serving more as a trim tab for fine-tuning.[28] Speth also uses the term "trim tab" to explain the rudder's role to new crew on his ships.[29] Dounce of *Kalmar Nyckel* concurs: "We fine tune our sail plan and trim from time to time, and can get it balanced enough that we don't need to use any helm to keep the ship on course …" This would have greatly increased ease of operation on a voyaging vessel with no automatic self-steering. Dounce says the rudder is for starting a tack but once the sails on the foremast come aback, the sails take over the whole process—yet more evidence of the effectiveness of backed sails on a square-rigger in tacking the ship.

Speth notes of his largest Jamestown vessel, *Susan Constant*, that she only has 12 degrees of rudder throw, so proper sail balance and trimming are critical.[30] Of the whipstaff, Dounce says "it's easy to know how much helm you have" unlike with a wheel, where all you know is how many times you have turned the wheel—even a new helmsperson can tell how much rudder angle they have with a tiller and whipstaff.[31] That advantage might be offset by decreasing mechanical advantage of the tiller and whipstaff as rudder angle increases, if

26 McGowan, *Tiller and Whipstaff*, 42.
27 Hocker corr. 8/15.
28 Reynolds interview.
29 Speth interview.
30 Speth interview.
31 Dounce, telephone interview, 23 April 2015—hereafter Dounce interview; and pers. comm. 9/19. Also see Steele, 49–50, and 332, note 39; Harland, "The Early History of the Steering Wheel," 41–68; and McGowan, *The Century Before Steam*, 16.

the forces involved are great enough. Hocker points out "that the mechanical advantage of the wheel is constant at all rudder angles, while the advantage of the whipstaff decreases as rudder angle (and thus force) increases, regardless of rudder size."[32] Hocker cautions that replica captains' anecdotes

> [tend] to maintain an older fiction that pre-wheel ships were steered with the sails rather than the rudder, but I believe that this is a misunderstanding. You CAN tack a ship with the sails alone, but that was equally possible in the 19th or 20th century. Both the primary literature (handbooks and ships' logbooks) from the seventeenth century and the practical experience [show] that the narrow rudder and whipstaff [are] more than adequate to steer a ship, and some operations (tacking and heaving to) require the rudder to operate in opposition to the sails. What is important is sail and helm balance (in ANY sailing ship)—choosing the right sails to set makes a large difference in the workload at the helm and course stability.[33]

It is clear that sail balance, sail trim, a functioning rudder, and an adequate steering apparatus were all necessary for satisfactory steering.

Much of the evidence points toward increasing vessel size as the imperative driving the installation of wheel steering for mechanical advantage. Dounce and Reynolds pointed out, respectively, however, that the Swedish flagship *Vasa* and the great Dutch East Indiamen of the period were tiller-and-whipstaff-steered.[34] Hocker says that seventeenth-century mariner and dictionary author Henry Mainwaring

> believed that the whipstaff was not useful on large ships, but the historical and archaeological evidence does not bear this out. Our calculations show that the whipstaff on *K*[*almar*] *N*[*yckel*] should take MORE effort to handle than that on *Vasa*, yet *KN* can be steered in a storm by a small woman in her 70s …[35]

---

32 Hocker corr. 8/15.

33 Hocker corr. 8/15. At that time, he and others involved with *Vasa* were working on a report on steering systems and what they have learned from the recovered Swedish wreck and from replica trials.

34 Dounce interview; Reynolds interview. This issue is still under investigation.

35 Hocker corr. 8/15. This comment also brings up the interesting complication of how to trust expert primary sources. Mainwaring (or Manwayring) was one of the most experienced and accomplished seamen of his day. As Hocker put it, "[t]his shows that there are

We have no reason to think that the steering of these larger ships was ineffective, but what if wheel steering on later, larger ships allowed for larger rudders? Burningham disposes of that question. "[I]n general large sailing ships had small rudders irrespective of the steering system."[36] That supports Hocker's comment that what was true of seventeenth-century ships was true of eighteenth-, nineteenth-, and twentieth-century ships as well in that regard. So, did wheel steering make larger vessels easier to steer than tiller-and-whipstaff systems? Hocker complicates it further by presenting a caveat to the "conventional wisdom … that bigger ships require bigger steering effort and more mechanical advantage." While generally true, the problem is actually "governed by tiller size … the larger vessel will probably not have a tiller proportioned as well to its length."

> By careful choice of geometry, it is possible to cheat a little. *Vasa* gets away with relatively small steering effort due to a narrow rudder and a very long tiller.… *Nyckel* has higher relative steering efforts because it has a proportionally wider rudder and much shorter tiller. This also means that it gets away with a much shorter and more manageable whipstaff …[37]

What Hocker is getting at here is that we have another issue of one component getting in the way of others. Look at any accurate model of a square-rigged sailing ship and you need know nothing at all about how any of it works to be struck by how complicated and crowded it is. A large ship cannot have as proportionally long a tiller as a smaller one. There is no room for it. This brings us back to considering size as the primary—but not exclusive—imperative toward wheel steering in its early years.

We have established that multiple factors always exist for any specific technological continuity or change on these ships. Given that, is there any merit to McGowan's idea that we must consider the wheel and the jib together? If so, is there a correlation between increasing size and the triangular jib? Or is Burningham entirely correct to dismiss that connection?

Speaking of triangular jibs, we read that their adoption improved windward ability and ease of sail handling over the clumsy *spritsail* and *sprit-topsail* of earlier vessels (see Figure 15 of the *Kalmar Nyckel* replica—while the sail itself is not set in this photograph, the mast and yard for the sprit-topsail are clearly

---

areas where replica sailing may not only challenge the received wisdom of the secondary sources, but may reveal errors in perception in primary sources" (Hocker corr. 8/15).

36 Burningham, personal e-mail, 4 May 2015.

37 Hocker corr. 8/15.

visible on the end of the bowsprit). Yet replica crews report that, once they learned to use those earlier sails, they served to control the head of the ship and assist the rudder in steering her handily. This goes back to Alan Villiers and his transatlantic voyage in *Mayflower II* in 1957—the only insight from a period replica I have found in a maritime history secondary source. John Harland refers to Villiers' experience in *Seamanship*, as well as in a *Mariner's Mirror* article.[38] Villiers' experience is corroborated by Burningham from his experience aboard *Duyfken*. Of *Kalmar Nyckel*'s spritsail, Dounce says it is difficult to balance the sail plan without it.[39]

It is clear that much of the specific skill set used every day on a seventeenth and early eighteenth-century vessel was lost even before the end of commercial sail. John Harland tacitly acknowledges this when he makes a refreshing break from the dismissive presumption regarding seventeenth-century designs and rigs that creeps into ship history by rejecting the idea that the sprit-topsail remained in use as long as it did for reasons other than usefulness. To him,

> … it is inconceivable that the seamen and shipbuilders of the day, as practical men, would have tolerated the consequences of poor design unless they saw, or thought they saw, some immediate advantage in so doing. However that may be, the old-fashioned spritsail topsail remains, in my view, pretty much an enigma.[40]

McGowan, too, writes that we do not know enough "about the positive qualities of the spritsail topsail, nor of exactly how and when it was used to the best advantage."[41] Only experimental archaeology is likely to help with that enigma at this point. *Kalmar Nyckel* has a sprit topsail, and Dounce reports that it does contribute to fine-tuning sail plan balance in light air (light wind), but that, while they have experimented with different ways to rig it, it is skill- and labor-intensive to use and requires care to set and furl, and that they do not use it very often.[42] Burningham wrote that the sprit-topsail was rarely set, and that it would have "probably contributed almost nothing to speed through the water under any circumstance."[43] Burningham would no doubt agree that, if the sail

38 Harland, *Seamanship*, 85; and "The Early History of the Steering Wheel," 65, n. 9. Also see Villiers, "How We Sailed the New *Mayflower* to America," 632.

39 Burningham, "Learning to Sail the *Duyfken* Replica," *International Journal of Nautical Archaeology* 30:1 (April 2001): 78–79; Dounce interview.

40 Harland, *Seamanship*, 89.

41 McGowan, *Tiller and Whipstaff*, 43.

42 Dounce interview.

43 Burningham corr. 5/15.

were intended as a balancing and/or maneuvering sail, a contribution to speed might not have been expected.

Dounce thinks *Kalmar Nyckel*'s sprit-topsail is undersized, as they have been able to bend a larger sail to the top yard in an experiment. A larger sail would more effectively accomplish the purpose. While *Kalmar Nyckel* makes day-sails and short coastal trips, the original was a large, heavy voyager, and prolonged light winds can be just as dangerous to a voyaging ship as storms, for different reasons. Any sails that allowed a ship to take full advantage of a light breeze to move a little faster, wallow a little less, or be easier to steer could have been valuable to a voyager carrying fatigued humans and limited, perishable stores.

Still, even after ships were carrying forward staysails—inner jibs, one could say—they were also still carrying spritsails. It would be worth further experiments with that combination, and perhaps worth keeping in mind the general axiom about square versus fore-and-aft sails: that each has its advantages and drawbacks, which is why ocean-going ships carried both. Modern sailors would expect a ship equipped with jibs to go to weather better than the same ship not so equipped. It would be worth coming as close as we can to making such a comparison with operational vessels. While conducting sea trials with the current *Godspeed* in Penobscot Bay, Maine, Speth took some of the builders and shipyard staff out. They were accompanied by some of the similarly sized late 19th- and early 20th-century schooners still common in those waters. People were surprised, he remembered, at how *Godspeed* sailed, how she carried sail, and how weatherly she was, with the qualification that she was a square-rigger and thus could not stay on a windward tack as long as a schooner. They were surprised that she could tack. It is common to assume that square-riggers do not tack, but always wear (jibe). "I do find them to be quite handy," Speth said, speaking of both *Godspeed* and the smaller *Discovery*, "probably because of its size and its two-masted rig, the *Discovery* is delightful to sail and is very, very maneuverable ... I can sail it backwards at times.... and *Godspeed*, there are times I can sail it backwards if the wind and the conditions are right." He once had *Maryland Dove* out in the Chesapeake Bay and sailed her with the extreme schooner—'Baltimore clipper'—*Pride of Baltimore*, and in the right conditions could keep up with her, even with 30–40% less waterline and much less sail area.[44]

Speth's anecdotes introduce the distinct possibility that the performance of these early seventeenth-century fuller-bodied ships was at least roughly comparable to ships with sharper lines, less capacity, and more labor-intensive and

44 Speth interview. He is referring to the original, lost *Pride of Baltimore*, sunk in a squall in 1986.

damage-prone rigs—at least on typical points of sail. If *Maryland Dove* could keep up with *Pride of Baltimore* in some conditions, that also helps us to understand why we see as little variance in passage times in the age of sail as we do. What these vessels could do was, generally, dictated by conditions, rather than by their own intrinsic qualities.[45] Those qualities, indeed, are best shaped to give the vessel the best chance of surviving those conditions and getting her crew and cargo where they are going. The sharp, tall-rig schooners and sloops, like the Baltimore clippers, were specialists. Their speed and windward ability could help them evade capture, which was one reason they were popular smugglers.

More controlled side-by-side trials with different rigs might reinforce what Speth, and Cook's experience with *Grenville*, suggest, and thus reveal something contrary to the accepted wisdom of modern sailing. If so, it would not be the first time. On *Duyfken*, the crew discovered by trial and error that, in winds strong enough to require reefing sails, or even lowering upper sails, in modern vessels, their best results were achieved by half-lowering their topsails, letting them belly out in the breeze. This is not done with modern sails; to a modern sailor, it contravenes a basic tenet of seamanship; but it worked with these.[46] Reynolds also reports positive results with the same technique on *Half Moon*.[47] Speth reports that he routinely employs this method of 'reefing' on the two smaller Jamestown replicas, and that it works well "from off the wind to a beam reach"—any point of sail where the wind is from perpendicular to the centerline of the ship to dead astern, which is, as we know, the range of angles within which all these ships performed best.[48]

Dounce of the *Kalmar Nyckel*, however, is not as sanguine about the rig of her vessel vis-à-vis that which would offer more versatile options for sail combinations. She remembered an experience in heavy weather in which they half-hoisted a topsail, but it was not a stable set—there was too much chafe, the sail blew around too much, the sail would fill and spill, contributing to the rolling and jerking the ship was enduring in the quartering sea (waves hitting the ship on her stern quarter—not dead astern, but just forward of it). A smaller lower topsail, as would be found on an eighteenth-century ship, would

---

45 The most important performance benefit of viable fuel-powered ships was not an increase in inherent speed. It was freedom from the tyranny of the fickle wind.

46 Modern sailors are taught to flatten, rather than belly, a sail to reduce its power in stronger wind. When a modern sail is reefed, it is shortened and drawn tight to create as flat a shape as possible. See Burningham, "Learning to sail the *Duyfken* replica."

47 Reynolds interview.

48 Eric Speth, personal e-mail, 11 February 2016. Unfortunately, he writes, they have not sailed the larger *Susan Constant* enough recently to have experimented with this.

have been better than the baggy half-hoisted larger topsail, she says. (Compare Figures 13 and 15.) She was concerned about strain on the topmast "due to the rig's jerky movement as the ship rolled and the backstays came tight against a gust in the billowing topsail, combined with the ship riding down into a wave trough." When the bow reaches the bottom of a trough, the ship's forward motion slows or even stops momentarily, and that can exacerbate the shock to the rig. Dounce attributes the later division of sails both to making it easier to handle, which she thinks it would be, and also to offer more options for setting sail in different conditions. She did say, however, that half-hoisting the topsails definitely gave the ship more power forward—just at a significant cost "in some conditions."[49] Her assessment of this experience on a ship larger than *Duyfken*, and thus with larger sails, might support Burningham's view of the merit for split lower topsails for larger ships—he calls double topsails "a great improvement for big ships."[50] The caveat here is that we need to take into account any changes in the cut of the sails themselves. How a sail behaves with a certain degree of tension on it has much to do with how flat or full it is cut. Another important element is the fiber type, weight, quality, and weave of the sailcloth. The closer a replica ship's sails are to what the originals would have been, the more weight we can assign to the results of experiments with those sails.[51]

Harland writes that, over the course of the seventeenth century, topsails evolved from "adjuncts to the courses [mainsails]" to the principal sails of the ship. Their height gave them the advantage of cleaner air—less affected by surface friction and the turbulence caused by the lower rig—even in the hollows between big waves. As the topsail's role grew in importance, it grew in size. So, by the late 1600s, even the same size ship had bigger topsails.[52] This development may have helped pressure for splitting the topsails in the eighteenth century, even disregarding the modest increases in vessel size that occurred at the same time.

If the later split topsails were adopted primarily as a labor-saving modification, did handier sail-shortening in heavy air come along as a bonus benefit? Or was more flexibility in shortening sail the driving factor behind the change? Future experiments comparing the manpower requirements of single versus

49 Dounce interview and pers. corr. 9/19.

50 Burningham corr. 5/15.

51 All sailors know the importance of properly-cut sails, made from quality sailcloth, and how the distortion of a sail's shape from being too long in service has a significant negative effect on its performance. The history of ship technology would likely benefit greatly from more research on period sail-making.

52 Harland, *Seamanship*, 82.

split topsails, on similarly-sized ships, would help answer those questions. We need to ascertain if there is a labor-efficiency advantage, and if so, if it is related to vessel size.

Complicating the consideration of rig changes as related to vessel size is the fact that, beginning in the 1740s and persisting until the end of the period, more and more mid-sized ocean-going square-riggers were two-masted rather than three-masted. Davis tells us that from 1680 to 1720, ships normally went from two-masted to three-masted at around 50 or 60 tons, and that this was true for both English and foreign vessels. By 1750, this number was around 150 tons, and would approach 200 tons by the 1770s.[53] That would seem to contradict the hypothesis that further division of the sail plan was a labor-saving adaptation allowing the operation of larger vessels with the same size crews. However, it may be that the division of the sail plan on each mast, and the handiness of the snow rig (the most popular two-masted square rig on the largest vessels in this class) made the overall work load easier, with the sails on two rather than three masts to handle. Laboratory experiments using marine-engineering software to compare the same hull with a three-masted and a two-masted rig, calculating total sail area and the loads on each sail, may help move us forward on that question. For that matter, comparing the most popular forms of two-masted square-rigger—brig and snow—should also prove instructive.

While the two-masted square-rigger gained popularity in the mid-size ocean-going merchantman, the smaller end of the spectrum saw an increase in the popularity of the two-masted schooner rig over the single-masted sloop. We know that sloops tended to have huge sails, and the schooner rig would have been more versatile, and could project the same sail area with each sail being easier to handle—the overarching advantage of dividing the sail plan. It would be worth delving into this development, using archival evidence on manning, wages, building and rigging costs, as well as experimental comparisons of the work involved on both types, since examples of both are currently operational. The results should yield further insight as to the relationship between vessel size, labor requirements, and rig.

All of this technical discussion and hypothesizing is directly, even intimately, related to the experiences of the humans who worked on these ships. The ship as a contrivance capable of traveling over water utilizing wind power was entirely dependent on the sailors' familiarity with the myriad components of the ship's rig, the layout of the decks and the location of equipment on them, how to use that equipment, the physical demands required to use that equipment, and the physical strength and stamina available to do the work. Conversely,

53 Davis (2012 ed.), 73–74.

whether or not a ship had two masts or one, larger sails or smaller sails, guns or no guns, and a crew adequate or inadequate to perform the work needed to sail her well, had a real effect on what his work was like for each sailor aboard. These "technical" matters affected how safe he was, how exhausted he was, and what he had to do for his wage. Given that ships carried the materials and tools needed to make not only repairs but modifications to their rigs, masters and crews could and did experiment with adjustments while underway. The technology of the wooden sailing ship lent itself well to improvisation.[54] It is not surprising, and indeed fitting, that we do not know the names of those who first came up with the adaptations we have considered here. We cannot study this technology by reading the notes and correspondence of inventors. We can only understand it by understanding the priorities of owners and shipwrights, the advantages and limitations of the materials and techniques available, and the skills and experience—some of which we must try to recreate ourselves, as best we can—of those who worked aboard.

54 The rig of the wooden sailing ship as improvisational technology would make a worthwhile study.

CONCLUSION

The Merchant Ship in the British Atlantic, 1600–1800

The English were latecomers to Atlantic expansion, and there was nothing original or distinctive about their early efforts at reconnaissance and exploration.[1] Spurred by suspicion, rivalry, and conflict with Spain, the Netherlands, and France, English adventurers combined experience gained from fishing and trading near the home islands and in the Mediterranean, their brutal conquest of Ireland, and their probing and raiding in Spanish America to launch their earliest attempts to defy the Spanish claim to the entire North American coast. Both the French and the English learned painfully that the Spanish were prepared to defend their claim with violence as far north as the Chesapeake, but both also learned that the Spanish could not keep up the pressure north of what later became Georgia. Landing in small, general-purpose galleons and pinnaces, English cohorts well-suited to war but ill-suited to settlement set about to find gold and the Northwest Passage to Asia. A colony on what is now Roanoke Island, North Carolina, died out before 1600. Small, struggling settlements of disappointment hung on in the Caribbean, Virginia, Bermuda, and Newfoundland in the first two decades of the seventeenth century, thanks to transatlantic re-supply from the companies sponsoring them. Colonies established by profit-seekers and religious dissenters took root in New England in the 1620s, starting with the landing of the *Mayflower*.

The English Atlantic of the seventeenth century did not survive and grow because English merchant ships were better than those of their rivals, but because their merchant ships were good enough to compete, and because English Atlantic shipwrights—on both sides of the ocean—built enough of them. Some politico-geographic aspects of the English and later British Atlantic proved fortuitous. With no prospects for plantation agriculture of any sort in New England, those who settled there could nevertheless grow and catch their own food, and a healthy population grew up to succeed in fishing, whaling, shipbuilding, and maritime commerce. While it was the Dutch who established themselves as the best carriers and merchants of the Atlantic world during their long struggle against Spain, English and French determination to cut the Dutch out of their imperial business led to restrictive protectionism

1 See Elliott, *Empires of the Atlantic World*.

 | DOI:10.1163/9789004426344_010

and a series of wars that, while never able to drive out the Dutch, curtailed them enough that English and French ships, merchants, and markets could grow in their own right. Too, both the English and French, it would seem, consciously adopted elements of Dutch ship design—elements that had proved overwhelmingly successful in economizing the enterprise, although the Dutch practice of sailing unarmed in the North and Baltic Seas would prove impracticable in the predatory Atlantic shipping environment.

With Spanish resources and attention focused heavily on settling and exploiting South and Central America, their original settlements, the Caribbean islands, suffered neglect. A trading vacuum opened up, to be filled in large part by Dutch, English, and French interlopers. When sugar spread from Portuguese Brazil to the West Indies, the sugar islands needed ever-increasing supplies of building materials, fish, salt, meat, and labor. New England shipping supplied that demand—and not only for the English islands. Despite legal restrictions, trade with the Dutch, Spanish, and French islands constituted an important part of New England's maritime commerce. With no domestic market for it, the French islands were especially dependent on New England ships to carry their molasses for rum distilling. Thanks to protectionist duties and price supports encouraged by the powerful English sugar lobby, French sugar and molasses were frequently cheaper for British Americans to buy. The Dutch gave up their Eastern Seaboard colony, New Netherlands, to the English as part of the settlement of the Anglo-Dutch Wars of the mid-1600s. The French presence in Atlantic North America was restricted to Canada; the rest of their efforts were concentrated around the Great Lakes and down the Mississippi. That left the British American Eastern Seaboard colonies with no competition in supplying ships and supplies to the West Indies from America, and allowed a thriving commercial maritime economy to develop in New England and the Mid-Atlantic, bolstered by access to England's credit networks and financial resources, which surpassed those of the Dutch in the eighteenth century. As the southern Eastern Seaboard colonies expanded their plantation economies based on chattel slavery, their needs were also served by the ships of the northern colonies.

This put British American shipbuilding in competition with the home-island industry, but with Britain's dependence on naval power commanding its biggest dockyards and remaining forests, and the seemingly-endless stands of oaks on the American Eastern Seaboard, this was to be expected. It cost too much to ship timber across the ocean, but it did not cost too much to build ships for sale and send them across. The successful British American shipbuilding industry is as obvious an example of the limitations of mercantilist principle as any.

Even before the yards of Boston and Philadelphia were challenging those of the Thames and the Mersey, American-built craft were fishing the Grand Banks, transporting Chesapeake tobacco, and calling on Caribbean islands. The sloops built in Bermuda of its deservedly-famous cedar won a reputation as the finest of their type. The schooners of the Merrimack River in New England achieved similar status in the eighteenth century as fishing vessels and coastal traders, although the type would prove itself capable of transatlantic passages too. Ships built in Britain enjoyed a reputation for quality and longevity that those built in America could not attain in this period, however, and British-built ships commanded a higher price accordingly. That higher price, on the other hand, encouraged more and more merchants to trust the craft of New England and Delaware Valley shipwrights.

Shipwrights building ordinary merchant ships, whether in England or Massachusetts or Jamaica, loosely followed accepted rules of proportion which we can find today in surviving published treatises, but practiced significant idiosyncratic variation within those parameters, as the archaeological record has shown. While some general preferences can be ascribed to specific European traditions, they shared more than they reserved for themselves. Wholesale captures of a rival's merchantmen during the chronic conflicts of the period ensured that no one's design or construction practices would remain a mystery to others. Perhaps the most important hybridization was, fittingly enough, the adoption of Dutch design elements by English and French builders. Having failed to beat the Dutch at merchant shipping, their rivals saw fit to join them, and the incorporation of Dutch *vlieboot* beam, draft, and hull shape into the "frigate-built" designs of *Princess Carolina* and *Henrietta Marie* seems to be one way they did so. This is an important example of a central trope in technological history: the incorporation of some aspects of one group's technology into another group's technology, to serve the adopting group's purposes.

This Dutch-influenced hybridization may well be one way designs changed to increase capacity-for-size, the trend McCusker infers from the changes in tonnage rules. Another may be the decrease in tumblehome. In the eighteenth century, the British Atlantic was not the same operating environment it had been in the early 1600s. The small, stout, general-purpose galleons like *Susan Constant* were replaced by merchantmen more optimized for commerce. This trend became more pronounced in the mid-eighteenth century and after, as the security offered by the Royal Navy and the insurance industry offered owners strong complements, and perhaps in some cases alternatives, to defensive armament. As superstructures came down and hull sides went more vertical, the same size ship could carry more cargo and stores lower down in the hull.

Susan Constant's hull shape, however, was not the only aspect of her design that changed as the British Atlantic developed. While ships increased modestly in size, they did increase, and the larger merchant ships of the established English and British Atlantics would have made these small galleons seem a bit quaint. While simple and handy, the rig of a *Susan Constant* or *Mayflower* went by the board, in favor of split topsails, staysails, jibs, and a mizzen sail and yard that terminated at, not before, the aftermost mast. While more complex, the later rig was also versatile, and it was likely primarily a response to the desire to economize on labor while allowing ships to grow in size and capacity. Ton-per-man ratios increased, as a general trend, over the course of the period. With smaller sails, larger ships could sail under a two-masted, rather than three-masted, rig. Wheel steering systems appeared and spread on medium-sized and larger ships, providing mechanical advantage and improved visibility from the helm.

Strong continuities persisted, however. Passage times remained consistent. The fish-form hull carried on, despite the decrease in tumblehome and superstructure and the increase in capacity-for-size. The core concept of a mix between square and fore-and-aft sails on most ocean-going ships was retained, although modified. Smaller vessels, which could carry their tillers on the weather deck like a modern yacht, and thus did not require whipstaffs, continued to employ this simple yet effective steering device. Merchant ships tended to remain general-purpose rather than specialized, although it was common for many to serve out their careers in the same trade. That trade likely meant carrying the same produce—tobacco or sugar or rice—to its destination seasonally, and then loading sundries ("market goods," as they were often referred to) for the return leg. Tobacco ports needed manufactured goods, tropical produce, indentured labor, and enslaved labor. Slave ships, as we have seen, were not a distinct type, but could be outfitted for a slaving voyage and then reconfigured for other work. With seasonal, unpredictable markets and populations small enough that an entire ship full of one product was rarely a viable proposition, there was more pressure against specialization than for it.

Predation also remained a fact of life throughout the period. Folded into the general irony presented by the timing of the American Revolution, coming as it did so shortly after the British Atlantic had triumphed over its rivals, is the irony that this conflict perpetuated maritime predation just when the Royal Navy, the insurance industry, and the marketplace had finally made the Atlantic much safer for ships than it had been earlier. Be that as it may, however, naval protection was more effective, as the Royal Navy had built up its infrastructure and administration so effectively by that point. Now, American ships were no longer protected by that Navy, but vulnerable to it. The British American economy, that robust product of a successful British Atlantic empire, collapsed. American delegations went begging to France and the Netherlands.

After 1779, when the War of American Independence expanded into a global conflict, even the mighty British Empire was stretched thin. Naval forces had to be kept close to home for defense. Admiral Rodney gorged himself on the spoils of St. Eustasius, long-defiant Dutch hub of free trade in the Caribbean, while French Admiral DeGrasse escaped north, surprising the British squadron in the York River. Cut off and besieged, General Lord Cornwallis surrendered to U.S. and French forces. By 1783, the civil war was over, too, the last British troops were evacuating, and thousands of loyalist Americans went into exile.

The American economy was no better off, however, and now saddled with massive war debt. The British government refused to allow U.S. ships to resume trade with the British sugar islands. That hurt the British sugar interest more than it did U.S. merchants and masters, who could still trade to the French islands, but both parties entreated Whitehall and Westminster for redress—to no avail, as the establishment in London was unmoved by their distress (and by the mass starvation of enslaved workers on British sugar plantations). While both the U.S. and the West Indies planters worked around British policy as best they could, the situation was never resolved to either party's complete satisfaction before the abolition of slavery in the Empire in 1833. Meanwhile, British manufactures trounced nascent U.S. alternatives in the markets of the new republic. Severe depression held down the U.S. economy throughout the 1780s.

Despite the hardships, however, the U.S. economy strengthened in the 1790s, and the British economy continued to thrive even through the wars of the French Revolution, with the Navy protecting the merchant fleet at sea as it carried on with the maritime commerce that had made the Empire. The same ships that had been carrying goods and people before the revolutionary disruptions continued to do so. Commercial interests in both the U.S. and Britain recognized the benefits of resuming normal trading relations to go along with normal political relations. The latter, however, would prove most difficult to accomplish, given the impossible situation the U.S. found itself in, batted about between the warring empires of Britain and France. U.S. merchant ships were stopped, searched, and sometimes seized by predators on both sides, touching off conflict at sea in 1798, with France, and 1812, with Great Britain.

Owners and crews, then, continued to face the same risks of doing business on the Atlantic as they always had. There were more owners, more crews, and more ships for them to sail. More of those ships were larger than they had been in earlier times. More of them sailed unarmed, although to be sure,

British ships stood a much better chance of sailing safely than U.S. vessels, as the former enjoyed a singular degree of naval protection, while the latter enjoyed precious little of it; big navies are for rich countries. The thirteen Eastern Seaboard colonies had given up much to leave the Empire, and the crews of their merchant ships were keenly aware of one major cost of that decision. The Atlantic was still British.

•••

While a summary like this serves a useful purpose at the end of a book, it can also be misleading—in the same way that maritime and Atlantic history can be, and have been, unwittingly misleading to some extent about the development and use of the ordinary merchant ship and its role in creating and sustaining the new world brought about by connecting Europe, the Americas, and Africa via the Atlantic. The primary misperception we risk is that we have answered all the important questions about how these people went about designing, building, and using these ships, and why they did it the way they did. We have not answered those questions. We have developed some promising hypotheses, presented evidence supporting them, and pointed out how those hypotheses have, in some cases, called into question some long-standing assumptions in the historical literature, primarily that merchant ship technology was for all intents and purposes static, and that any technological innovations that occurred were inconsequential. It is time to draw some conclusions, tentative though they may be, and lay out the work that still needs to be done.[2]

Credit is due maritime economic historians for deducing important trends from the historical record. As important as any of those is that ton-per-man ratios generally tended to increase over the course of the period. Ralph Davis demonstrated that in 1962. John McCusker put forth a strong argument a few years later that increased capacity-for-size was at least a major factor in that trend. Davis and McCusker also presented strong evidence for a general modest increase in average size over the course of the period, but that presents its own set of questions. If ships got bigger, even a little bit, and crews did not, did that mean that sailors were working harder later in the period than earlier? Only a careful, sophisticated examination of the work aboard ship can answer

2 I presented earlier versions of the questions and problems that follow here, and which are revisited in the Conclusion, in Phillip Reid, "The Ordinary Merchant Ship in the British Atlantic, 1600–1800: A Call for Further Research," *International Journal of Maritime History* 29:4 (November 2017): 911–26. This book represents some of that "further research"; there is much more to be done.

that question, which is one compelling reason to study continuity and innovation in ships' rigs. Did the adaptations to rigs that we know occurred decrease or increase crew workload? We see a clear trend toward division of sail plans into smaller sails, on both square-rigged and fore-and-aft-rigged vessels. That tells us that each of those sails would be manageable even with an increase in the size of the vessel, but that there would also be more sails to work. So, it is not clear that sailors would have had less work to do on vessels with the later rigs, but that the work was feasible for the same-size crew. Did the workload remain roughly the same, although the actual tasks performed differed somewhat? Comparative experimental archaeology, using replicas of both earlier and later vessels, should help us answer that question.

Did an increased use of fore-and-aft sails, even on primarily square-rigged vessels, allow the same crew to work a larger ship with the same expenditure of effort? If so, was there a cost to the ship's downwind performance? We know that fore-and-aft sails are less effective off the wind than square sails; that they did not, contrary to common misconception, make one of these vessels more maneuverable, nor did they necessarily allow a merchant vessel to sail closer to the wind, contrary to another common misconception. However, we also know they are easier to operate. We need to determine the comparative workloads on similar vessels with different ratios of square to fore-and-aft sails, as well as comparative downwind *and* upwind performance, to determine the cost-benefit relationship of all three. This is not a matter of determining primary causality, which is almost certainly an unattainable research quarry. If we can determine what the costs and benefits were in multiple important respects, however, we can make informed guesses as to why certain technological choices were made, carried forward, or abandoned. Meanwhile, continuing work in the archives, in the papers of ships' masters especially, may get us closer to a clear sense of that elusive primary causality.

What about the decrease in armament in the later eighteenth century, which Shepherd and Walton demonstrated in 1972? I have shown here that it takes more men to work guns and operate the vessel effectively than it does just to operate the vessel. Can we correlate that to what we find on crew lists from the period? Shepherd and Walton made a good start; it would be worth expanding their data considerably, to make sure the trend holds with a larger sampling. At the same time, we have to keep in mind that required manning complements and the actual numbers of crew listed cannot be assumed to be synonymous.

What about a decrease in sailor mortality? Compiling data on mortality rates on merchant ships over time could tell us how likely it is that any trend in those rates influenced ton-per-man ratios. A decrease in typical mortality rates could have been a contributing factor, if such a trend in fact existed. We should also consider the availability or scarcity of labor. We know that the supply of skilled seamen in the British Atlantic was never adequate, especially in wartime—hence the stubborn persistence of impressment despite the general loathing of it. Was a scarcity of skilled labor, and the consequent price of that labor, pressuring for labor-saving adaptations aboard? We know that building costs did not decrease, so whatever factors were pressuring for slightly larger ships must have been considered more important.

Questions of mortality, the actual manning of actual ships, wages, and building costs can be pursued and answered in the historical record. Official documents and merchants' papers have survived well enough for that, although it is certainly easier to research the eighteenth century than the seventeenth, given the growth of the British Atlantic that occurred in the eighteenth century and the fact that more regulations generated more documents during and after the 1690s.

As for understanding the work aboard and how the human component integrates with other components of the ship, however, we must turn to experimental archaeology. If we can study a replica based on clear, sound historical evidence, with documentary sources on the original still available, we can compare the number of crew needed to operate the replica with the number the records tell us were required to operate the original. Ideally, we would have records on this vessel in both armed and unarmed configurations. Even better would be a vessel that had been rigged two different ways, with extant detailed records on both ways. If we could compare the same vessel both armed and unarmed—including any and all extra costs of armament, in money and stowage, and that vessel with one rig to the same vessel with another, we might get closer to answering all the questions above by studying one ship. Fortunately, there is a vessel that may be able to answer more of these questions than any other: the schooner *Sultana*. Built at Boston by Benjamin Hallowell on speculation and sent to England in 1768, she was sold to the Royal Navy, who took her lines off and drew up plans and specifications on her, as was standard practice when buying a ship into the service. Intending to use her as an armed patrol craft-interceptor for customs enforcement on the Eastern Seaboard, the Navy re-rigged her for speed and power, armed her with swivel guns, and sent her back across the Atlantic, where she served in that capacity until 1772.[3]

3 See Drew McMullen with Captain Robert Brittain, "The Schooner *Sultana*," in Bennett, *Sailing Into the Past*, 150–61. For context, see Sarah Kinkel, "The King's Pirates? Naval Enforcement of Imperial Authority, 1740–76," *The William and Mary Quarterly* 71:1 (January 2014): 3–34; and *Disciplining the Empire*.

FIGURE 16 Replica of HM Schooner *Sultana*, 1767
PHOTOGRAPH BY MICHAEL WOOTTON, COURTESY OF THE SULTANA EDUCATION FOUNDATION

The records left by the Navy, and all the ship's logbooks, survived. We know, then, how she was originally designed and built, how she was rigged and re-rigged, what she carried, and how she performed in use configured as both a merchant ship and a small makeshift warship.

That depth of knowledge and treasure chest of information made her a perfect candidate for a replica, and the current *Sultana* was launched in 2001 after four years of construction. We can compare her speeds and performance characteristics with those recorded for the original. *Sultana* provides a rare opportunity to cross-reference replica performance to original performance. That in turn could take much of the speculation out of doing the same with other replicas we have reason to be confident are accurate, such as *Duyfken* and the Swedish Ship *Götheborg*. The relationship between replica and original performance of *Sultana* might serve as an otherwise-unattainable rule of thumb for hypothesizing that relationship in cases where the historical record is nowhere near as complete.

As an American-built ship purchased by owners in England, *Sultana* was by no means exceptional. While we know that shipbuilding for export was a successful enterprise for New England shipwrights as early as the 1640s, and for Delaware Bay yards in the eighteenth century, whether the data consistently, let alone conclusively, support the notion that American-built ships were shorter-lived than those built in Britain is a question still open for research. We should continue to test the anecdote relayed in de Tocqueville's *Democracy in America*, wherein the shorter lifespan of American ships is taken for granted and attributed to American proclivities for innovation, planned obsolescence, and quantity over quality. Considered in aggregate, the evidence does not inspire confidence in the broad applicability of the anecdote, and is likely more illustrative of an early Federal-period American cultural self-image than the state of American shipwrightry.

As we consider specific technical questions concerning hull design, construction, and rigging, we need to keep in mind that technological choice always has an aesthetic component, whether it is realized or acknowledged or not. The evidence makes abundantly clear that aesthetics mattered to owners, builders, and mariners—and not just in superfluous decoration. The "fancy" bow and stern work of the frigate-built ship offered no functional advantages over the plainer—and cheaper—alternatives offered by the bark and the flute. Specialists in seventeenth- and eighteenth-century aesthetics may be able to offer insight as to how taste, fashion, and the sense of what looked proper likely influenced design choices. The history of modern technology, drawing on material culture studies, has done much to situate technology in culture in this way, providing a good example to follow as we expand the history of premodern technology.

Those early seventeenth-century hulls, with their high, tapering stern castles forming tall, teardrop-shaped sterns, and their tumblehome exaggerating the roundness of their hulls, look quaint to us. Their brightly-painted decorations and streaming pendants make them seem perhaps more like amusement-park props than stout, capable ocean conveyances. Yet that was what they were, and we have good evidence that some were handy, weatherly, and reasonably fast when compared to more modern counterparts. On the other hand, we have other evidence that some had inherent stability problems. To be clear on what I mean by "inherent," I would remind the reader that all these ships had to be loaded or ballasted properly to be stable, as they carried no built-in hard ballast, as modern yachts do, or water-tank ballast, as modern ships do. Some, however, needed so much ballast, or cargo, just to be acceptably stable, that they did not float at their designed waterline, compromising their sailing ability and reducing their *freeboard*—the amount of vertical hull side between the waterline and the edge of the upper deck—more than was ideal for safety in rough seas.[4] *Mayflower II* and the ketch *Nonsuch* (1650, 1968), built with Matthew Baker's tangent-arc system in mind, suffered from this problem, as did other replicas built to that specification. We know from the historical record that English ships of this era were commonly furred or girdled to improve their inherent stability.

On the other hand, we know from Adams' laboratory experiments with the virtually-reconstructed *Sea Venture* that she proved quite stable. His ongoing experiments are important. The more such experiments are performed, the better; they are likely to yield a markedly more complete understanding of this issue than we have had since these ships disappeared from the seas. It is already clear that we cannot assume all early-seventeenth-century English vessels to have demonstrated the stability problems that seem to be inherent to Matthew Baker's tangent-arc method. Analysis of the midship bend of more early-seventeenth-century hull remains, should we be fortunate enough to find them, will give us a better sense of what design approaches were actually in use, and to what extent.

When we do experiments on inherent stability, we should include tumblehome and castling, as we know that changes in those two important aspects of hull form—which we have established are interrelated—both decreased over time. In the laboratory, it is possible to test the same virtual hull with varying

4 Modern Atlantic cargo ships have a set of marks and numbers painted on their hulls to indicate safe load waterlines for different expected sea conditions.

degrees of tumblehome and castling. By considering the weight of hull material, taking into account typical scantlings, we can test the hypothesis that decreasing tumblehome pressured for decreased castling to reduce the weight of upper works, lowering the center of gravity for better inherent stability—which is where the evidence points. Builders could have reduced castling without decreasing tumblehome, but they did not. That does not disprove the traditional notion that decreased castling was the manifestation of a de-escalating arms race, as high fighting platforms lost relevance in ship-to-ship combat, but if we can demonstrate that decreasing tumblehome without also decreasing castling would not have worked, we can add another compelling explanation for the simultaneous change, adding further weight to the more general argument that significant technological decisions are usually the result of multiple pressures.

Closely related to the connection of tumblehome to superstructure is that of superstructure to length-to-beam ratio. It may be that building a longer hull, relative to its beam, also pressured to reduce castling. Too, we must consider not just the weight of castling, but windage; recall that the windage of the stern castle on *Mayflower II* helped her lie a-hull in a severe Atlantic storm. Would the effect be altered on a longer hull? Would the ability to lie a-hull as well as *Mayflower II* be a necessary sacrifice of reduced superstructure, which itself was necessary in order to lengthen the hull? Laboratory analysis may be able to sort out answers to these questions, as well as how they may relate to one another.

Does a decrease in tumblehome help explain the increase in capacity-for-size that McCusker found so strongly implied in the changing laws for calculating tonnage for duties, and if so, to what extent? We need to calculate block coefficients for as many period ships as possible, wherever we have reliable records. If we do find such modifications, we need to consider their implications for stability and speed. We cannot assume that a fuller hull was necessarily a slower one, even if waterline length remained the same; there are too many variables. Even if it was a bit slower, that disadvantage might well have been moot in the Atlantic, given the preeminence of its fickle sea and weather conditions in determining actual passage times. Besides, if ships were getting modestly larger, and thus a bit longer, the increase in hull speed inherent in increased length may have offset whatever cost in speed a fuller hull form imposed. Laboratory experiments on resistance with virtual hulls should help sort this out. As it stands, the aggregate of evidence points to fuller hulls, with less tumblehome, slightly larger but also more capacious for their size, as the primary design innovation in hull design adopted over the course of the period.

While considering capacity-for-size, and how shipwrights may have increased it, we can compare fish-form hulls—the standard throughout the period—with a more box-form hull, whose midship bend, and thus maximum beam, was amidships rather than forward of it. Speth's experiences sailing replicas of early seventeenth century ships with nineteenth-century counterparts suggest that whatever penalties to speed the fish-form hull may have imposed—and we do not know that it did—may not have been sufficient to pressure for change, especially with the counter-pressures of capacity and buoyancy in the bow section. Virtual experiments should prove helpful here as well.

Recall, too, that there was a construction imperative to bluff bows—buoyancy, to bear upper weight and resist hogging. We cannot consider aspects of design in the abstract; we have to consider structural considerations, since ships had to be assembled from pieces of wood that had to stay together under multidirectional strains. Some of those strains were caused by the weight and recoil of guns, and we know what scantlings had to be increased and what structural reinforcements were necessary for carrying a complement of deck guns.

While we have treatises on period ship design and construction, the problem of determining to what extent those treatises were descriptive and prescriptive forces us to rely on archaeology to tell us how ships were actually built. Given the parts of ships usually missing from archaeological remains, that requires skilled conjecture based on a compilation of evidence from the archives and the seabed to virtually reconstruct what is not there. The more evidence we find, preserve, and analyze, the more confident we can be in our interpretation. In 2017 and 2018, archaeologists in Alexandria, Virginia found and excavated the remains of four deliberately-buried late-eighteenth-century merchant hulls.[5] The virtual reconstruction of the first one has just now been produced (see Figures 8 and 9). Analysis of these hulls will add materially to what we know about design and construction at the end of the period. We cannot predict when we will find such remains, but with archaeological surveys required for waterfront construction projects, our chances are excellent of adding several more vessels to our database in the near future. It is to be hoped that funding to cover the significant costs of conservation will be forthcoming. The more widely-disseminated an understanding of the potential advancement of human technological history offered by these remains, the more likely individuals and groups will be willing to contribute financial support.

5 On the analysis and documentation of the first Alexandria vessel, see http://nautarch.tamu.edu/CRL/Alexandria/index.html, accessed 24 Jan. 2019.

Conservation is not the only cost worthy of such support. Modern software allows actual archaeological remains to be digitized in full detail, piece by piece, such that those pieces can be virtually disassembled and re-assembled, in a process called "reverse naval architecture." The process allows archaeologists to figure out how a vessel was put together, down to each extant fastening, without having to handle the actual remains, which are fragile, cumbersome, and irreplaceable. It is thus much easier now to determine the proportions used in the design, compare those to the archaeological record and to the published treatises, and to study comparative construction techniques. Archaeologists have looked for patterns in construction techniques in the British Atlantic, but those have proven elusive. We may soon have a better chance to advance that search—wherever it might lead—than we ever have, and we certainly have better tools with which to do it. We should examine archaeological finds for any correlation between date and place of construction and evidence of any adaptation of construction techniques indicative of scarcer materials, such as fashioned pieces where we might expect to find compass timber, or less-than-ideal wood species used in certain applications. We should also note, as Warren Riess did, any similarity to Dutch methods, in which, for example, framing timbers were shifted out of uniform patterns, to some extent, in order to fit the hull shape with minimal trimming. It would stand to reason that British Atlantic shipwrights were influenced by that Dutch technique, just as they were influenced by Dutch design principles, but that does not prove such a correlation. We know that building techniques, such as using fire to hollow a log for a canoe, developed independently in different places.

The Dutch influence on design, and any correlations between Dutch and British Atlantic construction techniques, need to be considered for every vessel we investigate, as the archaeological evidence so far backs up the historical literature claiming that influence.

One possibility worth looking for is that the Dutch influence, with its tendency toward wider beam and shallower draft, allowed ships to grow in size without growing proportionally in draft. Deep draft posed problems all around the British Atlantic, just as it did in Northern Europe. If we were to find a pattern of shallower draft for hull length in the eighteenth century, that might be an important element of the Dutch influence. We could expect to find a similar trend on French ships, and we should look for it; if David Moore was correct that *Henrietta Marie* was French-built, and his evidence is strong, then he may well have discovered the same Dutch influence on a French vessel as Riess found on *Princess Carolina*.

As ships grew in size, they tended to adopt wheel steering systems in the eighteenth century. We should continue investigating whether that is a causal correlation. It would be helpful to locate period sources that commented on the adoption of the wheel—or on the use of the earlier tiller-and-whipstaff system on larger vessels, although caution is required, as anecdotes and opinions can be highly idiosyncratic, might have been hotly debated at the time, and might even be disprovable. We have seen that consensus in primary sources is not to be expected; take for example the disparate opinions on the fish-form hull expressed in the proceedings of the Society for the Improvement of Naval Architecture. The treatises are another good example.

As for experiments on steering, Fred Hocker is pursuing his theory of tiller length being a limiting factor in mechanical advantage on larger ships, based in part on his experience sailing with the crew of *Kalmar Nyckel*. It would seem that an examination of period models known to be accurate might further that inquiry, especially if the models were digitized so that exact proportions and loads could be calculated. We are fortunate that maritime museums hold fine collections of such models, and we can use them for such purposes without harming them.

If we can compare the labor required to steer a ship by wheel and a similar ship by tiller-and-whipstaff, we can mount an argument as to whether saving labor was a likely pressure to make the change; or, if we can find no such advantage, we can make the case that other considerations, such as visibility from the helm and the ability of the helmsman to respond more quickly and accurately to steering orders from the officer of the deck, were the primary driving factors. As for a possible connection between the adoption of the jib as a headsail and the adoption of wheel steering, it would be preferable to test such a connection as best we can rather than either accept it or dismiss it out of hand. It may be that, once it was technologically-feasible, increased size was the primary factor driving the adoption of wheel steering, but better visibility for the helmsman, making it easier for him to steer accurately, may have been just as important—or more. Until we experiment further with steering systems, and their relationship to choice of headsails, we cannot accept at face value McGowan's claim about the relationship of wheel steering and jibs, nor the claims made by Steele and Menard based on it.

Given the relative strengths and weaknesses of fore-and-aft versus square sails, it stands to reason that the spritsail, the square sail mounted under the bowsprit, survived into the era of the jib; neither would have interfered with the other. It would be worth asking, however, whether extended bowsprits, with their substantial support rigging, including *dolphin strikers* and *whisker stays* mounted beside and below the bowsprit, precluded the use of such a sail.

What about the sprit-topsail—the smaller square sail mounted on top of the bowsprit? Harland confessed it to be an enigma to him. Dounce said it provided some balance, but was labor-intensive to rig and adjust. It may well be that sailors and masters eventually decided it was not worth the bother, perhaps because the jib did a good enough job as a substitute. More experiments aboard appropriately-equipped operating replicas can test that hypothesis. Another potentially-fruitful avenue of inquiry would be to compare its use aboard merchant ships versus warships. If we were to find that it persisted longer on warships, it may be that the labor demands were the deciding factor, as the saving of labor was so much less a priority aboard warships.

The saving of labor cost—or perhaps more accurately, the avoidance of increased labor cost—probably served as the primary pressure for splitting the upper square rig into more but smaller sails. This too can be tested by comparing labor demands on similarly-sized ships with each type of rig. Calculating sail areas and thus the power generated by specific sails in specific winds may help as well. Since both wage data and shipbuilding costs are available to us, we can compare the cost of the rig adaptation to the saving of labor, should we prove able to discern the latter.

Before taking leave of the earlier single topsails, however, with their taller aspect ratios (height to width), we should conduct further experiments on the technique of half-lowering them as a form of reefing (reducing sail area aloft in higher winds). The experiential evidence we have suggests that the technique, so counter-intuitive to modern sailors, was more satisfactory with smaller sails, on smaller vessels, but it would be useful to know how well this suggestion holds up under further trial. We may find that a change in how typical sails were cut or their cloth woven may have played a part as well.

As for the mizzen, which on the average British Atlantic merchant ship morphed from the lateen mizzen, found on Columbus' ships, to the cut-back lateen, to the gaff-rigged spanker or driver, we know from Speth's experience that more labor is required to operate a larger lateen mizzen, and we know how much. We can then calculate how much the labor savings might have been at a given time, but we need to investigate the costs of converting a lateen yard to a gaff, especially the means by which the gaff was joined to the mizzen mast, to compare the labor savings to the cost of refit. We also need to conduct experiments to compare labor demands of similarly-sized and -rigged vessels with lateen mizzens and gaff spankers or drivers.

Speth also suggested that the increasing adoption of staysails may have pressured for the abandonment of the lateen mizzen. We can test this on accurate scale models, but we might be able to do so on certain replicas as well. The fact that staysails and lateen mizzens existed concurrently does not necessarily disprove Speth's hypothesis, as such changes never occurred immediately.

The final quarter of the period considered here saw a significant increase in the number of medium-sized and larger square-rigged ships rigged with two masts instead of three—snows and brigs. A detailed study of the costs and labor demands associated with these alternatives, compared to the quarter immediately preceding, should ultimately explain why this was so. It is likely that the split rig made sails easy enough to handle that the third mast could be dispensed with. If so, the change would have saved the cost of the third mast and all its spars, sails, and rigging, and saved the crew the tasks of working those. This would be an important stand-alone study in both maritime economic and maritime technological history. The *Lady Washington*, operating out of Grays Harbor, Washington, is a replica of an eighteenth-century brig rigged for merchant service. An investigation of her operational requirements would be valuable.[6] The U.S. Brig *Niagara*, though rigged as a warship, is a hybrid "snow-brig," which was not uncommon; it would be worth interviewing her crew about this rig.[7]

Another promising study of a shift in popularity in the other direction—more masts rather than fewer—would be that of sloops versus schooners, in which the popular single-masted coasting vessels in service since the beginning of the period ceded pride of place to the two-masted schooner by the late eighteenth century—or so the archival evidence suggests. Here, the case may be that the sails of a schooner were so much easier to handle than those of a sloop of similar size and capacity that the costs of the second mast and its rigging were easy to bear. It should be reasonably straightforward to determine that, again with a combination of experiment and archival work. We have adequate information on the now-retired Great Lakes replica Armed Sloop *Welcome* and her rough contemporary, the schooner *Sultana* on the Chesapeake Bay in Maryland. There is also an operational sloop relevant to the period, *Friends Good Will*, at the Michigan Maritime Museum. As we investigate sloops versus schooners, we must keep in mind the possibility of increasing size as a driving factor, so we should note the relative sizes of the vessels we are considering, not just their rigs and numbers of crew. We also need to understand the schooner as a counterpoint to the ketch, so popular and common as an ocean-going fishing and merchant vessel in the earlier part of the period but largely absent

6 See https://www.historicalseaport.org/ships/lady-washington/, accessed 27 November 2019.

7 See Figure 13. A promising place to start a virtual-modeling comparison may be found in David MacGregor's *Merchant Sailing Ships*, in which he discusses plans he found for the same vessel rigged as both a snow and a brig. See 74–75, 80–81.

by the end of it. Both were two-masted, but the schooner was more thoroughly fore-and-aft rigged. Was the increasing reliance on fore-and-aft rigs for smaller vessels the primary pressure driving that shift?

Since our own experiments can only suggest why past people made the decisions they did, researchers must continue to seek the technological opinions of shipowners and shipwrights, insurance underwriters and government officials, in the archives. It is a laborious process, as such opinions are rarely expressed in the course of ordinary business correspondence and even in building contracts, but no other source will substitute for them. It would be especially helpful to examine collections of merchants' and shipwrights' papers in UK repositories, which I was only barely able to do.

While the questions and proposed investigations laid out above make clear that there is still much to sort out, it is already clear that those who paid for, built, sailed, and maintained the British Atlantic merchant ship made decisions about technological retention and adaptation that met the needs of their world just as well as the ships of the next era would meet theirs, without the need of the "scientific" naval architecture that would later assume some of the predictive powers based earlier on cumulative experience, thus reducing the risk somewhat of launching ships with significant adaptations made in the relatively short term. Understanding how a technology remained viable over such a long period, without drastic change on the one hand, or becoming a drag on the growth and development of its world on the other hand, is of serious importance now, with the pace of change in digital technology constantly increasing, and with that technology inserting itself into more and more other technologies. We entered our own technological age still heavily under the influence of the Enlightenment, with its strong bias toward the idea of linear progress and concomitant technological functionalism and determinism, borne along through at least two centuries of technological imperialism.[8] Ironically, perhaps, given that we also live in the "Information Age," it is likely more difficult, not easier, for us to apprehend the sweep of human technological history. To correct that, we need to study the technological histories of other peoples at other times, as well as those of other places, both past and present. Doing so, given the limitations of the written record, requires historical inquiry deeply informed by archaeology, anthropology, and the study of material culture. The payoff for that, however, promises to be greater than

8 On technology and imperialism, see Adas, '*Machines as the Measure of Men*'; Daniel Headrick, *The Tools of Empire: Technology and European Imperialism in the Nineteenth Century* (Oxford: Oxford University Press, 1981); and John Ellis, *The Social History of the Machine Gun* (London: Century Hutchinson, 1975).

better pre-industrial technological history, for such an approach brings about a more sophisticated, culturally-situated history of technology in our own time as well.[9]

The ship is one of the oldest technologies there is. Paleoanthropologists have traced the use of ocean-capable watercraft so far back into the past that the continents and islands were not the same as they are now when early humans began to move from one to the other by water.[10] Most peoples have used watercraft to transport themselves and their goods across substantial bodies of water, and we still do. The operating environments provide strong consistencies against which to compare the different technological approaches people have employed to make use of, and survive, sojourns upon the element that comprises most of the surface of our planet—an element upon which, without the aid of relatively sophisticated and costly technology we cannot survive. So, the ship lends itself particularly well to comparative technological history. The approach used in this book, and the questions raised in it, have application beyond the British Atlantic in its formative two centuries. They can help us understand how any people, at any time and in any place, have taken their own particular technological approach to their sea ventures.

---

9 See Schatzberg, *Wings of Wood, Wings of Metal* for an outstanding example.

10 See Atholl Anderson, et al., eds., *The Global Origins and Development of Seafaring* (Cambridge, UK: McDonald Institute for Archaeological Research, 2010).

EPILOGUE

Ann & Hope in Canton—Beyond the British Atlantic

A few days before Christmas, 1798, after a voyage from Rhode Island of five months and ten days, the ship *Ann & Hope* reached Canton, China. She had been commissioned by Brown & Ives of Providence, and built by Benjamin Tallman, who had built other substantial vessels for the Browns and their partners. Substantial she was, too, at 550 tons, when a middling-sized Atlantic merchantman was 200 or so. The records of various iterations of the Brown family shipping business survive, in the research library of the university that bears their name. In those records, we find detailed records of their humble beginnings in sloops and schooners, in the coasting trade, in the ubiquitous trade from New England to the Caribbean to exchange stores, food, and sundries for sugar and molasses, to brigs and snows, success in the transatlantic slave trade, and finally to the building and launching of East Indiamen like *Ann & Hope*, as the new United States struggled out of a severe postwar depression and her shipping magnates sought, not just to re-establish their old trades, but to launch themselves into new ones.

Ann & Hope was not built for owners in Britain. She was not to be fractionally-owned by a large group of investors. The Browns had accumulated the kind of money necessary to commission an East Indiaman, and they were not the only American owners who had. From the 1790s until the implosion of the U.S. Civil War in the mid-1860s, the British Empire would face real and persistent maritime competition from the young breakaway republic whose existence represented both that Empire's great success and the limits thereof. Everything that British Americans had learned about ships, sailing, and the shipping business in two centuries under the imperial umbrella, they would now put to use as an independent power for their own sole benefit. No longer, however, would they enjoy the substantial privileges afforded by membership in the most successful maritime empire in history. Without privileged access to the ports and markets of the British world, the U.S. shipping industry would have to compete with that of France, the Netherlands, Spain, Portugal, and any other European or American maritime interest, with no special advantage. The former alliance with France was no consolation; indeed, in the same year *Ann & Hope* made her maiden voyage to China, U.S. ships were exchanging fire

 | DOI:10.1163/9789004426344_011

with French ships in the Caribbean in what became known as the Quasi-War. No recourse to Royal Navy protection was to be expected, naturally.

Steering a course of total neutrality between the still-warring imperial titans proved untenable, putting enormous strain on the republic's nascent and fragile political system. By 1812, the U.S. had gone from a Quasi-War with France to all-out war with Great Britain. That internally-divisive war cost the young country dearly, but it was a cost deemed necessary by those who supported it to protect the future of U.S. high-seas shipping.

Ultimately, however, the commercial and political interests of both Britain and her former colonies pressured strongly for a rapprochement. It turned out that the benefits to both parties of the old British Atlantic system were too important to discard completely. The ban on the transatlantic slave trade after 1807 was the result of an Anglo-American agreement, and the Royal Navy enforced it seriously. Small, fast ships—especially schooners—from the U.S. and Spanish America continued an illegal trade as best they could, although on nothing like the scale of the eighteenth century. Nevertheless, the U.S. did not oppose the British effort. Its slave population was self-reproducing, and the bulk of the slaves brought from Africa had always gone to the sugar islands of the Caribbean, of which the U.S. had none. In 1833, Great Britain abolished slavery throughout the empire. The reign of sugar was over. The rise of cotton would not depend on a steady supply of fresh African captives as the sugar economy had.

Both Great Britain and the U.S. turned their attention to opportunities in Spanish America. Despite the Monroe Doctrine, it was in both their interests for the U.S. to stay quiet about British incursions as the Empire moved to take advantage of the independence movements sweeping through South America. Meanwhile, southern slaveowners contemplated expanding the cotton kingdom into the Caribbean, as they shipped more cotton to the mills of industrializing Britain. Both powers still viewed the Atlantic Basin as a place for their ships to do the business they had long done, whether they were trading with each other or letting each other trade separately. Both, in important ways, were operating amicably in a British Atlantic.

That is not to suggest there was no competition. The arrival of *Ann & Hope* in Canton presaged the fierce but peaceful rivalry between U.S. and British tea clippers, a technological and literal race that transcended commerce to become a matter of national pride—as did the schooner races between New England and Canadian fast fishing vessels, which persisted into the twentieth century. In both cases, a niche market dependent on ships getting a highly-perishable, valuable commodity to market as fast as possible drove the technology of the fast ship forward. As both powers expanded—Britain into Asia and the Pacific and the U.S. across the continent—they needed fast ships to carry passengers and cargoes valuable enough to justify the high freight rates required to operate such specialized vessels, with their large crews and high riggers' bills. Only the Suez Canal and the U.S. Transcontinental Railroad, both completed in 1869, gave steamers a real chance to compete with long-distance sailing ships.

Still, the plank-on-frame wooden sail- and muscle-powered merchant ship remained in service, the last ones not disappearing from regular commercial use until the mid-twentieth century. We have photographs of ships that were built and put into service during the eighteenth century. Some were still in use, although in limited service and in many cases heavily modified from their original form. This technology proved to be exceptionally durable. Although the ships of one epoch would have surely seemed strange to mariners of another, those mariners would have understood perfectly well how they worked. The continuities are there, and they are important. Nevertheless, these ships never remained exactly the same for long. Adaptations were necessary as operating conditions changed with demographic and market shifts, and those adaptations were executed. Too, given the ease with which skilled mariners could improvise while underway, adaptations were bound to suggest themselves through routine trial and error. Perhaps too much has been made of the innovations that took place in the sixteenth century, and in the nineteenth, relative to those that accumulated during the intervening two centuries. If we make an attempt to understand these continuities and innovations, rather than accept them as given or gloss over them with too-easy explanations, we will further our understanding of the human relationship to one of the most world-altering technologies we have ever devised.

Glossary of Terms

Note: For online access to a comprehensive glossary of shipbuilding terms, see Steffy, "Illustrated Glossary of Ship and Boat Terms;" for more general works from the period, see Mainwaring's and Falconer's dictionaries. Italicized terms in this Glossary are themselves defined here.

Adze	a wood-shaping tool similar to an axe, but with the blade set perpendicular to the handle rather than parallel to it
Aloft	in the rig, above the main (*weather*) deck
Amidships	the longitudinal mid-point of the *hull*
Anchor stock	the heavy wooden bar set perpendicular to the anchor's flukes, to encourage the anchor to turn fluke-down when it hit the bottom
Athwartships	transverse; perpendicular to the *keel*
Backstay	a piece of *standing rigging*—a taut line—running from the top of a *mast* to an after point, to keep the mast from pitching forward; could be adjusted to adjust mast rake and thus sail plan balance
Badge	a decorative piece attached to the *hull* near the *stern*, containing a window or a carving of one, and decorated with ornament
Ballast	weight placed low in the *hull* for stability; could be heavy cargo, such as pig iron, or stone, gravel, or shingle
Bark	in this period, not a specific rig, but a *hull* type emphasizing simplicity and capacity over speed and aesthetic flourish
Beam	the breadth of a *hull*
Beat	to sail as close to *weather* (as high upwind) as possible; see also *Close-hauled*
Bilge	the lowest part of the inside of the *hull*, where water tends to collect
Bill of lading	a document, detailing cargo loaded, provided by the carrier to the owner or consignor of the cargo, acknowledging receipt of the cargo on board and acknowledging its value
Bitt	a short stout post, with a short crossbar set into its top, to which anchor or mooring lines are secured
Block	an assembly of wooden cheeks and wooden or metal sheaves, used for pulleys or, in nautical terminology, *tackles*
Block coefficient	the amount of a rectangle defined by a *hull's* length, breadth, and depth taken up by that hull, expressed as a ratio

 | DOI:10.1163/9789004426344_012

Boom	a *spar* to which the *foot* of a *fore-and-aft* sail may be attached; the front end of a boom is attached to the *mast*, while the after end is free to move side-to-side, controlled by a *sheet*
Bow	the forward end of a ship or boat
Bowsprit	a *spar* extending forward from the end of the *bow*, to which sails can be attached to project them past the *hull* itself
Brace	1. on a *square sail*, a line used to control the end of a *yard*; 2. To turn a yard by means of a brace
Breast hook	a v-shaped timber connecting longitudinal timbers as they run into the *stem* (or *stern*); helps to form the structure of the ship's end—a type of *knee*; if used astern, the piece could also be called a *crutch*
Brig	a rig description, denoting a *square-rigged* vessel with two *masts*, in which the *main mast* was aft of the *foremast*, and slightly taller (contrast to *Ketch* and *Snow*)
Brigantine	generally considered interchangeable with *Brig*
Broach (or Broach-to)	to be pushed broadside to the waves by a strong following sea, placing the ship in immediate danger of being rolled over
Broadside	along the side of the hull; typically applied to the arrangement of carriage guns carried on the main or lower deck(s) and deployed through hinged gunports cut into the side of the *hull* or *rail*
Bulkhead	the structural equivalent of a wall in a building, providing compartmentalization and structural rigidity
Camber	a slight convex curve, in a deck or an individual timber, to discourage the pooling of water and encourage fast drainage
Cant frame	a *frame*, usually in the *bow*, whose angle to the keel is less than 90 degrees; cant frames are used to form bow structures—contrast to *Square frames*
Capstan	a winch, mounted vertically, used to wind up anchor rode (anchor chain or line); stout wooden bars inserted in holes in its drum gave the crew levers to push on as they walked around the capstan in a circle, weighing (raising) the anchor—contrast to *Windlass*
Careening	leaning the ship way over, on a gently-sloping shore or against a wharf, to clean, repair, and *pay* her bottom
Cargo manifest	a detailed list of the vessel's cargo for presentation to customs officials
Cargo tonnage	a measure (or approximation) of how many tons of cargo a vessel can carry

Carling	a transverse (perpendicular to the beams) reinforcing member incorporated into the deck structure, used to reinforce openings in the deck such as *hatchways* and mast *partners*
Carpenter's measure	the tonnage measurement calculated by the shipwright, and used as the basis for the price of construction; same as *measured tonnage*; it was based on the agreed-upon length of the straight part of the *keel*
Cathead	a stout post protruding at an angle from the upper deck structure at the *bow*, used for securing bow anchors
Ceiling	the planks covering of the inside of the sides of the *hull*; analogous to a wall covering in a building (contrast to *Overhead*, which is analogous to the ceiling of a building)
Center of buoyancy	the *center of gravity* for the volume of water the *hull* displaces; for the ship to be stable, this point should be below the center of gravity, and when it moves relative to the center of gravity, because the vessel tilts, this creates a *righting moment*, a force which tries to bring the two points back into a vertical line, righting the ship; this only works if the center of gravity remains fixed, which it will unless the hull changes form or the cargo moves
Center of effort	the point at which all the force exerted upon the sails by the wind is focused; its location determines whether the sail plan is balanced or not, as it acts as a pivot point
Center of gravity	the point at which all the torque (rotational) force exerted by gravitational pull vanishes; the point from which the weight of a body acts; in a ship, the higher this point, the more unstable the ship is; the lower this point, the more stable she is
Center of lateral resistance	the center of pressure of the hydrodynamic forces acting on the *hull*
Channel	small platform on the outside of the hull near the *gunwales*, around which the *shrouds* (the standing rigging that held up the *masts* side to side) were bent
Chaser	a carriage gun mounted in the bow or stern, pointing directly fore or aft through a port cut into the *hull* or secured on deck
Chip log	an apparatus for measuring speed through the water, comprised of a reel, a long length of cord in which knots have been tied at regular intervals, and a wooden chip at the end to provide drag in the water as the line is let out
Chock	an angular piece of timber used to fill empty areas between other timbers and thus reinforce those areas

Clamp	thick *ceiling strakes*, used as longitudinal support, placed on the inside opposite *wales* on the outside, in a sandwich arrangement, analogous to that of *keel* and *keelson*
Cleat	a t-shaped or horn-shaped piece affixed to a *spar* or structural member for tying off lines
Close-hauled	any *point of sail* in which the wind is coming from forward of the *beam*
Companionway	an opening in a deck equipped with steps by which crew move between decks
Compass timber	a naturally-curved piece of wood used to make a curved timber in ship construction
Counter	1. The transverse section between the bottom of the *stern* and the highest, main *transom*, called the *wing transom*; 2. The entire transverse area between the top of the *sternpost* and the *rail* or *taffrail*; these are somewhat contradictory definitions, and the sense in which the term is used depends on the source
Counter-bracing	using a *sheet* or *brace* to pull or hold a sail or its *spar* to the *windward* or *weather* side of the ship, for maneuvering
Crosstree	a straight wooden piece attached *athwartships* to the *mast*, to help spread the *shrouds* for better mast support.
Crutch	1. Forked or notched vertical timbers used to cradle a spare *spar* on deck; 2. One forked or notched vertical timber used to support the end of a *boom* at rest; 3. A bracing timber used for lateral support of a *mast step*; 4. A bracing timber serving the same purpose as a *breast hook* but used in the *stern* instead of the *stem*
Cutwater	the forwardmost part of the *stem*; the first piece of the *bow* to part the water
Dagger knee	a *knee* placed at any angle other than horizontal or vertical
Dead work	all *hull* structure below the waterline
Deadrise	the degree to which the *floor* rises from the *keel* to the *turn of the bilge*; a principal aspect of *hull* design defining how flat or deep the hull was
Deal	sawn plank; later, specifically softwood, but in this period, not necessarily
Depth in hold	distance between bottom of main deck and covering of *floor* timbers, measured at the midship *frame*
Douse	to strike or take down a fore-and-aft sail; to furl a fore-and-aft sail; contrast to *furl* and *hand*
Draft	the depth of the *hull* from the load waterline to the bottom of the *keel*; how much water the ship needs to clear the bottom

Drag	resistance to forward movement caused by friction; the opposite of thrust
Driver	in the mid-eighteenth century, the term *driver* became interchangeable with the term *spanker*, whereas earlier, the driver had been a sort of *studding sail* to augment the *mizzen* sail; the later driver or spanker was a large mizzen sail hung from a *gaff* and attached to a *boom*
Duck	canvas for sailcloth
Dugout	a method of boat construction in which a whole log is hollowed out into a hull, using fire, *adzes*, or both
Embayment	being pushed into and trapped in an indentation in the coast by wind, seas, or both, and unable to clear the headland at either end of the indentation due to the direction of wind and/or seas
Entrance	the forwardmost underwater part of the *hull*; the shape of this part of the hull is a principal aspect of the hull's design
Fish-form hull	also called "cod's head and mackerel's tail," a fish-form hull has its widest breadth (maximum *beam*) forward of *amidships*; the form does not necessarily include a bluff (broad, rounded) bow, but in this period, it typically did
Flat of floor	the portion of the *floor* that is flat, between the *keel* and the beginning of the *floor sweep*; this dimension is usually obtainable from a wreck, as this section tends to survive
Floor sweep	the portion of the *floor* between the *flat of floor* and the *turn of the bilge*; the radius of this portion is usually obtainable from a wreck, as this section tends to survive
Fluit (also *fluyt*)	a Dutch cargo vessel of the seventeenth century, characterized by pronounced *tumblehome*, a full *hull* shape, a narrow *stern*, and a relatively-small *weather* deck; the type was economical to build, usually unarmed or lightly armed, and highly successful in the early decades of the century; some of its characteristics were blended with other traditions at the end of the century and into the eighteenth
Following sea	waves moving in roughly the same direction as the vessel, so that they approach her from astern
Foot	the bottom edge of a sail
Footrope	a line run just underneath a *yard* for crew to stand on when handling the sail on that yard
Fore-and-aft rig	a sailing *rig* whose primary sails are rigged such that their leading edges are fixed to a *mast*, or *stay*, so that they move from the longitudinal centerline of the vessel to perpendicular (or close to it) on one side or the other; contrast with *Square rig*

Forecastle — frequently shortened to "fo'c'sle," and pronounced as such; originally, a high superstructure built over the main deck in the forward section of the ship; once castling decreased and even disappeared, the term was retained for the crew's quarters in this section of the ship; it is intimately associated with the sailors on a ship, as opposed to the officers and passengers

Foremast — the forward-most *mast* on a vessel

Foresail — 1. The fore-course, or the lowest driving sail on the *foremast* of a *square-rigged* vessel; 2. All of the sails forward of the mainmast considered together for purposes of balancing the sail plan and its effects on the ship's longitudinal motion and its ability and tendency to pivot about its CLR (*Center of Lateral Resistance*)[1]

Foul — 1. When describing the wind, a wind from a direction that hinders the vessel's progress toward her intended destination; 2. When describing the weather in general, unsettled or stormy weather, with reduced visibility; 3. When referring to a line, to wrap or snag a line unintentionally around another line or an object, so as to hinder the intended function of that line; 4. When referring to a ship's bottom, encumbered with significant growth of weed, algae, and/or barnacles

Frame — the rib of the skeleton of the ship in plank-on-frame construction; composed of multiple pieces, frames define the shape of the *hull*

Freeboard — the distance from the waterline to the *rail*; the portion of the side of the ship that is out of the water

Frigate-built — a design that incorporates elements of the frigate—a small warship—into a merchant *hull*, providing some speed, weatherliness (see *Weatherly*), and more ornate *bow* and *stern* construction than the simplest merchant hulls

Furl — to secure a square sail by tying it up to its *yard*; also see *Hand*

Furring — also see *girdling*; adding an additional layer of planking around the waterline to add stability to an unstable *hull*

Futtock — one of the timbers making up the upper part of a *frame*, as opposed to the *floor* timbers, which made up the lower part

Futtock shrouds — *shrouds* which jutted out from the *mast* around the *top*, such that climbing them required the sailor to be angled backward from the vertical

1 On the latter, see Harland, *Seamanship*, 51.

Gaff	a *spar* to which the top edge of a quadrilateral, *fore-and-aft* sail was attached; the gaff joined the *mast* at its throat; its use became more common in the middle of the eighteenth century and later
Galleon	a stout multi-purpose blue-water three-masted ship used in the fifteenth and sixteenth centuries by all European maritime powers, most famously by Spain, whose large well-armed galleons carried most of the treasure expropriated from the New World
Galley	while more commonly used to refer to sleek, fast warships propelled by oars and used in the Mediterranean, here the term refers to small early seventeenth-century sailing warships
Girdling	also see *furring*; adding an extra layer of planking around the waterline of a ship to increase stability
Grating	latticed wooden structure placed over a *hatch* to allow light and air to pass through
Gudgeon	one part of the hinge assembly that attached the *rudder* to the *stern* and allowed it to pivot side-to-side; the gudgeon was attached to the stern and received the *pintle*, which was attached to the rudder
Gunwale (gunnel)	the top *rail* along the side of a watercraft
Halyard	from "haul-yard"; the line used to raise and lower a *yard*, or a yardless sail
Hance	a curved or arched piece of deck structure
Hand	an older term for *furl*; the traditional definition of an able seaman was one who could "hand, reef, and steer"
Hanging knee	a *knee* attached in the vertical plane, used as a supporting piece
Hard bilges	refers to a hull shape in which the *turn of the bilge* is more abrupt than gradual, forming more of a box shape and less of a round one; contrast to *slack bilges*
Harpin	a *wale* around the *bow*, for increased strength
Hatch	an opening in the deck, usually fitted with *gratings* in clean weather and with tarpaulins and battens in foul weather (hence the expression "batten down the hatches")
Hawsehole	an opening in the *bow* for passage of the anchor cable
Hawse-piece	a reinforcing framing timber in the forwardmost part of the *bow*, through which the *hawseholes* were partially cut
Head	1. a structure built onto the *bow* of a vessel, slightly below the main or *weather* deck, that typically contained latrine seats for the crew; the term became generic for any such toilet on a vessel, regardless of where it was located; 2. The top end of something on board, e.g. a sail or a *mast*

Headsail	any sail deployed on the *foremast* or ahead of it, used for driving power, balance, maneuverability, or any combination thereof
Heave-to	to set the *headsails* and after-sails against each other, with the rudder put over so that it is trying to turn the vessel into the wind, such that the vessel remains virtually motionless; used as a heavy-weather tactic, but requires sail up—contrast to *lie a-hull*
Heel	1. To lean over under the pressure of the wind in the sails; more pronounced the closer to the direction of the wind the vessel points; 2. the bottom end of a *mast*
Helm	the steering station; also, the part of the steering apparatus directly operated by the helmsman (crewman steering the vessel)
Hogging	longitudinal structural deformation of the *hull*, in which the ends droop relative to the mid-section—contrast to *sagging*
Hoist	to raise a sail
Hull	the body of a vessel
Hull speed	the maximum speed a displacement *hull* (one that cannot plane on top of the water) can achieve; $1.34\sqrt{\text{LWL}}$ (waterline length); based on the principle that a displacement hull cannot climb out of the wave it creates with its own *bow*
Initial stability	the degree to which a hull initially resists *heel*—contrast to *Ultimate stability*; a hull with high initial stability is said to be *stiff*
Jackline	a line running along the deck, longitudinally, to which a sailor wearing a harness can clip on to for secure movement along the deck in heavy weather; a modern safety feature, these were not typically employed on period vessels
Jib	a triangular *headsail* attached to a *stay*, the *foremast*, and the *bowsprit*
Jibe	to move through the wind, from one side of the wind direction to the other, while sailing away from, rather than toward, the wind; in a *square-rigged* vessel, also called *wear* or *wear ship*
Keel	the principal structural member of any vessel; the longitudinal member making up the spine of the *hull*
Keelson	a longitudinal structural member fastened on top of the *frames*, forming the inside layer of a sandwich of the frames, with the *keel* being the outside layer
Ketch	a two-masted rig in which the main *mast* is the forwardmost, and the after mast is significantly shorter
Knee	a structural-reinforcement piece filling a 90-degree intersection of two straight pieces, to keep them from flexing; knees can be *hanging knees*, if they are placed in the vertical plane; *lodging*

	knees if they are placed in the horizontal plane; and *dagger knees* if they are placed at any other angle
Knockdown	a ship-threatening event that occurs when the wind or waves push the vessel onto her *beam* ends, or onto her side, so that at least part of her rig is in the water, and she may or may not recover her upright posture
Larboard	the left side of a vessel when the observer is facing forward; an older synonym for *port*; contrast to *starboard*
Lateen	refers to a triangular sail and its *yard*, in use in the Mediterranean since ancient times and incorporated into European rigs in the Middle Ages; the yard descends from high to low, with its center attached to the *mast*; to move the sail from one side of the wind to the other requires shifting this yard from one side of the mast to the other; most ocean-going European ships of this period utilized a lateen *mizzen* until the mid-eighteenth century
Ledge	along with *carlings*, *clamps*, and *lodging knees*, ledges were intermediate deck-support pieces set parallel to the deck beams
Lee	downwind; *in the lee of the …* means "downwind of …"; a *lee shore* is a shore downwind of the vessel, toward which the wind and seas may well be pushing the vessel, which is dangerous; *leeward* (pronounced "LOO-ard") means on the side opposite to that from which the wind is blowing—contrast to *Windward* or *to weather*
Lie a-hull	to rest at sea, with the *rudder* hard over, and no sail up; only certain vessels, with the proper *windage*, can do this; others must *heave to*
Lodging knee	a *knee* mounted in the horizontal plane, to reinforce a 90-degree joint
Loose-footed	refers to the bottom edge (*foot*) of a sail that is not laced to a *boom*, but left unsupported; a boom might be present, but if so, the sail's foot is attached to it only at the *tack* and *clew*
Lubber	a pejorative term used by experienced seamen to describe a novice deckhand, or a landsman when the intent was to point out ignorance of nautical matters
Lubber's hole	an opening in the *top* through which a sailor may pass; so-called because it is easier and safer, but slower, to use than climbing the *futtock shrouds*; see *Lubber*
Luff up	in modern parlance, to point the ship toward the wind enough that the sails are barely filling, or even beginning to flap some as they lose smooth airflow

Mainmast	the tallest, largest *mast* on the vessel, on which its *mainsail* is mounted
Mainsail	the largest driving sail of the rig; also called *main course* on a *square-rigged* vessel
Mast	a vertical *spar* upon which sails are mounted
Measured tonnage (see also *Carpenter's measure*)	the ship's internal volume, estimated by the shipwright using measurements of her *hull* dimensions; the shipwright's fee was usually based on this figure (i.e. so much per ton as so measured)
Metacenter	the point at which a vertical line through the vessel's *center of buoyancy* while at rest intersects another vertical line through a new center of buoyancy when the vessel is tilted; this point must be above the *center of gravity* for the vessel to be stable; calculating this point was a primary goal of eighteenth-century mathematicians interested in the physics of ship design
Midship bend	the shape of the widest *frame* of the *hull*, which determines the shape of the hull at its maximum *beam*; techniques for creating the midship bend help define different approaches to hull design
Mizzen	the aftermost *mast*, driving sail, or associated *spar* of a three-masted *square rig*; the only relevant exception being the *bonaventure* (also spelled *bonadventure*), a small fourth mast behind the mizzen on some large ships c. 1600; an analogue to the bonaventure is the *ringtail*, which, in that sense of the term, meant a small light-weather sail rigged on the *taffrail* flagpole, and this persisted through the eighteenth century
Mizzenmast	see *Mizzen*; the aftermost *mast* on a three-masted ship
Naval stores	pine products necessary for ship maintenance; *tar*, *pitch*, and *turpentine*; varnish was also derived from pine sap
Oakum	unlaid rope fibers forced into the joints between hull planks, then covered with *pitch* to seal the joints
Overhang	the extension of the *bow* or *stern* at deck level past the forward or after end of the *keel*; a *hull* whose bow or stern is directly over the end of the keel is said to have a "plumb bow" or "plumb stern"
Overhead	the underside of a deck; analogous to the ceiling of a building but different from the *ceiling* of a ship
Partner	the reinforced opening in a deck through which a *mast* passes

Pay	to coat the bottom of a ship with an antifouling agent such as *white stuff*; more generally, to apply a protective coating, such as varnish to a ship's sides
Periauger	an open boat, propelled by sail, used by Europeans on the American coasts throughout the period; the type was a *dugout*, perhaps copied from native watercraft; as large cypress trees dwindled, sawn wings were added to the center dugout section to form the *hull*
Pink	a relatively sharply-built *hull* form, with a narrow *stern*—related to "pinch"
Pinnace	a loose term applied both to open work boats and small ships built for good speed and maneuverability, especially in the seventeenth century; the term was also applied to vessels propelled by both sails and oars
Pintle	the male component of the hinge assembly securing the *rudder* to the *stern*; contrast to *gudgeon*
Pitch	1. Viscous resin derived from pine sap; more viscous than *tar*—see also *naval stores*; 2. Longitudinal rocking motion (contrast to *Roll* and *Yaw*)
Point	1. To sail upwind; a *weatherly* vessel may be said to "point high"; 2. *point of sail*, referring to the direction in which the vessel is sailing relative to the wind direction; 3. A division of the compass card less precise than the degree, which was not used in this period; on a 32-point compass, the point equals 11.25 degrees; the cardinal points are North, East, South, and West, the inter-cardinal points Northeast, Southwest, etc.
Poop (deck)	a high deck constructed over the *stern* section, reserved for the master and other officers
Port	the left side of a vessel when the observer is facing forward; contrast to *starboard* and compare to *Larboard*
Protest, crew	a legal deposition, sworn before a notary or other official, by a delegation of a ship's crew after an accident or loss, which records the crew's version of the events leading up to the incident and the actions taken by the crew during and after the incident
Quick work	all of the *hull* constructed above the waterline; contrast to *Dead work*
Rail	a bulwark constructed above deck level
Rake	1. The distance between a plumb hung from the furthest point of the *bow* or *stern* and the end of the *keel* at that end; measures extent of *overhang*; 2. An inclination of a *mast* from the vertical, either fore or aft, used to adjust the balance of the sail plan

Ratline	a short line tied between two *shrouds* used for climbing the rigging
Reach	a *point of sail* on which the wind is perpendicular to the ship's direction of travel (beam reach, in modern parlance), or between perpendicular and dead astern (broad reach); most vessels performed at their best on a broad reach
Reefing	reducing the amount of sail presented to the wind by partially-lowering the sail and tying off the loose part
Registered tonnage	legal tonnage—the figure written on all official documents pertaining to the vessel; taxes and duties were based on that figure
Rig	the *spars*, sails, supporting lines (*standing rigging*), and control lines (*running rigging*) of a sailing vessel
Righting moment	a torque force created when the *center of buoyancy* moves out of the vertical line shared with the *center of gravity* when the hull is fully-upright; this force tries to bring the two points back into line, restoring an upright posture; this is key to stability
Ringtail (or "ring-tail")	1. A small light-weather sail set on the *taffrail* flagpole of eighteenth-century ships, analogous to the earlier *bonaventure* (see *mizzen*); 2. A scudding-sail added as an extension to a *gaff*-rigged *mainsail* for *scudding*, or *running* fast before the wind
Roadstead	an area off a beach or shore at which ships can ride at anchor; some roadsteads are well-sheltered and others are only usable in calm weather or with wind and sea from certain directions
Roll	side-to-side motion (contrast to *Pitch* and *Yaw*)
Rudder	the blade hung from the *stern* and turned side-to-side to steer the ship
Run	1. A *point of sail* on which the wind is directly astern; 2. "the aft-most or hindmost part of a ship's bottom, where it grows extremely narrow, as the floor approaches the stern-post"[2]
Rung head	the tops of the *floor* timbers
Running rigging	rigging that moves, used to control sails; contrast to *standing rigging* which is used to provide static support to the *masts* and other fixed *spars*
Sagging	longitudinal *hull* deformation in which the middle of the hull droops relative to the ends—contrast to *hogging*
Scantling	breadth and thickness of a timber

2 Falconer, *An universal dictionary.*

Scarf	a method of joining the ends of two timbers in which both ends are angled for maximum overlap, providing much more strength than a butt-joint
Schooner	a two-*masted fore-and-aft*-rigged vessel whose main *mast* is abaft the fore *mast* and either the same height or slightly taller
Scuttle	opening in the deck equipped with a lid
Sea chantey	a rhythmic work song used by sailors performing repetitive tasks, especially *weighing anchor* by turning the *capstan* or *windlass* or hauling on a *halyard*; like other such work songs, became a form of group entertainment and frivolity
Sextant	a navigational instrument invented and diffused in the eighteenth century, used to measure the angle of a celestial body to the horizon, or to measure horizontal angles between points of land when turned on its side; so-named because its arc was one-sixth of a circle; the sextant was derived from the earlier octant, which was based on a smaller arc
Shallop	a loose term for a large boat, one- or (perhaps more commonly) two-*masted*, sometimes open and sometimes partially-decked, used throughout the period as a ship's tender, coastal transportation and trade, and coastal exploration
Sheer	the curve made by the *gunwale*, helping to define the profile of the *hull*
Sheet	a line—or, more commonly in this period, a tackle—part of the *running rigging* attached to a corner of a sail, and used to pull in or let out that sail, and thus adjust its angle to the wind and the amount of concavity, or belly, it presented to the wind
Ship	1. A generic term for any ocean-going watercraft larger than a boat; 2. A *square rig* of three *masts*
Shroud	a line, part of the *standing rigging*, used to provide support for a *mast*; analogous to a guy wire on a utility pole
Sirmark	mark made to note where a flat section was to meet a curve, or two curves were to meet
Slack bilges	refers to a *hull* shape in which the *turn of the bilge* is more gradual than abrupt, forming more of a round shape and less of a box; contrast to *Hard bilges*
Sloop	a vessel rigged with a single *mast*, whose primary sails are *fore-and-aft*
Snow	a medium-sized ocean-going vessel with two primary *masts*, like a *brig*, except with a short *trymast* set just behind the main (after) mast, upon which the *mizzen* sail is raised and lowered; this was a

	smaller sail, than the mizzen of a brig, usually *boom*-less, and the snow was considered by many to be a handier vessel; it became very popular in the mid-eighteenth century
Soundings	measurement of the depth and sampling the composition of the sea bottom, taken simultaneously by lowering a lead weight attached to a long, knotted line—called "heaving the lead;" the concave bottom of the weight was filled with tallow or wax, so that it picked up a sample of the bottom material—sand, mud, gravel, etc.; a vessel was said to be *on soundings* when the water was shallow enough for the depth to be measured with the lead line
Spanker	see also *driver*; in the mid-eighteenth century, the term *driver* became interchangeable with the term *spanker*, whereas earlier, the driver had been a sort of *studding sail* to augment the *mizzen* sail; the later driver or spanker was a large mizzen sail hung from a *gaff* and attached to a *boom*
Spar	a stout wooden pole used for suspending sails; spars included *masts*, *bowsprits*, *yards*, and *booms*
Speed	1. Speed through the water: how much distance the vessel travels in a certain time relative to the water, which is assumed to stand still; in this period, measured with a *chip log*; 2. Speed over the ground (or over the bottom): how much distance the vessel travels in a certain time relative to the bottom, which is fixed; in this period, could only be estimated by celestial navigation or coastal piloting using known landmarks; also see *Velocity made good*
Spritsail	a *headsail* hung underneath the *bowsprit*; used throughout most of the period
Sprit-topsail	a small *headsail* mounted on a short *mast* placed at the end of the *bowsprit*; disappeared in the early eighteenth century
Squall	a thunderstorm, with strong, gusty winds, heavy rain, and drastically-reduced visibility
Square frame	a *frame* mounted perpendicular to the *keel*; contrast to *Cant frame*
Square rig	a rig in which the primary sails are *square* (actually trapezoidal), hung from *yards*, and rotated around the mast from their center-points; contrast to *fore-and-aft* rigs; square-rigged vessels were the primary ocean-going merchant ships of the European world from the fifteenth through the twentieth centuries
Square sail	a sail of a trapezoidal shape, hung from a yard (see *Square rig*); on a *schooner*, such a sail flown on the foremast in lieu of the *fore-and-aft foresail*

Stanchion	a short pole or pillar mounted to a deck or *rail* to support a railing on deck, netting or awning
Standing rigging	rigging used to provide static support to the *masts* and other fixed *spars*; as opposed to *running rigging*, which moves, and is used to control sails
Starboard	the right side of a vessel when the observer is facing forward; contrast with *port*, which is synonymous with *larboard*
Stay	a line, part of the *standing rigging*, that helps support a *spar*; for a *mast*, stays are fore-and-aft supports, as opposed to *shrouds*, which are side-supports
Staysail	a sail whose leading edge is attached to a *stay*
Stem	the central longitudinal member making up the spine of the *bow*
Step	1. To place the *heel* of a *mast* on the *keel*, or whatever other structure it will rest on; 2. The reinforced, notched structure built on the keel (or on a deck) to receive the heel of the mast—also called *mast step*
Stern	the after section of the *hull*—contrast to *bow*
Stern castle	the superstructure built over the *stern* section; contrast to *Forecastle*; as an actual structure, the stern castle survived longer than the forecastle on European ships
Stiff	said of a vessel with high *initial stability*—one resistant to *roll*; if excessively-stiff, a vessel tended to snap-roll, in which she would come upright from a *heel* with a jerking motion, which was extremely uncomfortable for the crew and stressful to the *rig*
Strake	a plank, either continuous or made up of joined lengths, extending the entire length of the *hull*
Studding sail	frequently shortened to "stuns'l"—an auxiliary sail deployed as an extension to a larger sail for increased driving power in light to moderate air; used to increase speed
Swivel gun	a small smoothbore cannon, frequently brass but also cast in iron or bronze, mounted on a swiveling support on the *rail* or on deck, usually loaded with scatter-shot, and used either to repel boarders or kill enemies—or, in the case of a slave ship, resisting prisoners—on deck; analogous to a large shotgun
Tack	1. To turn the *bow*, rather than the *stern* through the eye of the wind; to shift course such that the vessel goes from having the wind on one side of the bow to having it on the other (contrast to *jibe* and *wear*); 2. A reference to which side of the ship the wind is coming from as she sails—either *starboard* or *port/larboard*;

	3. The front lower corner of a *fore-and-aft* sail, equipped with a grommet for the attachment of a *sheet* or other control line; 4. A line or *tackle* used to hold the foremost lower corner of a *square* (see *Square rig*) sail in a fixed position when the wind crosses the sail at an oblique angle
Tackle	a pulley arrangement made up of *blocks* and line, to provide mechanical advantage for adjusting sails or moving heavy loads
Taffrail	also spelled "taffarel"; the upper part of the *stern*, made of a curved piece, and frequently ornamented
Tangent-arc method	a technique for forming the shape of the *midship bend* using intersecting arcs of circles
Tiller	a long, stout handle attached to the head of the *rudder*, used to turn it to steer the ship
Top	a small platform at the top of a primary *mast*—a lowermost mast—used for observation
Top timber	the uppermost timber in a *frame* assembly
Topgallant	in a *square rig*, the third square sail on a *mast* from the bottom; the lower-most sail is the *main* or *course*, the second is the *topsail*, and the third is the *topgallant*
Topmast	the second, upper *mast*, set at the *top* of the *main mast*; the masts of larger vessels were comprised of at least a main mast and a *topmast*
Topsail	in a *square rig*, the second square sail on a *mast* from the bottom; the lower-most sail is the *main* or *course*, the second is the topsail; square topsails were also used as auxiliary driving sails for light and moderate air on *fore-and-aft*-rigged vessels such as *schooners*; a schooner so equipped was called a "topsail schooner"
Transom	one of the *athwartships* members of the *stern* structure, affixed to the *sternpost*, helping to define the shape and size of the stern, and to strengthen it; also used more generally for the entire stern section formed by these transom timbers
Treenail (trunnel)	a round wooden peg, driven through a hole just large enough to accept it, used to join pieces together; usually, a small wedge of wood or metal was driven into the head of the trunnel to fix it more firmly in the hole
Trestle tree	a straight wooden piece mounted at the *top* of a *mast* to support the top and accept the heel of the *topmast*; masts on these vessels were not all one piece
Trunnel (treenail)	see "treenail"

Trymast	a short *mast* placed just behind the *main mast* on a *snow*, from which the vessel's *mizzen* was flown
Tuck	where the planks forming the bottom of the hull terminated under the *stern* or *counter*; if this area was of a convex shape, the tuck was said to be "round"; if the bottom and stern sections were perpendicular here, the tuck was said to be "square"
Tumblehome	the curving-inward of the sides of a *hull* from the waterline to the deck
Turn of the bilge	where the *floor* transitions into the side of the *hull*; depending on how abrupt or gentle this curve was determined whether the hull had *hard bilges* or *slack bilges*
Ultimate stability	the ability of a vessel to recover from extreme angles of *heel*, returning to an upright position rather than capsizing; contrast to *initial stability*
Velocity made good (VMG)	the speed a vessel averages toward her destination over the course of the voyage, no matter her course changes and the changes in wind speed and direction
Vlieboot	Dutch term whose meaning, like so much of nautical etymology, is difficult to pin down; for our purposes, the term (translated into English as "fly-boat") connotes typically Dutch hull characteristics: relatively shallow *draft*, wide *beam*, and high capacity-for-size, at the expense of maximum speed and *weatherliness* (see *Weatherly*)
Wale	a thick *strake*, or belt of strakes, run along the side of the vessel for strengthening
Watch	The portion of the crew on duty at a given time, traditionally divided into two, the *starboard* and *port* (or *larboard*) watches.
Waterway	a "timber or gutter along the side of a deck whose purpose was to prevent the deck water from running down between the frames and to divert it to the scuppers [openings in the rail for water to run overboard]"[3]
Wear	to change course through the wind with the wind astern, so that the wind is coming from the opposite side of the *stern*; see also *jibe*; contrast to *tack*, which refers to changing course through the eye of the wind, with the wind coming over the *bow*; in general, the fuller a hull form a vessel had, the more her master would have preferred to wear or jibe rather than tack if he could, as tacking risked "missing stays," in which the bow failed to come all the

3 Steffy, "Illustrated Glossary."

	way through the wind, leaving the ship "in irons"—dead in the water and likely to start moving backward if the wind or sea were strong enough
Weather	1. in the nautical sense, the direction from which the wind is blowing; "to weather" means toward that direction; contrast to *leeward* (see *Lee*) which is the direction toward which the wind is blowing; 2. When referring to a deck, the term refers to the main deck, exposed to the elements
Weatherly	able to *point* relatively high
Weigh anchor	to retrieve the anchor by winding in the anchor cable until the anchor is secured to the outside of the *hull*
Whipstaff	a vertical rod, attached to the head of the *tiller*, and run through an opening in the deck to the *helm*; used to move the tiller on larger seventeenth-century vessels
White stuff	a mixture of whale oil (or tallow), rosin, and brimstone (sulfur), commonly used to *pay* the ship's bottom to retard weed growth and repel shipworms; the use of copper sheathing for this purpose was only just coming into widespread use at the end of the period
Whole moulding	a method of determining the shape of a hull by using molds of principal *frames* and shifting them to different positions; the technique did not require using a compass or transferring a set of lines from a plan to full-scale
Windage	vertical surface area presented to the wind; the higher a vessel's superstructure, the more windage she will present, and the more pronounced the effect of the wind will be on her upper works
Windlass	a winch, equipped with pawls and notches to prevent the load from slipping backward when turned, used to wind a cable to move a load, whether an anchor cable or a hoist attached to heavy cargo or a piece of artillery, to use common examples; unlike the *capstan*, a windlass is mounted horizontally, such that the line it pulls is in a vertical rather than horizontal direction
Windward	the direction from which the wind is blowing; synonymous with *to weather*; contrast to *leeward* (see *Lee*)
Yard	a *spar*, attached to a *mast* or *bowsprit*, and at an angle to it from perpendicular to nearly parallel, from which a sail is flown; on a *square-rigged* vessel, sailors must climb out onto the yards to hoist and *douse* the main square sails
Yaw	to move side-to-side horizontally around the central pivot point; contrast to *Pitch* and *Roll*

APPENDIX 1

Basic Sails on a Square-Rigged Ship, a Sloop, and a Schooner

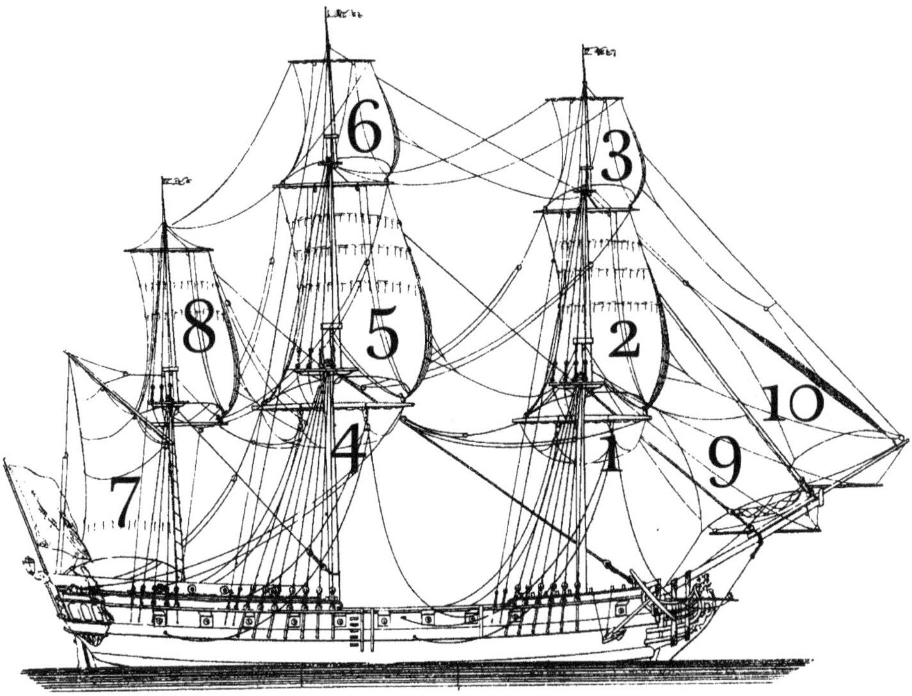

FIGURE 17 Basic sails on an eighteenth-century square-rigged ship
FROM ARCHITECTURAN LXII, FREDRIK HENRIK AF CHAPMAN, SJÖHISTORISKA MUSEET, STOCKHOLM

1 Foresail or forecourse
2 Fore topsail
3 Fore topgallant
4 Mainsail or main course
5 Main topsail
6 Main topgallant
7 Spanker or driver
8 Mizzen topsail
9 Fore staysail
10 Jib

 | DOI:10.1163/9789004426344_013

FIGURE 18 Basic sails on a sloop
FROM ARCHITECTURAN LXII, FREDRIK HENRIK AF CHAPMAN, SJÖHISTORISKA MUSEET, STOCKHOLM

1 Flying jib
2 Jib
3 Fore staysail
4 Mainsail
5 Topsail
6 Topgallant

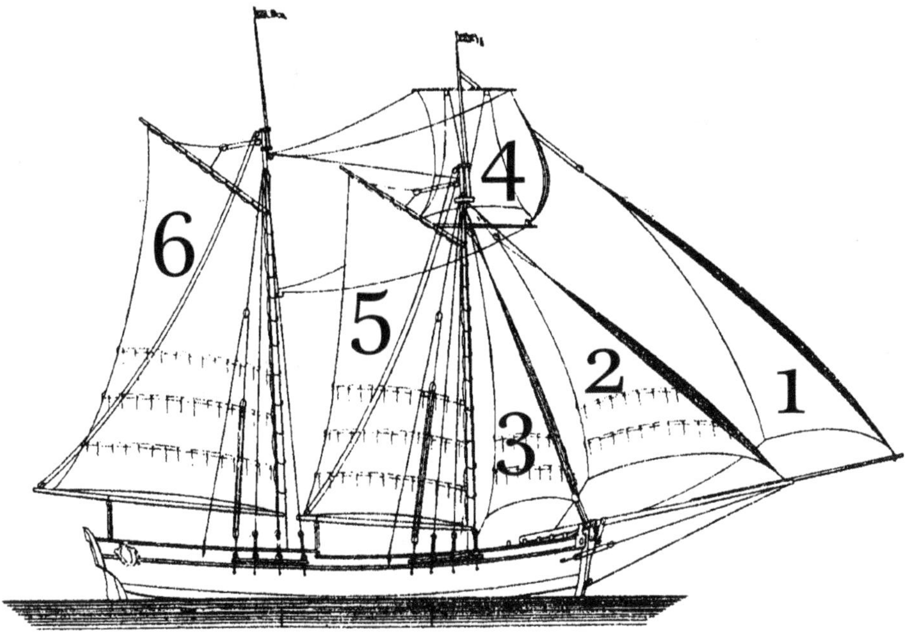

FIGURE 19 Basic sails on a schooner
FROM ARCHITECTURAN LXII, FREDRIK HENRIK AF CHAPMAN, SJÖHISTORISKA MUSEET, STOCKHOLM

1 Flying jib
2 Jib
3 Fore staysail
4 Topsail
5 Foresail
6 Mainsail

APPENDIX 2

Full Transcription of Winne & Hawksworth Letter to William Jones, 17 January 1733

Bristol 17 January 1733

Dementions for a New Ship, to be Built at Boston in New=England of 58 Foot Keel & 21 Feet Beame, 12 Foot Rake by the Stem & 5 by the Post; 9 1/2 Foot in the Hould from under the Lower Deck Beame to the Ceiling & 3 Foot 10 Inches between Decks from under the upper Deck Beames to the Lower Deck; & to Carry her Breadth well forward even to the after part of the fore Channell Vizt. [viz.] Length of Keele & Breadth of Beame Measur'd as Customary: the Breadth of her Floor from Surmark to Surmark 11 Foot & 7 or 8 Inches Dead Riseing; the Floor Timbers to be 12 Inches Square.
The Stem to be a Good Crooked Piece & one that will Tread well in proportion. The Stern Post Likewise to be a Good Piece & both pieces not Less than 10 Inches. The Keele 12 Inches deep & 10 Inches thick with a Good false Keele on it of 3 Inches thick & Nail'd well on.
Transom 15 1/2 Foot high & 16 Broad & 12 Inches Square, the Harpin to be so many Foot high as the Builder thinks fit & that will give the Ship a Good Sheare not too much nor too Little but middling in proportion to most they Build.
12 Inches high in the Waste with a Rise for a forecastle another abaft at the Main Mast for a halfe Deck both of 10 or 12 Inches with 2 Hances proper for each Rise with a Plank Raile from Hance to Hance sett on 4 Timbers left for that purpose, the 2 Bulk Heads for the forecastle & halfe Deck to be 4 Inches thick at Least & Timbers left up at each Hance with Crutches of a proper Heighth fitt to put Spare Topmast in.
11 Flat Timbers not to Varey in the Least but placed on the apron of the Keele of one Heighth & Length but so as to make a Fair Body & Good Work & 10 Inches a sunder & thickness as above 12 Inches & 8 Inches at the Rung Heads the Rest of the Floor Timbers of the Same Thickness both before & abaft & to Carry their Length well so as to give a Good Long Floor & to rise in proportion from the Middle of the Keele aft on & from the foremost Flatt forward to Content so as to make a Cleane Sail & Good Entrance.
The Lower Futtocks 9 Inches Broad & 8 Inches thick at the Rung Heads & Grov'd Rounding fitt for their Work. The upper Futtocks 8 Inches Broad & 7 Inches thick & not Less than 6 Inches at the Bend. The Topp Timbers

 | DOI:10.1163/9789004426344_014

7 Inches thick at Foot & not Less than 4 1/2 Inches on the Topp; all these Timbers before Mentioned to be of Good White Oak & every Timber & Futtock to Butt & Butt with a Square Butt & no Less Scarf than 4 1/2 Foot or 4 at Least. A Good Kelson to Run from Stem to Stern & a Good Knee fixed against the Stern Post; the Foots of the Stern & Bow Timbers to be well Chock'd & Boulted & a Good broad apron on the Stem not Less than
[new page]
then 6 1/2 Inches thick with 2 Breast Hooks in the Hould one pretty low down the other to Lodg the Lower deck on & to Come well Round the Bows 2 Breast Hooks between decks, one in the middle the other to Lodg the Forecastle Deck on to Come well Round the Bows & arm'd [?] so as it may take the nuckle & some other Timbers in. The Quarter & Stern Timbers to Rake well according to Pleasure with a Good Transom athoart [athwart] them & well kneed 2 good Hass peices [hawsepieces] forward of a Good thickness, good Timber & Long Enough to Run Low down in the Bows so as they may be firm [?] & the Catt Heads to Turn from the Hass peices & not to Lodg on the Deck; Lower Deck to Lodg on the Transom abaft & the Beams to be 11 Inches thick & 12 Broad to Round 7 or 8 Inches all Double Kneed with Knees Grown to their Work & the Beams 4 1/2 or 5 Foot a sunder with Ledges & Carlings in proportion. The Beams of the upper Deck to Round according to discretion & to be 7 1/2 Inches thick & 10 Inches Broad all Double Kneed with Lodging Knees Grown to their Work Except a Good Hanging Knee of each side at the Bulk Head of the forecastle & halfe Deck abaft the Beams 4 1/2 or 5 Foot a sunder with Ledges & Carlings in proportion. The halfe Deck & forecastle deck Beams to be 4 1/2 Inches thick & 7 Inches Broad to Round according to discretion all Double Kneed with Knees Grown to their Work. The Ship to be Plank'd with Good White Oak from the Keele to the Bend of 2 1/2 Inch thick Except 2 Streaks of 3 Inches & one of 3 1/2 Inch at the point of the Buldg. [bilge] one Streak of 3 Inch & one of 3 1/2 Inches under the Bend, 2 Bends one on Top of the other 9 Inches Broad & 5 1/2 Thick one Streak of 3 Inch Plank upon Top of the Bends & one on Top of that of 2 1/2 Inch the Rest 2 Inch Plank to the Gunil Except a Channell Bend of 8 Inches Broad & 3 1/2 Inches thick to Carry a Round side & to Tumble in at the Channell Bend & then turn off to the Gunill so as to make a Handsom side every Streak of Plank to be Work'd fair & none Broader than 9 or 10 Inches at most. The Ceiling in the Hould & between decks to be 2 Inch White Oak Plank Except 2 Streaks of 2 1/2 Inch & one of 3 Inch Plank on the Rung Heads or point of the Buldg & one of 2 1/2 Inch the other of 3 Inch upon Top of that for Riseing under the Lower Deck one of 2 1/2 Inches for Riseing under

the upper Deck & one of 2 1/2 Inch for Spurcut [spirket] riseing on the Lower deck all the planks both within & without to be free from Sapp—All the Decks to be Laid with full 2 or 2 1/4 Inch Deale Plank free from Sapp & a Good Length at Least 30 Foot Long Except a 2 or 2 1/4 Inch Oak Plank for Waterway & 4 Inch Oak for Partners on the Lower Deck. The upper deck Waterway must be Thick Oak & [two words partially obscured]
[new page]
so as to make a Good Seam above the Deck. The Partners of Oak Plank 2 1/2 Inch thick both before & abaft & the halfe Deck to Run aft with a Fair Sheare so as it may be 5 Foot 9 Inches in the Cabin under the Beame & the Quick Work on Top of the Deck 6 or 7 Inches high & 7 or 8 on the Fore Castle & all the upper deck Plank both for Main & halfe Deck to Run the full Length of the Deck so as there may be no Butts.
A Good false Stern Post & Rudder & the Counter to be Plank'd with good 2 1/2 Inch Oak Plank the Stern with 2 1/2 Inch Deale Plank a Good Drum headed Capson for 6 or 8 Bars & fix'd in his Place, a Good Substantiall Windlist & Bitts of 6 Inches thick & 18 or 19 Broad & fix'd up high Enough from the Deck so as it mayn't hinder the Loop Holes in the Bulk Head of the Fore Castle under A Good Head CuttWater & Badges with Topsail Sheet Bitts afore & abaft & those before to go down to the Lower deck & be so Substantiall & Strong so as to make a Gallows to Step the Bowsprit in which must be well Chock'd both above & below. To find Crosstrees Trusstletrees anchor Stocks, Cleets for the Yards & Masts as many be Wanting.
A Main Hatch Way with Hatches an after Hatch way with Hatches & Grateing for the halfe Deck & Likewise a Grateing with high Combings [coamings] for the forecastle, a Small Scuttle to go down out of the fore Castle, another abaft in the Cabin to go down in the Powder Roome 2 Ports Cutt out & Hanged through the Counter for Stern Chase one of each Side in the Cabbin 2 of each side in the Steridge [steerage], these Ports to be of an Equal Distance one from the other as neare as Can be & to be between 20 & 24 Inches from the Lower Deck all of the Same heighth as the Gun Carriges may serve the one as well as the other if need be all Hanged well & Lind & so as they Shut Close & Contrive so as they dont Cutt into the Plank thats below them on the Outside But that the 2 Streaks above the Bends may Run fair alongt, one Port to be Cutt out Hang'd & Lin'd of each side of the fore Castle & the Lower Sill to be so as it may be Clear of the 2 planks above the Bends, a Tiller fitted to the Rudder a Well Built in the Hould to find Oakham & be well Caulk'd all over to find Pitch & a Coate of White Stuff for her Botome. To be Trunld with good dry White Oak well Seasoned Trunels all the Timber & Plank to be well Season'd & that Plank

thats above the Bends must be through Dry Sufficient Timbers Left to Come through the Gunils of both Sides fitt to belay to with Stantions & Rails for a Breast Work at the Bulk head of the halfe Deck & Double Timbers fix'd at [?]
[new page]
through the Gunil of each side the halfe Deck so as to make Round ports & 2 in the Stern through the Taferell [taffrail] which Timbers of each side to be Strong Enough to bare a Cable if it should be put out through the Stern as it will sometimes be with Good Substantiall Rails fix'd on them, breast high, with 2 Chestrees & Sundry other things as maynt be heare thought of & may be Wanting all finished & done in a Workmanlike manner & Ship Built to Content. We would have her to be Built so as she may be a Floaty Ship & not to draw above 11 1/2 Foot of Water when Loaden if Less the better, to Sail well in order [short smudged-out word] she must have a Cleane Tail carried Fair without Hollows & a Good Entrance with a Round Harpin to keep her up & a Round Taught Body so as she may Carry Sail; & be stiff not only when she is Loaden but at her LIght Water mark, as she may Shift without Ballast if Possible to be done, for we shall want her to Ship in Wild Roads & with as Little Ballast as possible, so it will Suite us to to have a Stiff Shifting Ship which will be no hindrance to her Sailing. Wittness our Gallys in the Last Warr was so, & Sail'd as well as most Ships in England which we should be glad if this did, dont forgett to rise at the Middle of the Keele afton Gradually & Fair not Quick up at once abaft; & as we shall send a ship over with all materialls so we shall not want all the Masts & yards as is Common but you may agree for to have them as Customary & what we don't take oblidge the Builder to alow for. Mind to agree that 2 Foot of the 58 Foot Keele to be given in as if it was out of the Rake, which is Common & so as we may pay ba[ck?] for 56 Foot & that she must be Finished & Launched in all August at farthest, but sooner if Possible, & Call'd the Bristol merchant. If there is any thing amiss in these Dementions which may be mended & for our Interest please to Lett it be done & it will oblidge

[signed by] John Winne
Walter Hawksworth]

to Mr. William Jones in Bristol to be forwarded to his friend in Boston at first opportunity to New England-------

Bibliography

Interviews

Note: With the permission of the interview subjects, I have retained audio and/or written files of all these interviews. They may be accessed with the permission of the subject(s).

Dounce, Sharon. Captain, *Kalmar Nyckel*. Skype-to-phone interview, recorded and stored as MP3 audio. 23 April 2015.

Graczyk, James. Former volunteer crew and interpreter, Armed Sloop *Welcome*. Written questionnaire exchange via e-mail, filed. 5 November 2014.

McMullen, Drew. Executive Director, *Sultana* Education Foundation. Skype-to-phone interview, recorded and stored as MP3 audio. 8 January 2015.

Reynolds, William T. ("Chip"). Master, *Half Moon*. Skype interview, recorded and stored as MP3 audio. 8 January 2015.

Rybka, Walter. Senior Captain and Site Administrator, U.S. Brig *Niagara*. Skype-to-phone interview, recorded and stored as MP3 audio. 21 April 2015.

Severinson, Joakim. Master Shipwright, Swedish Ship *Götheborg*. Written questionnaire exchange via e-mail, filed. 19 March 2015.

Speth, Eric. Maritime Program Manager, Jamestown-Yorktown Foundation. Skype-to-phone interview, recorded and stored as MP3 audio. 14 August 2015.

Personal Correspondence

Burningham, Nick. Archaeologist and mariner, author of published work on *Duyfken* replica and experimental maritime archaeology. Exchange of electronic messages regarding *Duyfken* sail handling and Dutch vernacular ship design. 5–8 April 2015. Substantive comments on earliest draft of work on replica experiments, via e-mail. 4–5 May 2015. Remarks on performance of fore-and-aft versus square rigs, Cook's experience with *Grenville*, 30 August–2 September 2019.

Dounce, Sharon. Captain, *Kalmar Nyckel*. E-mail detailing experience with lateen mizzen in operation. 24 April 2015; e-mails clarifying earlier conversations, 4 May 2015. Comments on earlier quotes in manuscript, 5 September 2019.

Fontenoy, Paul. Curator of Maritime History, North Carolina Maritime Museum. E-mails recommending published articles on Dutch ship design by Albert Hoving, relaying of lengthy query of mine to Ab Hoving and his response. 2 April 2015.

Graczyk, James. Follow-up discussion via e-mail of design and performance characteristics of Armed Sloop *Welcome*. 30 October–25 November 2014.

Hocker, Fred. Archaeologist, Research Director, Ship Unit, Swedish National Maritime Museums. E-mail correspondence and full review of first replica analysis paper. 18–20 August 2015.

Hoving, Albert (Ab). Formerly of the Rijksmuseum, Amsterdam, expert ship modeler and historian of Dutch shipbuilding and seafaring, author. E-mail via Paul Fontenoy. 2 April 2015; e-mail 4 May 2015.

Klomps, Steve. Controller, Peabody Essex Museum, historian. E-mails. June 2015–December 2018.

Reynolds, William T. ("Chip"). Follow-up discussions of sail handling and rig of *Half Moon* and seventeenth-century ships in general. 5–8 February 2015.

Severinson, Joakim. Comments on structural reinforcements and ventilation for carriage guns. 1 November 2018.

Speth, Eric. E-mail correspondence on proposed replica experiments. 11 February 2015.

Manuscript Collections

Abbreviations for repositories used in the notes follow their full names below.

Bristol Archives

19835/7/h Schedule of deeds and abstract of title to two houses in Lamb Street. Walter Hawksworth 1706–40

AC/JS/53/3a-b Summonses from the Mayor of Bristol against William Jones and Joseph Farrell, owners of the 'Stockton' and the 'New Kent', for bringing their ships into the Key without license. 1736 August 28

AC/JS/53/4a-k Robert Smith, William Jones, … owners of the 'Boston Merchant', v. Thomas Whitchurch … in a plea of trespass. Court papers: Summonses; writ for arrest; deposition re theft of anchor; pleas and demurrer; counsels' opinion; argument; bill of costs

BCC/F/E/398 772/20 Ancient lease: Corporation of Bristol to Walter Hawksworth and others Property-Castle Orchard

BCC/F/E/399 772/21 Ancient lease: Corporation of Bristol to Walter Hawksworth and others Property-Castle Orchard

BCC/F/E/6432 2059/7 Ancient lease: Corporation of Bristol to Captain John Winne

P/St.Aug/Ch/4 Bond in £200 Philip Baker of Stowick, Henbury, yeo, to John Winne, merchant, and Henry Combe, esq.

Historical Society of Pennsylvania (HSP)

Boyer, Brooke & Wilson Account Book, 1783–1790.

Coates and Reynell Family Papers, 1729–1764.

George (ship) logbook, 1805–1806.
Gregory Marlow Account Book, 1676–1703.
Jones and Clarke Papers.
Joshua Humphreys Papers, 1660–1931.
Log of the Brig *Lydia*, 1808.
Maitland Family Papers, 1729–1896.
Wharton Family Papers, 1679–1891.

John Carter Brown Library (JCB)

Brown Family Business Records

The National Archives of the UK (TNA)

C 11/1141/26 Jones v Jones. Bill only. Plaintiffs: William Jones … 1729
C 11/1627/1 Missing v Nicholas. Bill and answer. Defendants: … Walter Hawksworth … 1747
C 11/1652/23 Winne v Hawksworth. Bill and answer. Plaintiffs: William Winne; Defendants: Walter Hawksworth and others, 1752
C 11/1666/39 Winne v Hawksworth. Bill and two answers. Plaintiffs: William Winne, mariner of London; Defendants: Walter Hawksworth, merchant of Bristol … 1750
C 11/369/51 Hunt v Allen … Bill only. Plaintiffs: … Walter Hawksworth … 1716
C 11/491/13 Jones v Ruddock. Bill and plea. Plaintiffs: William Jones, … 1727
C 11/491/15 Jones v Merewether. Answer only. Plaintiffs: William Jones, … 1727
C 11/500/28 Jones v Brickdale. Answer only. Plaintiffs: William Jones, … 1730
C 11/591/33 Duff v Hawksworth. Bill and two answers. 1752
C 11/667/43 Winne v Lees. Bill and two answers. Plaintiffs: John Winne … 1726
C 11/930/1 Jones v Ruddock. Depositions. Plaintiffs: William Jones, … 1729
E 134/3Geo2/East14 Margaret Goldney, widow. V. Henry Goldney, gent.: Last will and testament of John Goldney, … The names and possessions of … Walter Hawksworth …
HCA 26/15/4 Commander: Joseph Hiscox. Ship: Westbury Galley. Burden: 120 tons. Crew: 24 … Owners: … Walter Hawksworth … 1710 June 12
PROB 31/181/617 Exhibit: 1738/617. William Jones of Bristol, Somerset. Probate inventory, or declaration of the estate of the same, deceased, with account. 1738 October

New Hampshire Historical Society (NHHS)

George Boyd Letter Book, 1773–1775.
Langdon Family Papers.
Shipping Papers (a topical collection), 1707–1886.
Tredick Family Papers, 1753–1835.
William Hale Correspondence, 1791–1839.

Phillips Library, Peabody Essex Museum (PEM-*Phillips*)

Abner Dole Account Book, 1730–1760.
Account Book of David Lowell, 1781–1801.
Account Book of the Horton Family, 1781–1828.
Account Book of Samuel Coker, 1749–1774.
Crowninshield Family Papers, 1727–1891.
Derby Family Papers, 1716–1925.
Joshua Coffin Papers, 1647–1862.
Orne Family Papers, 1719–1899.
Shipbuilding Agreement, 1739.
William Hacket Papers, 1739–1808.
Woodwell Family Papers, 1752–1884.

Published Primary Sources

Anderson, R.C., ed. *Journals and Narratives of the Third Dutch War. Publications of the Naval Records Society, Vol. LXXXVI*. London: Navy Records Society, 1946.

Annals of Lloyd's Register. London, 1884. Digital facsimile.

Barker, Robert. *The Unfortunate Shipwright: Or Cruel Captain*. London, 1795. Google Books digital facsimile.

Barlow, Edward. *Barlow's journal of his life at sea in king's ships, East & West Indiamen & other merchantmen from 1659 to 1703*. Two volumes. Transcribed by Basil Lubbock. London: Hurst & Blackett, 1934.

Boardman, Timothy. *Log-Book of Timothy Boardman*. Albany, NY: Joel Munsell's Sons, 1885.

Bushnell, Edmund. *The Compleat Shipwright*. Sixth edition. London: For Richard Mount, 1699.

Chapman, Frederik H. af. *Architectura Navalis Mercatoria: The Classic of Eighteenth-Century Naval Architecture*. Mineola, NY: Dover Publications, 2006.

Charnock, John. *An History of Marine Architecture*. Volume II. London: 1801. ECCO facsimile.

Churchill, Awnsham and John Churchill. *Collection of Voyages and Travels, Some Now First Printed from Original Manuscripts, Others Now first Published in English, In Six Volumes, With a General Preface, giving an Account of the Progress of Navigation, from its first Beginnings*. Six Volumes. Third Edition, London, 1747. Google Book, accessed 16 February 2015 at https://books.google.com/books?id=oFVEAAAAcAAJ&lpg=PA343&ots=knNjzh2Iyu&dq=Charles%20May%20An%20Account%20of%20the%20Wonderful%20Preservation%20of%20the%20Ship%20Terra%20Nova%20of%20London&pg=PP1#v=onepage&q=Charles%20May%20An%20Account%20

of%20the%20Wonderful%20Preservation%20of%20the%20Ship%20Terra%20Nova%20of%20London&f=true.

Dassié, F. *L'Architecture Navale, contenant la maniere de construire des navires ...* Paris: Chez Jean de la Caille, 1677. Google e-book via archive.org.

Elliot, Jonathan. "Remarks and Observations on the Use of Cotton Canvas, as Preferable to Hemp, for the Sails of Ships of War or Merchant Vessels, Supported by the Testimony of Naval Officers, Ship Owners, Captains, &c." Washington, DC: Jonathan Elliot, 1829.

Euler, Leonhard. *A compleat theory of the construction and properties of vessels: with practical conclusions for the management of ships made easy to navigators*. Trans. Henry Watson. London: J. Sewell, 1790.

Falconer, William. *An universal dictionary of the marine: or, a copious explanation of the technical terms and phrases employed in the construction, of a ship. Illustrated with variety of original designs of shiping The second edition, corrected*. London: 1771. ECCO facsimile.

Faraday, Michael. "On the practical prevention of dry rot in timber." London: John Weale, 1838.

Guest, Henry. "Observations on sheathing vessels: seasoning timber; the proper time to fall timber; the nature and what force it is that causes the sap to rise; with a number of other valuable observations." New Brunswick, NJ: A. Blauvelt, 1805.

Hamer, Philip M., ed. *The Papers of Henry Laurens*, Volumes 1–5. Columbia: South Carolina Historical Society, 1968–1976.

Hemphill, James. "Leaves from the Log-Books and Letters of James Hemphill, Mariner and Merchant of Wilmington, 1793–1797." *Papers of the Historical Society of Delaware*, LXIV, 1914.

Hutchinson, William. *A Treatise on Naval Architecture*. Fourth (1794) Edition. London: Conway Maritime Press, 1970.

Kelly, Samuel. *Samuel Kelly: An Eighteenth Century Seaman*. Edited by Crosbie Garstin. New York: Frederick A. Stokes Company, 1925.

Lavery, Brian, ed. *Deane's Doctrine of Naval Architecture, 1670*. Annapolis, MD: Naval Institute Press, 1981.

Mainwaring, Henry. *The Sea-mans Dictionary : or, an exposition and demonstration of all the parts and things belonging to a shippe* (1644). In *The Life and Works of Sir Henry Mainwaring*, Volume 2. Edited by G.E. Mainwaring. London: The Navy Records Society, 1920. E-reader edition published by The Internet Archive, 2007.

May, Charles. "An Account of the Wonderful Preservation of the Ship Terra Nova of London." In Churchill, Awnsham and John Churchill. *Collection of Voyages and Travels, Some Now First Printed from Original Manuscripts, Others Now first Published in English, In Six Volumes, With a General Preface, giving an Account of the Progress of Navigation, from its first Beginnings*. Six Volumes. Third Edition, London, 1747.

Google Book, accessed 16 February 2015 at https://books.google.com/books?id=0FVEAAAAcAAJ&lpg=PA343&ots=knNjzh2Iyu&dq=Charles%20May%20An%20Account%20of%20the%20Wonderful%20Preservation%20of%20the%20Ship%20Terra%20Nova%20of%20London&pg=PP1#v=onepage&q=Charles%20May%20An%20Account%20of%20the%20Wonderful%20Preservation%20of%20the%20Ship%20Terra%20Nova%20of%20London&f=true.

Murray, Mungo. *Treatise on Shipbuilding, 1765. Includes translation of Du Hamel du Monceau's treatise of the same year.* https://archive.org/stream/treatiseonshipbu00murr#page/n0/mode/2up.

Ollivier, Blaise. *18th Century Shipbuilding: remarks on the navies of the English & the Dutch from observations made at their dockyards in 1737*. Trans. and ed. David H. Roberts. Rotherfield, UK: Jean Boudriot Publications, 1992.

Salisbury, W. and R.C. Anderson, eds. *A Treatise on Shipbuilding* and *A Treatise on Rigging Written About 1620–1625*. The Society for Nautical Research Occasional Publications Number 6. London: The Society for Nautical Research, 1958.

Saunders, Edmund. "An Address to the Owners of Ships, &c., in particular, and in general to all who have an Interest in preserving Timber under Water." Plymouth, UK: M. Haydon & Son, 1790.

Smith, John. *The Sea-man's Grammar and Dictionary, Explaining all the difficult Terms in Navigation, and the Practical Navigator and Gunner: In Two Parts*. London: 1691 edition. Google e-book facsimile.

Society for the Improvement of Naval Architecture. "An address to the public …" London: Society for the Improvement of Naval Architecture, 1791.

Society for the Improvement of Naval Architecture. *Collection of Papers on Naval Architecture*. 3 volumes. London: The European Magazine, 1791–1793.

Society for the Improvement of Naval Architecture. "Some account of the institution, plan, and present state of the society, of the Society for the Improvement of Naval Architecture: together with the premiums offered by the Society, list of subscribers, rules and orders, of the Society: and 3 original letters, addressed to the committee, from Sir Charles Middleton, Bart. and Messrs. Gore and Patton, on subjects of naval architecture." London: Society for the Improvement of Naval Architecture, 1792.

Stalkartt, Marmaduke. *Naval Architecture or the Rudiments and Rules of Ship Building*. London, 1781. Google Books facsimile.

Steel, David. *The ship-master's assistant and owner's manual: containing complete information, as well to merchants, masters of ships, and persons employed in the merchant-service*. Sixth Edition. London: 1795. ECCO facsimile.

Steel, David. *The Elements and Practice of Rigging, Seamanship, and Naval Tactics. Volume III—Rigging*. London: 1806. Cambridge Library Collection facsimile.

Sutherland, William. *The Ship-builder's Assistant: Or Some Essays Toward Compleating the Art of Marine Architecture*. London, for Richard Mount, 1711.

Thorpe, David. "The *Mayflower II* Story—Reflections of a Sailor." Part 1. *The Mayflower Quarterly* 78:2 (June 2012): 117–21.

Thorpe, David. "The *Mayflower II* Story—Reflections of a Sailor." Part 2. *The Mayflower Quarterly* 78:3 (September 2012): 216–19.

Thorpe, David. "The *Mayflower II* Story—Reflections of a Sailor." Part 3. *The Mayflower Quarterly* 78:4 (December 2012): 324–27.

Thorpe, David. "The *Mayflower II* Story—Reflections of a Sailor." Part 4. *The Mayflower Quarterly* 79:1 (March 2013): 35–38.

Thorpe, David. "The *Mayflower II* Story—Reflections of a Sailor." Part 6. *The Mayflower Quarterly* 79:3 (September 2013): 286–90.

United States Bureau of Labor Statistics. *History of Wages In the United States From Colonial Times to 1928*. Washington, DC: United States Govt. Print. Off., 1934), 21. Accessed at https://babel.hathitrust.org/cgi/pt?id=uc1.32106007458745;view=1up;seq=35 3 October 2018.

Villiers, Alan. "How We Sailed the New *Mayflower* to America." *National Geographic* 92:5 (November 1957): 627–74.

Villiers, Alan. *The New Mayflower*. Leicester, UK: Brockhampton Press, 1959.

Wallis, John. "Cono-Cuneus: or the Shipwright's circular wedge." In *A treatise of algebra, both historical and practical…* London: John Playford, for Richard Davis, Oxford, 1685.

Secondary Sources

Abell, Sir Westcott. *The Shipwright's Trade*. Conway/Cornell Maritime, 1948, 1981.

Adams, Jonathan and Johan Rönnby, eds. *Interpreting Shipwrecks: Maritime Archaeological Approaches*. Southampton, UK: Highfield, 2013.

Adams, Jonathan. *A Maritime Archaeology of Ships: Innovation and Social Change in Medieval and Early Modern Europe*. Oxford: Oxbow, 2013.

Adams, Jonathan. "*Sea Venture*: A second interim report—part 1." *International Journal of Nautical Archaeology and Underwater Exploration* 14:4 (November 1985): 275–99.

Adas, Michael. *'Machines as the Measure of Men': Science, Technology, and Ideologies of Western Dominance*. Ithaca, NY: Cornell University Press, 1989.

Anderson, Atholl, James H. Barrett & Katherine V. Boyle, eds. *The Global Origins and Development of Seafaring*. Cambridge, UK: McDonald Institute for Archaeological Research, 2010.

Anderson, R.C. *The Rigging of Ships in the Days of the Spritsail Topmast 1620–1720*. London: Conway Maritime, 1982.

Anderson, R.C. "Comparative Naval Architecture, 1670–1720, Part I." *Mariner's Mirror* 7:2 (1921): 38–45.

Anderson, R.C. "Comparative Naval Architecture, 1670–1720, Part II." *Mariner's Mirror* 7:6 (1921): 172–81.

Anderson, R.C. "Comparative Naval Architecture, 1670–1720, Part III." *Mariner's Mirror* 7:10 (1921): 308–15.

Andrews, Charles M. *The Colonial Period of American History*. Four volumes. Yale, 1934–38.

Armitage, David, Alison Bashford, and Sujit Sivasundaram, eds. *Oceanic Histories*. Cambridge: Cambridge University Press, 2000.

Bailyn, Bernard. *The New England Merchants in the Seventeenth Century*. Cambridge, MA: Harvard University Press, 1955.

Bailyn, Bernard. *Soundings in Atlantic History: Latent Structures and Intellectual Currents, 1500–1830*. Cambridge, MA: Harvard University Press, 2009.

Bailyn, Bernard and Lotte Bailyn. *Massachusetts Shipping, 1697–1714: A Statistical Study*. Cambridge, MA: Belknap, 1959.

Baker, William A. *Colonial Vessels: Some Seventeenth-Century Sailing Craft*. Barre, MA: Barre Publishing Co., 1962.

Baker, William A. *The Mayflower and Other Colonial Vessels*. London: Conway Maritime Press, 1983.

Baker, William A. *The New Mayflower: Her Design and Construction*. Barre, MA: Barre Gazette, 1958.

Baker, William A. *Sloops and Shallops*. Barre, MA: Barre Publishing Co., 1966.

Banks, Kenneth J. *Chasing Empire Across the Sea: Communications and the State in the French Atlantic, 1713–1763*. Montreal: McGill-Queen's University Press, 2002.

Bankston, J. "Some Notes on Early Spanish Maritime Technology and Practices in the New World." In J.B. Arnold III, ed., *Beneath the Waters of Time: The Proceedings of the Ninth Conference on Underwater Archaeology*, 63–66. Texas Antiquities Committee Publication No. 6, Austin, Texas.

Barbour, Violet. "Dutch and English Merchant Shipping in the Seventeenth Century." *Economic History Review* 2:2 (January 1930): 261–90.

Barker, Richard. "A Manuscript on Shipbuilding, circa 1600, Copied by Newton." *Mariner's Mirror* 80:1 (February 1994): 16–29.

Barker, Richard. "Why tumblehome?" *Mariner's Mirror* 84:1 (1998): 95–97.

Barker, Rosalin. *The Rise of an Early Modern Shipping Industry: Whitby's Golden Fleet, 1600–1750*. Woodbridge, UK: The Boydell Press, 2011.

Barth, Jonathan. "Reconstructing Mercantilism: Consensus and Conflict in British Imperial Economy in the Seventeenth and Eighteenth Centuries." *William and Mary Quarterly* 73:2 (April 2016): 257–90.

Bartlett, J.V. "Why tumblehome?" *Mariner's Mirror* 83:4 (1997): 478.

Bass, George F., ed. *Beneath the Seven Seas*. New York: Thames & Hudson, 2005.

Beaglehole, J.C. *The Life of Captain James Cook*. Stanford, CA: Stanford University Press, 1974.

Bell, Whitfield J., Jr. "Adam Cunningham's Atlantic Crossing, 1728." *Maryland Historical Magazine* 50:3 (September 1955): 195–202.

Bennett, Jenny, ed. *Sailing Into the Past: Learning from Replica Ships*. Barnsley, UK: Seaforth, 2009.

Bernstein, Peter L. *Against the Gods: The Remarkable Story of Risk*. New York: Wiley, 1996.

Berry, Stephen R. *A Path in the Mighty Waters: Shipboard Life & Atlantic Crossings to the New World*. New Haven, CT: Yale University Press, 2015.

Bijker, Wiebe et al., eds. *The Social Construction of Technological Systems: New Directions in the Sociology and History of Technology*. Anniversary Edition. Cambridge, MA: MIT Press, 2012. Kindle edition.

Blake, John. *The Sea Chart: The Illustrated History of Nautical Maps and Navigational Charts*. Annapolis, MD: Naval Institute Press, 2009.

Blue, Lucy et al., eds. *Connected by the Sea: Proceedings of the Tenth International Symposium on Boat and Ship Archaeology, Roskilde 2003*. Oxford: Oxbow, 2006.

Bolster, W. Jeffrey. *Black Jacks: African American Seamen in the Age of Sail*. Cambridge, MA: Harvard University Press, 1997.

Boucher, Philip P. *France and the American Tropics to 1700: Tropics of Discontent?* Baltimore, MD: Johns Hopkins University Press, 2008.

Bowen, H.V. et al., eds. *Britain's Oceanic Empire: Atlantic and Indian Ocean Worlds, c. 1550–1850*. Cambridge: Cambridge University Press, 2012.

Breen, T.H. "An Empire of Goods: The Anglicization of Colonial America, 1690–1776." *Journal of British Studies* 25:4 (October 1986): 467–99.

Bruyns, Willem F.J. Mörzer. "Research in the History of Navigation: Its Role in Maritime History." *International Journal of Maritime History* 21:2 (December 2009): 261–87.

Burningham, Nicholas. "Experimental Maritime Archaeology." In Claire Smith, ed. *Encyclopedia of Global Archaeology*, 2716–18. New York: Springer, 2014.

Burningham, Nick and Adriaan de Jong. "The *Duyfken* Project: an Age of Discovery ship reconstruction as experimental archaeology." *International Journal of Nautical Archaeology* 26:4 (November 1997): 277–92.

Burningham, Nick. "Early Seventeenth-Century Ships." In Jenny Bennett, ed. *Sailing Into The Past: Learning from Replica Ships*, 100–119. Barnsley, UK: Seaforth, 2009.

Burningham, Nick. "Learning to Sail the *Duyfken* Replica." *International Journal of Nautical Archaeology* 30:1 (April 2001): 74–85.

Burningham, Nick. "Tallships and Replicas." *Maritime Studies* 3 (2000) 110 Maritime Studies 19. Accessed at http://www.austlii.edu.au/au/journals/MarStudies/2000/3 .html. 5 May 2015.

Butel, Paul. *The Atlantic*. Trans. Iain Hamilton Grant. London: Routledge, 1999.

Butel, Paul. *Histoire des Antilles françaises, XVIIe–XXe siècle*. Paris: Perrin, 2002.

Butler, Joseph T., ed. *The European Origins of the Small Watercraft of the United States and Canada*. Thibodeaux, LA: Center for Traditional Louisiana Boatbuilding and the Museum Small Craft Association, 1992.

Cann, B. "Notes on the effect of catharpings on shroud tension and mast displacement." *Mariner's Mirror* 98:2 (May 2012): 193–99.

Canny, Nicholas. "Writing Atlantic History; or, Reconfiguring the History of Colonial British America." *Journal of American History* 86:3, *The Nation and Beyond: Transnational Perspectives on United States History: A Special Issue* (December 1999): 1093–114.

Carlos, Ann M. and Frank D. Lewis. *Commerce by a Frozen Sea: Native Americans and the European Fur Trade*. Philadelphia: University of Pennsylvania Press, 2010.

Carrington, Selwyn H.H. "The American Revolution and the British West Indies' Economy." *Journal of Interdisciplinary History* 27:4 (Spring 1987): 823–50.

Carroll, Charles. "The Forest Society of New England." In Brooke Hindle, ed., *America's Wooden Age: Aspects of its Early Technology*, 13–36. Tarrytown, NY: Sleepy Hollow Press, 1985.

Castro, Filipe V. *The Pepper Wreck: A Portuguese Indiaman at the Mouth of the Tagus River*. College Station: Texas A&M University Press, 2005.

Castro, Filipe V. "Rising and Narrowing: 16th-century Geometric Algorithms used to Design the Bottom of Ships in Portugal." *International Journal of Nautical Archaeology* 36:1 (March 2007): 148–54.

Cecelski, David S. *The Waterman's Song: Slavery and Freedom in Maritime North Carolina*. Chapel Hill: University of North Carolina Press, 2001.

Chapelle, Howard I. *American Small Sailing Craft: Their Design, Development and Construction*. New York: Norton, 1951.

Chapelle, Howard I. *The Search for Speed Under Sail, 1700–1855*. New York: Norton, 1984.

Cheney, Paul. *Revolutionary Commerce: Globalization and the French Monarchy*. Cambridge, MA: Harvard University Press, 2010.

Chet, Guy. *The Ocean is a Wilderness: Atlantic Piracy and the Limits of State Authority, 1688–1856*. Amherst: University of Massachusetts Press, 2014.

Christopher, Emma. *Slave Ship Sailors and their Captive Cargoes*. Cambridge: Cambridge University Press, 2006.

Cipolla, Carlo M. *Guns, Sails, and Empires: Technological Innovation and the Early Phases of European Expansion, 1400–1700*. New York: Pantheon, 1965.

Classen, Albrecht, ed. *Handbook of Medieval Studies: Terms, Methods, Trends*. Berlin: De Gruyter, 2010.

Clowes, G.S. Laird. *Sailing Ships: Their History and Development*. Two volumes. London: Her Majesty's Stationery Office, 1952.

Clowse, Converse D. "Shipowning and Shipbuilding in Colonial South Carolina: An Overview." *American Neptune* 46 (1984): 221–44.

Coates, John et al. "Experimental Boat and Ship Archaeology: Principles and Methods." *International Journal of Nautical Archaeology* 24:4 (November 1995): 293–301.

Coclanis, Peter A. "*Drang Nach Osten*: Bernard Bailyn, the World-Island, and the Idea of Atlantic History." *Journal of World History* 111:3 (June 2006): 741–57.

Coclanis, Peter A. "In Retrospect: McCusker and Menard's *Economy of British America*." *Reviews in American History* 30:2 (June 2002): 183–97.

Coclanis, Peter A., ed. *The Atlantic Economy During the Seventeenth and Eighteenth Centuries*. Columbia: University of South Carolina Press, 2005.

Combs, Edwin L. III. "Trading in Lubberland: Maritime Commerce in Colonial North Carolina." *The North Carolina Historical Review* 80:1 (January 2003): 1–27.

Cottrell, P.L. and D.H. Aldcroft, eds. *Shipping, Trade, and Commerce: Essays in Memory of Ralph Davis*. Leicester, UK: University of Leicester Press, 1981.

Cowan, Ruth Schwartz. "The Consumption Junction: A Proposal for Research Strategies in the Sociology of Technology." In Wiebe Bijker et al., eds., *The Social Construction of Technological Systems: New Directions in the Sociology and History of Technology*, 253–72. Anniversary Edition. Cambridge, MA: MIT Press, 2012. Kindle edition.

Craig, Robin. "Printed guides for master mariners as a source of productivity change in shipping, 1750–1914." In *British Tramp Shipping, 1750–1914*. Research in Maritime History Number 24, 121–36. St. John's, Canada: International Maritime Economic History Association, 2003.

Creighton, Margaret S. and Lisa Norling, eds. *Iron Men, Wooden Women: Gender and Seafaring in the Atlantic World, 1700–1920*. Baltimore, MD: Johns Hopkins University Press, 1996.

Crittenden, Charles C. *Commerce in Colonial North Carolina, 1763–1789*. New Haven, CT: Yale University Press, 1936.

Crosby, Alfred W. *Ecological Imperialism: The Biological Expansion of Europe, 900–1900*. Cambridge: Cambridge University Press, 2004.

Crowhurst, Patrick. *The Defence of British Trade 1689–1815*. Folkestone, UK: Dawson, 1977.

Crowther, Simeon J. "The Shipbuilding Output of the Delaware Valley, 1722–1776." *Proceedings of the American Philosophical Society* 117:2 (April 10, 1973): 90–104.

Crumlin-Pedersen, Ole. "Experimental archaeology and ships—principles, problems and examples." In Lucy Blue et al., eds., *Connected by the Sea: Proceedings of the Tenth International Symposium on Boat and Ship Archaeology, Roskilde, 2003*, 1–7. Oxford: Oxbow, 2006.

Curtin, Philip. *The Rise and Fall of the Plantation Complex: Essays in Atlantic History*, 2nd ed. Cambridge: Cambridge University Press, 1998.

Daston, Lorraine J. "The Domestication of Risk: Mathematical Probability and Insurance, 1650–1830." In Lorenz Krüger, et al., eds., *The Probabilistic Revolution, Vol. 1: Ideas in History*, 237–60. Cambridge, MA: MIT Press, 1987.

Davis, Ralph. "Shipping Records." *Archives* 7 (April 1966): 135–42.

Davis, Ralph. *The Rise of the English Shipping Industry in the Seventeenth and Eighteenth Centuries*. London: Macmillan, 1962. Reprint Newton Abbott: David & Charles, 1972. Re-issue St. John's, Canada: International Maritime Economic History Association, Research in Maritime History Number 48, 2012.

Decencière, P. "Three French Sailing Ship Performance Trials." *Mariner's Mirror* 94:3 (August 2008): 276–84.

Deetz, James. *In Small Things Forgotten: An Archaeology of Early American Life*, Revised Edition. New York: Doubleday, 1996.

Devine, T.M. and Gordon Jackson, eds. *Glasgow, Volume I: Beginnings to 1830*. Manchester, UK: Manchester University Press, 1995.

Doerflinger, Thomas M. *A Vigorous Spirit of Enterprise: Merchants and Economic Development in Revolutionary Philadelphia*. Chapel Hill: University of North Carolina Press, 1986.

Donovan, Arthur. "The Impact of Containerization from Adam Smith to the 21st Century." *Review of Business* 25:3 (Fall 2004): 10–15.

Dunn, Richard S. *Sugar and Slaves: The Rise of the Planter Class in the English West Indies, 1624–1713*. Chapel Hill: University of North Carolina Press, 1972.

Earle, Peter. *Sailors: English Merchant Seamen 1650–1775*. London: Methuen, 1998.

Earle, Peter. *The Sack of Panamá: Captain Morgan and the Battle for the Caribbean*. New York: Thomas Dunne, 2007.

Ebert, Christopher. "Early Modern Atlantic Trade and the Development of Maritime Insurance to 1630." *Past & Present* 213 (November 2011): 87–114.

Edgerton, David. *The Shock of the Old: Technology and Global History since 1900*. Oxford: Oxford University Press, 2011 (reprint). Kindle edition.

Elliott, John H. *Empires of the Atlantic World: Britain and Spain in America, 1492–1830*. New Haven, CT: Yale University Press, 2007.

Ellis, John. *The Social History of the Machine Gun*. London: Century Hutchinson, 1975.

Elster, Jon. *Explaining Technical Change*. Cambridge: Cambridge University Press, 1983.

Endsor, R., W. Benbow and P. Quinn, "Notes: A Drawing of the Midship Bend of the Hampshire 1653." *Mariner's Mirror* 91:1 (2005): 67–100.

Englert, Anton. "Trial voyages as a method of experimental archaeology: The aspect of speed." In Lucy Blue et al., eds., *Connected by the Sea: Proceedings of the Tenth International Symposium on Boat and Ship Archaeology, Roskilde 2003*, 35–42. Oxford: Oxbow, 2006.

Enthoven, Victor. "That Abominable Nest of Pirates: St. Eustasius and the North Americans, 1680–1780." *Early American Studies: An Interdisciplinary Journal* 10:2 (Spring 2012): 239–301.

Eriksson, Niklas. "Lodging in a Fluitship: The Material Setting of Everyday Life on Board *Anna Maria* of 1694." *Journal of Maritime Archaeology* 10.1007/s11457-015-9134-3. http://dx.doi.org/10.1007/s11457-015-9134-3 (March 2015): 1–19.

Evans, Amanda M. "Defining Jamaica Sloops: A Preliminary Model for Identifying an Abstract Concept." *Journal of Maritime Archaeology* 2:2 (December 2007): 83–92.

Fawcett, Howard H., Jr. "Sailing the Jamestown Ships." *Transactions of the Society of Naval Architects and Marine Engineers* (1958): 46–66.

Feldbæk, Ole. "Danish North Atlantic Shipping, 1720–1814." *International Journal of Maritime History* 10:1 (June 1998): 105–20.

Ferguson, Eugene S. "Elegant Inventions: The Artistic Component of Technology." *Technology and Culture* 19:3 (July 1978): 450–60.

Ferreiro, Larrie D. "The Social History of the Bulbous Bow." *Technology and Culture* 52:2 (2011): 335–59.

Ferreiro, Larrie D. *Ships and Science: The Birth of Naval Architecture in the Scientific Revolution, 1600–1800*. Cambridge, MA: MIT Press, 2006.

Finamore, Daniel, ed. *Maritime History as World History*. Gainesville: University Press of Florida, 2004.

Fischer, David Hackett. *Albion's Seed: Four British Folkways In America*. Oxford: Oxford University Press, 1991.

Fischer, Lewis R. "Revolution without Independence: The Canadian Colonies, 1749–1775." In Ronald Hoffman et al., eds., *The Economy of Early America: The Revolutionary Period, 1763–1790*, 88–125. Charlottesville: University of Virginia, 1988.

Fischer, Lewis R., and Eric W. Sager. "An Approach to the Quantitative Analysis of British Shipping Records," *Business History* 22:2 (July 1980): 135–51.

Flatman, Joe. "Cultural biographies, cognitive landscapes and dirty old bits of boat: 'theory' in maritime archaeology." *International Journal of Nautical Archaeology* 32:2 (2003): 143–57.

Flatman, Joe, and Mark Staniforth. "Historical maritime archaeology." In Dan C. Hicks and Mary Beaudry, eds., *The Cambridge Companion to Historical Archaeology*, 168–90. Cambridge: Cambridge University Press, 2006.

Ford, Ben et al., eds. *The Oxford Handbook of Maritime Archaeology*. Oxford: Oxford University Press, 2013. Retrieved 8 Jan. 2019, from http://www.oxfordhandbooks.com/view/10.1093/oxfordhb/9780199336005.001.0001/oxfordhb-9780199336005-e-48.

Forestier, Albane. "Risk, kinship and personal relationships in late eighteenth-century West Indian trade: The commercial network of Tobin & Pinney." *Business History* 52:6 (October 2010): 912–31.

Forty, Adrian. *Objects of Desire: Design and Society since 1750*. London: Thames & Hudson, 1986.

Fox, Robert, ed. *Technological Change: Methods and Themes in the History of Technology*. Amsterdam: Harwood, 1996.

French, Christopher J. "The Longevity of Ships in Colonial Trade: Some Further Evidence." *International Journal of Maritime History* 3:1 (June 1991): 155–163.

French, Christopher J. "Productivity in the Atlantic shipping industry: a quantitative study." *Journal of Interdisciplinary History* 17:3 (Winter 1987): 613–38.

Friel, Ian. *The British Museum Maritime History of Britain and Ireland, c. 400–2001.* London: British Museum, 2003.

Games, Alison. *The Web of Empire: English Cosmopolitans in an Age of Expansion, 1500–1660*. Oxford: Oxford University Press, 2008.

Gardiner, Robert, ed. *Cogs, Caravels and Galleons: The Sailing Ship 1000–1650*. London: Conway Maritime Press, 1994.

Gardiner, Robert, ed. *The Heyday of Sail: The Merchant Sailing Ship, 1650–1830*. Edison, NJ: Chartwell, 2002.

Gardiner, Robert, ed. *The Line of Battle: The Sailing Warship 1650–1840*. London: Conway Maritime Press, 1992.

Geertz, Clifford. *The Interpretation of Cultures*. New York: Basic Books, 1973.

Gilfillan, S.C. *Inventing the Ship*. Chicago: Follett, 1935.

Gilfillan, S.C. *The Sociology of Invention*. Chicago: Follett, 1935. Reprint Cambridge, MA: MIT Press, 1970.

Gilje, Paul. *Liberty on the Waterfront: American Maritime Culture in the Age of Revolution*. Philadelphia: University of Pennsylvania Press, 2004.

Gillingham, Harold E. "Some Colonial Ships Built in Philadelphia." *The Pennsylvania Magazine of History and Biography* LVI (1932): 156–86.

Gillmer, Thomas C. *A History of Working Watercraft of the Western World*. 2nd ed. Camden, ME: International Marine, 1994.

Glassie, Henry H. *Folk Housing in Middle Virginia: A Structural Analysis of Historic Artifacts*. Knoxville: University of Tennessee, 1975.

Glassie, Henry H. *Material Culture*. South Bend: Indiana University, 1999.

Glassie, Henry H. *The Spirit of Folk Art*. New York: Abrams, 1989.

Glete, Jan. *Navies and Nations: Warships, Navies and State Building in Europe and America, 1500–1860*. Stockholm: Almqvist and Wiksell International, 1993.

Glover, William. "The Eighteenth Century Practice of Navigation As Recorded in the Logs of Hudson's Bay Company Ships." *The Northern Mariner/Le Marin du Nord* 26:2 (April 2016): 145–64.

Glover, William. "The Navigation of the *Nonsuch*, 1668–69." *The Northern Mariner/Le Marin du Nord* 11:4 (October 2001): 49–63.

Goldenberg, Joseph A. *Shipbuilding in Colonial America*. Charlottesville: University of Virginia Press for the Mariners' Museum, 1976.

Greene, Jack P. and Philip Morgan. *Atlantic History: A Critical Reappraisal*. Oxford: Oxford University Press, 2009.

Griffin, Patrick. *The Townshend Moment: The Making of Empire and Revolution in the Eighteenth Century*. New Haven, CT: Yale University Press, 2017.

Haggerty, Sheryllynne. "Risk, networks and privateering in Liverpool during the Seven Years' War, 1756–1763." *International Journal of Maritime History* 30:1 (February 2018): 30–51.

Hancock, David. *Citizens of the World: London Merchants and the Integration of the British Atlantic Community*. Cambridge: Cambridge University Press, 1997.

Hancock, David. *Oceans of Wine: Madeira and the Emergence of American Trade and Taste*. New Haven, CT: Yale University Press, 2009.

Harding, Richard. *The Emergence of Britain's Global Naval Supremacy: The War of 1739–1748*. Woodbridge, UK: The Boydell Press, 2010.

Harding, Richard. *Seapower and Naval Warfare, 1650–1830*. London: UCL Press, 1999.

Harland, John. "Catharpings, swifters, and Bentinck shrouds." *Mariner's Mirror* 95:3 (2009): 328–36.

Harland, John. "The Early History of the Steering Wheel." *Mariner's Mirror* 58:1 (1972): 41–68.

Harland, John. *Seamanship in the Age of Sail*. Annapolis, MD: Naval Institute Press, 1984.

Harms, Robert. *The* Diligent*: A Voyage Through the World of the Slave Trade*. New York: Basic Books, 2002.

Hattendorf, John B., ed. *Maritime History Volume 2: The Eighteenth Century and the Classic Age of Sail*. Malabar, FL: Krieger, 1997.

Hattendorf, John B., ed. *The Oxford Encyclopedia of Maritime History*. Four volumes. Oxford: Oxford University Press, 2007.

Headrick, Daniel. *The Tools of Empire: Technology and European Imperialism in the Nineteenth Century*. Oxford: Oxford University Press, 1981.

Hedges, James. *The Browns of Providence Plantations: Colonial Years*. Cambridge, MA: Harvard University Press, 1952.

Hicks, Robert. *Voyage to Jamestown: Practical Navigation in the Age of Discovery*. Annapolis, MD: Naval Institute Press, 2011.

Hindle, Brooke, ed. *America's Wooden Age: Aspects of its Early Technology*. Tarrytown, NY: Sleepy Hollow Press, 1985.

Hindle, Brooke. *Technology in Early America: Needs; Opportunities for Study*. Chapel Hill: University of North Carolina Press, 1966.

Hocker, Frederick M. "Bottom-Based Shipbuilding in Northwestern Europe." In Hocker and Cheryl A. Ward, eds., *The Philosophy of Shipbuilding: Conceptual Approaches to the Study of Wooden Ships*, 65–94. College Station: Texas A&M University Press, 2004.

Hocker, Frederick M. "The Clydesdale Plantation Vessel: Savannah River, South Carolina." In *Beneath the Seven Seas*, edited by George F. Bass, 192–93. New York: Thames & Hudson, 2005.

Hocker, Frederick M. "The Clydesdale Plantation Vessel Project: 1992 Field Report." *Institute of Nautical Archaeology Quarterly* 19:4 (Winter 1992): 12–16.

Hocker, Frederick M. and Cheryl A. Ward, eds. *The Philosophy of Shipbuilding: Conceptual Approaches to the Study of Wooden Ships.* College Station: Texas A&M University Press, 2004.

Hoffman, Ronald et al., eds. The *Economy of British America: The Revolutionary Period, 1763–1790.* Charlottesville: University Press of Virginia, 1988.

Holly, H.H. "Sparrow-Hawk: A Seventeenth-Century Vessel in Twentieth-Century America." *Pilgrim Society Booklet Series.* Boston: Nimrod, 1969.

Holmes, Sir George C.V. *Ancient and Modern Ships, Part I: Wooden Sailing Ships.* London: His Majesty's Stationery Office, 1906.

Hoppit, Julian. *Risk and Failure in English Business, 1700–1800.* Cambridge: Cambridge University Press, 1987.

Hornsby, Stephen J. *British Atlantic, American Frontier: Spaces of Power in Early Modern British America.* Lebanon, NH: University Press of New England, 2004.

Hornsby, Stephen J. *Surveyors of Empire: Samuel Holland, J.F.W. Des Barres, and the Making of the Atlantic Neptune.* Montreal: McGill-Queen's University, 2011.

Hoving, Albert. "Early Shipbuilding Literature in the Dutch Republic." *Nautical Research Journal* 58:3 (Autumn 2013): 163–74.

Hoving, Albert. "Early Shipbuilding Literature in the Dutch Republic, Part 2." *Nautical Research Journal* 58:4 (Winter 2013): 253–58.

Hoving, Albert. "The Fluit." *Nautical Research Journal* 42:1 (March 1997): 24–35.

Hoving, Albert. "Seagoing Ships of the Netherlands." In Robert Gardiner, ed., *The Heyday of Sail: The Merchant Sailing Ship, 1650–1830* in the Conway's History of the Ship series, 47–51. Edison, NJ: Chartwell, 2002.

Hoving, Albert. *17th Century Dutch Merchant Ships: Text, Photos, and Plans for the Ship Modeler.* Florence, OR: SeaWatch, 2014.

Howard, Mark. "Richard Mount, London Shipowner." *Mariner's Mirror* 103:1 (February 2017): 93–97.

Howse, Derek. *Greenwich Time and the Longitude.* Oxford: Oxford University Press, 2003.

Hughes, Thomas P. "The Evolution of Large Technological Systems." In Wiebe Bijker et al., eds., *The Social Construction of Technological Systems: New Directions in the Sociology and History of Technology.* Anniversary Edition, 45–76. Cambridge, MA: MIT Press, 2012. Kindle edition.

Huxtable, George and Ian Jackson. "Journey to Work: James Cook's Transatlantic Voyages in the *Grenville* 1764–1767." *Journal of Navigation* 63 (2010): 204–14.

Hyde, Francis E. *Liverpool and the Mersey: An Economic History of a Port, 1700–1970.* Newton Abbot, UK: David and Charles, 1971.

Jackson, Gordon. *The History and Archaeology of Ports.* Tadworth, UK: World's Work, 1983.

Jackson, Gordon. "Scottish shipping, 1775–1805." In P.L. Cottrell and D.H. Aldcroft, eds., *Shipping, Trade, and Commerce: Essays in Memory of Ralph Davis*, 117–36. Leicester, UK: University of Leicester Press, 1981.

Janzen, Olaf Uwe. "A Scottish Sack Ship in the Newfoundland Trade 1726–27." *Scottish Economic & Social History* 18:1 (May 1998): 1–18.

Jarvis, Adrian, ed. *Port and Harbour Engineering*. Aldershot, UK: Ashgate, 1998.

Jarvis, Michael J. "'The Fastest Vessels in the World': The Origin and Evolution of the Bermuda Sloop, 1620–1800." *Bermuda Journal of Archaeology and Maritime History* 7 (1995): 31–50.

Jarvis, Michael J. *In The Eye Of All Trade: Bermuda, Bermudians, and the Maritime Atlantic World, 1680–1783*. Chapel Hill: University of North Carolina Press, 2010.

Jarvis, Rupert C. "Sources for the History of Ships and Shipping." *Journal of Transport History* 3:4 (November 1958): 212–34.

John, A.H. "The London Assurance Company and the Marine Insurance Market of the Eighteenth Century." *Economica* n.s. 25 (1958): 126–41.

Johnson, Donald S. *Charting the Sea of Darkness: The Four Voyages of Henry Hudson*. Camden, ME: International Marine, 1993.

Jones, Michael Owen. *Craftsman of the Cumberlands: Tradition and Creativity*. Lexington: University Press of Kentucky, 1989.

Karras, Alan. "Smuggling and its Malcontents." In *Interactions: Transregional Perspectives on World History*, edited by Jerry H. Bentley, Renate Bridenthal, and Anand A. Yang, 135–49. Honolulu: University of Hawai'i, 2005.

Karras, Alan. *Smuggling: Contraband and Corruption in World History*. London: Rowman & Littlefield, 2010.

Karras, Alan. "Transgressive Exchange: Rewriting Atlantic Law in the Eighteenth-Century Caribbean." Paper presented at Seascapes, Littoral Cultures, and Trans-Oceanic Exchanges, Library of Congress, Washington D.C., February 12–15, 2003.

Kendrick, John. "The Evolution of Shipbuilding in the Eighteenth Century." In Stephen Haycox, James Barnett, and Caedmon Liburd, eds., *Enlightenment and Exploration in the North Pacific, 1741–1805*, 88–102. Seattle: University of Washington Press, 1997.

Kennedy, Cynthia M. "The Other White Gold: Salt, Slaves, the Turks and Caicos Islands, and British Colonialism." *The Historian* 69:2 (Summer 2007): 215–30.

Kennedy, Neil. "Impermanence and Empire: Salt Raking in the Turks and Caicos Islands." In Peter E. Pope with Shannon Lewis-Simpson, eds., *Exploring Atlantic Transitions: Archaeologies of Transience and Permanence in New Found Lands*, 80–92. Woodbridge, UK: The Boydell Press, 2013.

Kingston, Christopher. Review of Guy Chet's *The Ocean is a Wilderness: Atlantic Piracy and the Limits of State Authority, 1688–1856*. *Journal of Economic History* 75:3 (2015): 933–36.

Kinkel, Sarah. *Disciplining the Empire: Politics, Governance, and the Rise of the British Navy*. Cambridge, MA: Harvard University Press, 2018.

Kinkel, Sarah. "The King's Pirates? Naval Enforcement of Imperial Authority, 1740–76." *The William and Mary Quarterly* 71:1 (January 2014): 3–34.

Kline, Ronald and Trevor Pinch. "Users as Agents of Technological Change: The Social Construction of the Automobile in the Rural United States." *Technology and Culture* 37:4 (October 1996): 763–95.

Klooster, Wim. *Illicit Riches: Dutch Trade in the Caribbean, 1648–1795*. Leiden: KITLV, 1998.

Klooster, Wim. "Inter-Imperial Smuggling in the Americas, 1600–1800." In Bernard Bailyn and Patricia L. Denault, eds., *Soundings in Atlantic History: Latent Structures and Intellectual Currents, 1500–1830*, 141–80. Cambridge, MA: Harvard University Press, 2009.

Klooster, Wim. *The Dutch Moment: War, Trade, and Settlement in the Seventeenth Century Atlantic World*. Ithaca, NY: Cornell University Press, 2016. Kindle edition.

Knight, Roger. "Devil bolts and deception? Wartime naval shipbuilding in private shipyards, 1739–1815." *Journal for Maritime Research* 5:1 (April 2003): 34–51.

Koot, Christian J. *Empire at the Periphery: British Colonists, Anglo-Dutch Trade, and the Development of the British Atlantic, 1621–1713*. New York: New York University Press, 2011.

Kristiansen, Kristian. "Discourse and dialogue. Some observations and further reflections." *Archaeological Dialogues* 11:2 (December 2004): 114–32.

Kristiansen, Kristian. "Genes versus agents. A discussion of the widening theoretical gap in archaeology." *Archaeological Dialogues* 11:2 (December 2004): 77–99.

Krüger, Lorenz et al., eds. *The Probabilistic Revolution, Vol. 1: Ideas in History*. Cambridge, MA: MIT Press, 1987.

Kuhn, Thomas. *The Structure of Scientific Revolutions*. Chicago: University of Chicago Press, 1962. Kindle edition.

Kupperman, Karen Ordahl. "Before 1607." *William and Mary Quarterly* 72:1 (January 2015): 3–24.

Kurlansky, Mark. *Cod: A Biography of the Fish That Changed the World*. New York: Penguin, 1997.

Lane, Frederic. « Progrès technologique et productivité dans les transports maritimes de la fin du Moyen Age au début des Temps Modernes », *Revue Historique* 251 (1974): 277–302.

Lavery, Brian. *The Arming and Fitting of English Ships of War 1600–1815*. Annapolis, MD: Naval Institute Press, 1989.

Lavery, Brian. *The Colonial Merchantman* Susan Constant *1605*. Annapolis, MD: Naval Institute Press, 1988.

Leeson, Peter. *The Invisible Hook: The Hidden Economics of Piracy*. Princeton, NJ: Princeton University Press, 2009.

Leggett, Don and Richard Dunn, eds. *Re-inventing the Ship: Science, Technology and the Maritime World, 1800–1918*. Farnham, UK: Ashgate, 2012.

Lemmers, Alan and Ab Hoving. "Drawing the Line: Or Why the Rotterdam Shipwrights Would Not Adopt the Amsterdam (English) Drafting Techniques." *Nautical Research Journal* 52:2 (Summer 2007): 67–81.

Leonard, A.B. [Adrian]. "Underwriting Marine Warfare: Insurance and Conflict in the Eighteenth Century," *International Journal of Maritime History* 25:2 (December 2013): 173–85.

Leonard, Adrian and David Pretel, eds. *The Caribbean and the Atlantic World Economy: Circuits of Trade, Money and Knowledge, 1650–1914*. London: Palgrave-Macmillan, 2015.

Leonard, Adrian. Review of Guy Chet's *The Ocean is a Wilderness: Atlantic Piracy and the Limits of State Authority, 1688–1856. International Journal of Maritime History* 27:3 (August 2015): 583–84.

Lipman, Andrew. *The Saltwater Frontier: Indians and the Contest for the American Coast.* New Haven, CT: Yale University Press, 2017.

Loewen, Brad. "Ships and Shipbuilding / *Navires et Construction Navale:* Recent Advances in Ship History and Archaeology, 1450–1650: Hull Design, Regional Typologies and Wood Studies." *Material Culture Review/Revue de la culture matérielle* 48 (Fall/Automne 1998). Available at: http://journals.hil.unb.ca/index.php/MCR/article/view/17791/22170. Accessed 10 February 2015.

Long, Pamela O. *Artisan/Practitioners and the Rise of the New Sciences, 1400–1800*. Corvallis: Oregon State University Press, 2011.

Lucas, Adam. "Narratives of Technological Revolution in the Middle Ages." In Albrecht Classen, ed., *Handbook of Medieval Studies: Terms, Methods, Trends*, 967–90. Berlin: De Gruyter, 2010.

Maarleveld, Thijs. "Type or technique. Some thoughts on boat and ship finds as indicative of cultural traditions." *International Journal of Nautical Archaeology* 24:1 (1995): 3–7.

MacGregor, David R. *Fast Sailing Ships: Their Design and Construction, 1775–1875*. Lymington, UK: Nautical Pub. Co., 1973.

MacGregor, David R. *Merchant Sailing Ships 1775–1815: Their Design and Construction*. Watford, UK: Argus, 1980.

MacGregor, David R. *The Schooner: Its Design and Development from 1600 to the Present*. Barnsley, UK: Pen & Sword, 2003.

MacLeod, Christine. "Concepts of Invention and the Patent Controversy in Victorian Britain." In Robert Fox, ed., *Technological Change: Methods and Themes in the History of Technology*, 137–53. Amsterdam: Harwood, 1996.

Magra, Christopher P. *Poseidon's Curse: British Naval Impressment and Atlantic Origins of the American Revolution*. Cambridge: Cambridge University Press, 2016.

Magra, Christopher P. Review of Guy Chet's *The Ocean is a Wilderness: Atlantic Piracy and the Limits of State Authority, 1688–1856*. *American Historical Review* 120:3 (June 2015): 969.

Mallagh, Cris. "Some Aspects of the Life and Career of William Sutherland." *Mariner's Mirror* 100:1 (February 2014): 17–28.

Mandelblatt, Bertie. "A Transatlantic Commodity: Irish Salt Beef in the French Atlantic World." *History Workshop Journal* Volume 63 (2007): 18–47.

Marquardt, Karl Heinz. *Eighteenth Century Rigs and Rigging*. London: Conway Maritime Press via ANOVA, 1993.

Marquardt, Karl Heinz. *The Global Schooner: Origins, Development, Design and Construction, 1695–1845*. Annapolis, MD: Naval Institute Press, 2003.

Marzagalli, Silvia. "The French Atlantic." *Itinerario* 23:2 (July 1999): 70–83.

McConnell, David. *British Smooth-bore Artillery: A Technological Study to Support Identification, Acquisition, Restoration, Reproduction, and Interpretation of Artillery at National Historic Parks in Canada*. Ottawa: Minister of Supply and Services Canada, 1988. Accessed at https://sha.org/assets/documents/British%20Smooth-Bore%20Artillery%20-%20English.pdf, 3 October 2018.

McCusker, John J. "Colonial Tonnage Measurement: Five Philadelphia Merchant Ships as a Sample." *Journal of Economic History* 27:1 (March 1967): 82–91.

McCusker, John J. "The Demise of Distance: The Business Press and the Origins of the Information Revolution in the Early Modern Atlantic World." *American Historical Review* 110:2 (April 2005): 295–321.

McCusker, John J. "The Early History of 'Lloyd's List'." *Historical Research* 64:155 (October 1991): 427–31.

McCusker, John J. *Essays in the Economic History of the Atlantic World*. London: Routledge, 1997.

McCusker, John J. "The Pennsylvania Shipping Industry in the Eighteenth Century." Unpublished bound typescript, Historical Society of Pennsylvania, Philadelphia, 1973.

McCusker, John J. "The Rise of the Shipping Industry in Colonial America." In Robert A. Kilmarx, ed., *America's Maritime Legacy: A History of the U.S. Merchant Marine and Shipbuilding Industry since Colonial Times*, 1–26. Boulder, CO: Westview, 1979.

McCusker, John J. *Rum and the American Revolution: The Rum Trade and the Balance of Payments of the Thirteen Continental Colonies*. London: Routledge, 1997.

McCusker, John J. and Russell R. Menard. *The Economy of British America, 1607–1789*. Chapel Hill: University of North Carolina Press, 1991.

McDonald, Roderick A. and Richard B. Sheridan, eds. *West Indies Accounts: Essays on the History of the British Caribbean and the Atlantic Economy*. Kingston, Jamaica: University of the West Indies, 2002.

McGowan, Alan. *The Century Before Steam: The Development of the Sailing Ship 1700–1820*. London: Her Majesty's Stationery Office, 1980.

McGowan, Alan. *Tiller and Whipstaff: The Development of the Sailing Ship, 1400–1700*. London: National Maritime Museum, 1981.

McGrail, Seán. "Experimental Archaeology: Replicas and Reconstructions." In Jenny Bennett, ed., *Sailing Into the Past: Learning From Replica Ships*, 16–23. Barnsley, UK: Seaforth, 2009.

McGrail, Seán. "Replicas, reconstructions and floating hypotheses." *International Journal of Nautical Archaeology* 21:4 (November 1992): 353–55.

McIlvenna, Noeleen. *A Very Mutinous People: The Struggle for North Carolina, 1660–1713*. Chapel Hill: University of North Carolina Press, 2009.

McKee, Eric. *Working Boats of Britain*. London: Conway Maritime Press, 1983.

McMullen, Drew with Captain Robert Brittain. "The Schooner *Sultana*." In Jenny Bennett, ed., *Sailing Into the Past: Learning From Replica Ships*, 150–61. Barnsley, UK: Seaforth, 2009.

McNeill, J.R. *Mosquito Empires: Ecology and War in the Greater Caribbean, 1620–1914*. Cambridge: Cambridge University Press, 2010.

Meide, Chuck. "On the Rocks: A New Approach to Atlantic World Trade." In Adrian Leonard and David Pretel, eds., *The Caribbean and the Atlantic World Economy: Circuits of Trade, Money and Knowledge, 1650–1914*, 36–67. London: Palgrave-Macmillan, 2015.

Menard, Russell R. "Transport Costs and Long-Range Trade, 1300–1800: Was there a European 'transport Revolution' in the Early Modern Era?" In James D. Tracy, ed., *The Political Economy of Merchant Empires*, 228–75. Cambridge: Cambridge University Press, 1991.

Merrens, Harry Roy. *Colonial North Carolina in the Eighteenth Century: A Study in Historical Geography*. Chapel Hill: University of North Carolina Press, 1964.

Merton, Robert K. *Science, Technology and Society in Seventeenth-Century England*. New York: Howard Fertig, 1970.

Middlebrook, Louis F. "The Ship Mary of Philadelphia, 1740." *Pennsylvania Magazine of History and Biography*, LVIII (April 1934): 127–51.

Miller, Daniel, ed. *Car Cultures*. Oxford: Berg, 2001.

Mims, Stewart L. *Colbert's West India Policy*. New Haven, CT: Yale University Press, 1912. Google books facsimile.

Minchinton, Walter E. "Characteristics of British Slaving Vessels, 1698–1775." *Journal of Interdisciplinary History* 20:1 (Summer 1989): 53–81.

Mintz, Sidney W. *Sweetness and Power: The Place of Sugar in Modern History*. New York: Penguin, 1985.

Mitchell-Cook, Amy. *A Sea of Misadventures: Shipwreck and Survival in Early America*. Columbia: University of South Carolina Press, 2013.

Mokyr, Joel. *The Enlightened Economy: An Economic History of Britain, 1700–1850*. New Haven, CT: Yale University Press, 2009.

Mokyr, Joel. "Evolution and Technological Change: A New Metaphor for Economic History?" In Robert Fox, ed. *Technological Change: Methods and Themes in the History of Technology*, 63–83. Amsterdam: Harwood, 1996.

Mokyr, Joel. *The Lever of Riches: Technological Creativity and Economic Progress*. Oxford: Oxford University Press, 1990. Kindle edition.

Mokyr, Joel. *Twenty-Five Centuries of Technological Change: An Historical Survey*. London: Taylor & Francis, 2007.

Moore, David and Corey Malcom. "Seventeenth-Century Vehicle of the Middle Passage: Archaeological and Historical Investigations on the *Henrietta Marie* Shipwreck Site." *International Journal of Historical Archaeology* 12:1 (January 2008): 20–38.

Morgan, Edmund S. and Helen M. *The Stamp Act Crisis: Prologue to Revolution*. Chapel Hill: University of North Carolina Press, 1953.

Morgan, Kenneth. "Robert Dinwiddie's Reports on the British American Colonies." *William and Mary Quarterly* 65:2 (April 2008): 305–46.

Morgan, Kenneth. *Bristol & the Atlantic Trade in the Eighteenth Century*. Cambridge: Cambridge University Press, 1993.

Morgan, Philip D. "Maritime Slavery." *Slavery & Abolition* 31:3 (September 2010): 311–26.

Morris, E.P. *The Fore-And-Aft Rig in America*. New Haven, CT: Yale University Press, 1927.

Morris, John W., Gordon P. Watts Jr., and Marianne Franklin. "The Comparative Analysis of Eighteenth-Century Vessel Remains in the Archaeological Record: A Synthesized Theory of Framing Evolution." *Underwater Archaeology Proceedings from the Society for Historical Archaeology Conference 1995*, 125–33. Washington, DC: Society for Historical Archaeology, 1995.

Mukherjee, Rila, ed. *Networks in the First Global Age, 1400–1800*. Delhi: Primus, 2011.

Mulcahy, Matthew. *Hurricanes and Society in the British Greater Caribbean, 1624–1783*. Baltimore, MD: Johns Hopkins University Press, 2006.

Norling, Lisa. *Captain Ahab Had A Wife: New England Women and the Whalefishery, 1720–1870*. Chapel Hill: University of North Carolina Press, 2000.

North, Douglass C. "Sources of Productivity Change in Ocean Shipping, 1600–1850." *Journal of Political Economy* 76:5 (September–October 1968): 953–70.

Ojala, Jari. "Productivity and Technological Change in Eighteenth- and Nineteenth-Century Sea Transport: A Case Study of Sailing Ship Efficiency in Kokkola, Finland, 1721–1913." *International Journal of Maritime History* 9:1 (June 1997): 93–123.

Ortner, Sherry B. "Theory in Anthropology since the Sixties." *Comparative Studies in Society and History* 26:1 (January 1984): 126–66.

Paine, Lincoln. *The Sea and Civilization: A Maritime History of the World*. New York: Alfred A. Knopf, 2013.

Palmer, Colin. "Measuring Performance Under Sail." In Jenny Bennett, ed., *Sailing Into the Past: Learning From Replica Ships*, 24–32. Barnsley, UK: Seaforth, 2009.

Palmer, Colin. "Reflections on the Balance of Traditional Sailing Vessels." *International Journal of Nautical Archaeology* 38:1 (2009): 90–96.

Palmer, Michael A. *Stoddert's War: Naval Operations During the Quasi-War with France, 1798–1801*. Annapolis, MD: Naval Institute Press, 1987.

Pares, Richard. *Yankees and Creoles: The Trade Between North America and the West Indies Before the American Revolution*. Hamden, CT: Archon, 1956.

Parrott, Daniel S. *Tall Ships Down: The Last Voyages of the* Pamir, Albatross, Marques, Pride of Baltimore, *and* Maria Asumpta. Camden, ME: International Marine/ McGraw-Hill, 2003.

Parry, J.H. *Trade and Dominion: The European Overseas Empires in the Eighteenth Century*. London: Weidenfeld and Nicolson, 1971.

Parry, J.H. "Transport and Trade Routes." In E.E. Rich and C.H. Wilson, eds., *The Cambridge Economic History of Europe from the Decline of the Roman Empire, Vol. 4: The Economy of Expanding Europe in the Sixteenth and Seventeenth Centuries*, 155–219. Cambridge: Cambridge University Press, 1967.

Parthesius, Robert. *Dutch Ships in Tropical Waters: The Development of the Dutch East India Company (VOC) Shipping Network in Asia, 1595–1660*. Amsterdam: Amsterdam University Press, 2010.

Peters, Andrew. *Ship Decoration, 1630–1780*. Barnsley, UK: Seaforth, 2013.

Phillips, Carla Rahn. "The Caravel and the Galleon." In Robert Gardiner, ed., *Cogs, Caravels, and Galleons: The Sailing Ship 1000–1650*, 91–114. Annapolis, MD: Naval Institute Press, 1994.

Pitt, Ken and Damian M. Goodburn. "18th- and 19th-Century Shipyards at the South-East Entrance to the West India Docks, London." *International Journal of Nautical Archaeology* 32:2 (October 2003): 191–209.

Pomeranz, Kenneth. *The Great Divergence: China, Europe, and the Making of the Modern World Economy*. Princeton: Princeton University Press, 2001.

Pomeranz, Kenneth and Steven Topik. *The World That Trade Created: Society, Culture, and the World Economy, 1400 to the Present*. Armonk, NY: M.E. Sharpe, 1999.

Pope, Peter E. *Fish Into Wine: The Newfoundland Plantation in the Seventeenth Century*. Chapel Hill: University of North Carolina Press, 2004.

Pope, Peter E., with Shannon Lewis-Simpson, eds. *Exploring Atlantic Transitions: Archaeologies of Transience and Permanence in New Found Lands*. Woodbridge, UK: The Boydell Press, 2013.

Price, Jacob M. *Tobacco in Atlantic Trade: The Chesapeake, London, and Glasgow, 1675–1775*. Brookfield, VT: Variorum, 1995.

Price, Jacob M. "What Did Merchants Do? Reflections on English Overseas Trade, 1660–1790." *Journal of Economic History* 49:2 (June 1989): 267–84.

Pye, David. *The Nature of Design*. New York: Reinhold, 1964.

Rankin, Laird. *The Return of the Nonsuch: The Ship That Launched an Empire*. Winnipeg: Heartland Associates, 2004.

Rediker, Marcus. *Between the Devil and the Deep Blue Sea: Merchant Seamen, Pirates and the Anglo-American Maritime World, 1700–1750*. Cambridge: Cambridge University Press, 1987.

Rediker, Marcus. *The Slave Ship: A Human History*. New York: Viking, 2007. Kindle edition.

Rees, Gareth. "Copper Sheathing: An Example of Technological Diffusion in the English Merchant Fleet." *Journal of Transport History* New Series, 1 (1971–1972): 85–94.

Reid, Phillip. "Notes from a published treatise in an ordinary eighteenth-century shipwright's journal." *Mariner's Mirror* 104:1 (January 2018): 79–83.

Reid, Phillip. "The Ordinary Merchant Ship in the British Atlantic, 1600–1800: A Call for Further Research." *International Journal of Maritime History* 29:4 (November 2017): 911–26.

Reid, Phillip. "Something Ventured: The Ordinary Merchantman of the British Atlantic as a Technology of Risk Mitigation, 1600–1800." *Journal of Transport History* 38:2 (December 2017): 196–212.

Reid, Phillip. "The Time Machine? Using Replica Analysis to Understand Merchant Ships and the Development of the British Atlantic, 1600–1800." *The Northern Mariner/Le Marin du Nord* 26:3 (July 2016): 299–316.

Rhoden, Nancy, ed. *English Atlantics Revisited*. Kingston, ON: McGill-Queens University, 2007.

Rich, E.E. and C.H. Wilson, eds. *The Cambridge Economic History of Europe from the Decline of the Roman Empire, Vol. 4: The Economy of Expanding Europe in the Sixteenth and Seventeenth Centuries*. Cambridge: Cambridge University Press, 1967.

Ricoeur, Paul. *Lectures on Ideology and Utopia*. Ed. George H. Taylor. Princeton, NJ: Princeton University Press, 1975.

Riess, Warren C. with Sheli O. Smith. *The Ship That Held Up Wall Street*. College Station: Texas A&M University Press, 2015.

Riess, Warren. Angel Gabriel, *The Elusive English Galleon: Its History and the Search for Its Remains*. Bristol, ME: 1797 House, 2001.

Robb, John. "Agency. A personal view." *Archaeological Dialogues* 11:2 (December 2004): 103–7.

Rodger, N.A.M. *The Command of the Ocean*. London: Penguin, 2004.

Rodger, N.A.M. "Navies and the Enlightenment." In Pieter van der Merwe, ed., *Science and the French and British Navies, 1700–1850*, 5–23. Greenwich, UK: National Maritime Museum, 2003.

Rodger, N.A.M. "War as an Economic Activity in the 'Long' Eighteenth Century." *International Journal of Maritime History* 22:2 (December 2010): 1–18.

Rodger, N.A.M. *The Wooden World: An Anatomy of the Georgian Navy*. Glasgow: Collins, 1986.

Rosecrance, Richard. *The Rise of the Trading State: Commerce and Conquest in the Modern World*. New York: Basic Books, 1986.

Rosenberg, Nathan. "Factors Affecting the Diffusion of Technology." *Explorations in Economic History* 10:1 (Fall 1972): 3–33.

Rosenberg, Nathan and L.E. Birdzell. *How the West Grew Rich: The Economic Transformation of the Industrial World*. New York: Basic Books, 1986.

Rossano, Geoffrey L. "Launching Prosperity: Samuel Townsend and the Maritime Trade of Colonial Long Island, 1747–1773." *American Neptune* 48 (1988): 37–43.

Rozwadowski, Helen M. *Vast Expanses: A History of the Oceans*. London: Reaktion Books, 2018.

Russell, Andrew L. and Lee Vinsel. "After Innovation, Turn to Maintenance." *Technology and Culture* 59:1 (January 2018): 1–25.

Sahlins, Marshall. *Culture and Practical Reason*. Chicago: University of Chicago Press, 1976.

Scammell, Geoffrey. "British Merchant Shipbuilding, c. 1500–1750." *International Journal of Maritime History* 11:1 (June 1999): 27–52.

Schatzberg, Eric. *Wings of Wood, Wings of Metal: Culture and Technical Choice in American Airplane Materials, 1914–1945*. Princeton, NJ: Princeton University Press, 1998.

Shennan, Stephen. "Culture, society and evolutionary theory." *Archaeological Dialogues* 11:2 (December 2004): 107–14.

Shepherd, James F. and Gary M. Walton. *Shipping, Maritime Trade, and the Economic Development of Colonial North America*. Cambridge: Cambridge University Press, 1972.

Sheridan, Richard B. *Sugar and Slavery: An Economic History of the British West Indies, 1623–1775*. Kingston, Jamaica: University of the West Indies, 2000.

Smelser, Marshall and William L. Davisson. "The Longevity of Colonial Ships." *American Neptune* 30 (1973): 16–19.

Smith, Claire, ed. *Encyclopedia of Global Archaeology*. New York: Springer, 2014.

Solow, Barbara L., ed. *Slavery and the Rise of the Atlantic System*. Cambridge: Cambridge University Press, 1991.

Souza, G.B. "Convergence before Divergence: Global Maritime Economic History and Material Culture." *International Journal of Maritime History* 17:1 (June 2005): 17–27.

Spectre, Peter H. and David Larkin. *Wooden Ship: The Art, History, and Revival of Wooden Boatbuilding.* Boston: Houghton Mifflin, 1991.

Stammers, Michael. "Slipways and Steamchests: The Archaeology of 18th- and 19th Century Wooden Merchant Shipyards in the United Kingdom." *International Journal of Nautical Archaeology* 28:3 (August 1999): 253–64.

Starkey, David J. "Voluntaries and sea robbers: A review of the academic literature on privateering, corsairing, buccaneering and piracy." *Mariner's Mirror* 97:1 (February 2011): 127–47.

Staudenmeier, John. "Recent Trends in the History of Technology." *The American Historical Review* 95:3 (June 1990): 715–25.

Stedall, Jacqueline. "Notes made by Thomas Harriot (1560–1621) on ships and shipbuilding." *Mariner's Mirror* 99:3 (2013): 325–27.

Steele, Ian K. *The English Atlantic, 1675–1740: An Exploration of Communication and Community*. Oxford: Oxford University Press, 1986.

Steffy, J. Richard. "Illustrated Glossary of Ship and Boat Terms." In Ben Ford et al., eds., *The Oxford Handbook of Maritime Archaeology.* Oxford: Oxford University Press, 2013. Retrieved 8 Jan. 2019, from http://www.oxfordhandbooks.com/view/10.1093/oxfordhb/9780199336005.001.0001/oxfordhb-9780199336005-e-48.

Steffy, J. Richard. *Wooden Ship Building and the Interpretation of Shipwrecks*. College Station: Texas A&M University Press, 1994.

Stern, Jessica Yirush. *The Lives in Objects: Native Americans, British Colonists, and Cultures of Labor and Exchange in the Southeast.* Chapel Hill: University of North Carolina Press, 2017.

Stewart-Brown, R. *Liverpool Ships in the Eighteenth Century*. Liverpool and London: University Press of Liverpool/Hodder & Stoughton Ltd., 1932.

Swanson, Carl E. *Predators and Prizes: American Privateering and Imperial Warfare, 1739–1748*. Columbia: University of South Carolina Press, 1991.

Swanson, Carl E. "'To Procure a Good Freight:' Charles Town Merchants in the Eighteenth Century." *International Journal of Maritime History* 20:2 (December 2008): 111–42.

Taylor, David A. *Boat Building in Winterton, Trinity Bay, Newfoundland*. Ottawa: National Museums of Canada (Canadian Museum of Civilization, Mercury Series, Canadian Centre for Folk Culture Studies Paper Number 41), 1982.

Tracy, James D., ed. *The Political Economy of Merchant Empires*. Cambridge: Cambridge University Press, 1991.

Trimingham, Eldon. "The Development of the Bermuda Rig." *Bermuda Journal of Archaeology and Maritime History* 2 (1990): 127–42.

Truxes, Thomas M. *Defying Empire: Trading with the Enemy in Colonial New York*. New Haven, CT: Yale University Press, 2008.

Truxes, Thomas M. "Transnational Trade in the Wartime North Atlantic: The Voyage of the Snow *Recovery*." *The Business History Review* 79:4 (Winter 2005): 751–80.

Turner, A.J. "Advancing Navigation in Eighteenth-Century France: Teaching and Instrument-Making in the Port of Rochefort." *Mariner's Mirror* 91:4 (November 2005): 531–47.

Unger, Richard W. *The Art of Medieval Technology: Images of Noah the Shipbuilder*. New Brunswick, NJ: Rutgers University, 1991.

Unger, Richard W. "Early Modern Vessels," subentry under "Trading Vessels." In John B. Hattendorf, ed., *The Oxford Encyclopedia of Maritime History*, Vol. 4, 171. Oxford: Oxford University Press, 2007.

Unger, Richard W. "The *Fluit*: Specialist Cargo Vessels 1500–1650." In Robert Gardiner, ed., *Cogs, Caravels, and Galleons: The Sailing Ship 1000–1650*, 91–114. Annapolis, MD: Naval Institute Press, 1994.

Unger, Richard W. "Ship Design and Energy Use, 1350–1875." In Unger, ed., *Shipping and Economic Growth 1350–1850*, 249–68. Leiden: Brill, 2011.

Unger, Richard W. *The Ship in the Medieval Economy, 600–1600*. Montreal: McGill-Queen's University, 1980.

Unger, Richard W., ed. *Shipping and Economic Growth 1350–1850*. Leiden: Brill, 2011.

Unger, Richard W. *Ships and Shipping in the North Sea and Atlantic, 1400–1800*. Basingstoke, UK: Variorum, 1998.

Unger, Richard W. "Warships, Cargo Ships and Adam Smith: Trade and Government in the Eighteenth Century." *Mariner's Mirror* 92:1 (February 2006): 41–59.

Usher, Abbott Payson. "The Growth of English Shipping, 1572–1922." *Quarterly Journal of Economics* 42 (1928): 465–78.

Van den Belt, Henrik and Arie Rip. "The Nelson-Winter-Dosi Model and Synthetic Dye Chemistry." In Wiebe Bijker et al., eds. *The Social Construction of Technological Systems: New Directions in the Sociology and History of Technology*. Anniversary Edition, 129–54. Cambridge, MA: MIT Press, 2012. Kindle edition.

Van der Merwe, Pieter, ed. *Science and the French and British Navies, 1700–1850*. Greenwich, UK: National Maritime Museum, 2003.

Veit, Ulrich. "'Mission impossible!' Comment on Kristian Kristiansen, 'Genes versus agents. A discussion of the widening theoretical gap in archaeology.'" *Archaeological Dialogues* 11:2 (December 2004): 99–103.

Vickers, Daniel with Vince Walsh. *Young Men and the Sea: Yankee Seafarers in the Age of Sail*. New Haven, CT: Yale University Press, 2005.

Villiers, Patrick. "The Slave and Colonial Trade in France Just Before the Revolution." In Barbara L. Solow, ed., *Slavery and the Rise of the Atlantic System*, 210–36. Cambridge: Cambridge University Press, 1991.

Walton, Gary M. "Obstacles to technical diffusion in ocean shipping, 1675–1775." *Explorations in Economic History* 8:2 (Winter 1970): 123–40.

Weski, Timm. "The value of experimental archaeology for reconstructing ancient seafaring." In Lucy Blue et al., eds. *Connected by the Sea: Proceedings of the Tenth*

International Symposium on Boat and Ship Archaeology, Roskilde 2003, 63–7. Oxford: Oxbow, 2006.

White, Lynn Jr. *Medieval Technology and Social Change*. Oxford: Oxford University Press, 1962.

Whitewright, Julian. "The Potential Performance of Ancient Mediterranean Sailing Rigs." *International Journal of Nautical Archaeology* 40:1 (2011): 2–17.

Whitewright, Julian. "Technological Continuity and Change: The Lateen Sail in the Medieval Mediterranean." *Al-Masāq* 24:1 (April 2012): 1–19.

Wilde-Ramsing, Mark U. et al. *The Rose Hill Wreck: Historical and Archaeological Investigations of an Eighteenth-Century Vessel at a Colonial River Landing Near Wilmington, North Carolina*. Kure Beach, NC: North Carolina Dept. of Cultural Resources, 1992.

Williams, Alan F. "Bristol Port Plans and Improvements of the 18th Century." *Transactions of the Bristol and Gloucestershire Archaeological Society* Volume 81, 1962.

Williams, Justin. "English Mercantilism and Carolina Naval Stores, 1705–1776." *Journal of Southern History* 1:2 (May 1935): 169–85.

Wilson, Garth. Review of Maik-Jens Springmann and Horst Wernicke, eds., *Historical Boat and Ship Replicas: Conference Proceedings on the Scientific Perspectives and the Limits of Boat and Ship Replicas. Torgelow 2007*. *International Journal of Maritime History* 22:1 (June 2010): 27–29.

Wing, John F. "Shipping Productivity in Maryland's Tobacco Trade, 1689–1759." *International Journal of Maritime History* 20:2 (December 2008): 223–39.

Wingood, Allan J. "*Sea Venture*: An interim report on an early seventeenth century shipwreck lost in 1609." *International Journal of Nautical Archaeology* 11:4 (November 1982): 333–47.

Wingood, Allan J. "*Sea Venture* Second interim report—part 2: the artefacts." *International Journal of Nautical Archaeology* 15:2 (May 1986): 149–59.

Wood, Bradford J. *This Remote Part of the World: Regional Formation in Lower Cape Fear, North Carolina 1725–1775*. Columbia: University of South Carolina Press, 2004.

Wrike, Peter. "The Jamestown Replicas." In Jenny Bennett, ed., *Sailing Into the Past: Learning from Replica Ships*, 120–33. Barnsley, UK: Seaforth, 2009.

Zahedieh, Nuala. *The Capital and the Colonies: London and the Atlantic Economy 1660–1700*. Cambridge: Cambridge University Press, 2010.

Zwick, Daniel. "Conceptual Evolution in Ancient Shipbuilding: An Attempt to Reinvigorate a Shunned Theoretical Framework." In Jonathan Adams and Johan Rönnby, eds., *Interpreting Shipwrecks: Maritime Archaeological Approaches*, 46–71. Southampton, UK: Highfield, 2013.

Unpublished Theses and Reports

Bradshaw, Price. "The role of technology in the failure of the rigid airship as an invention." PhD dissertation, University of Florida, 1975.

Clifford, Sheila Alicia. "An Analysis of the Port Royal Shipwreck and its Role in the Maritime History of Seventeenth-Century Port Royal, Jamaica." Master's thesis, Texas A&M University, 1993.

Cook, Gregory D. "The Readers Point Vessel: Hull Analysis of an Eighteenth-Century Merchant Sloop Excavated in St. Ann's Bay, Jamaica." Master's thesis, Texas A&M University, 1997.

Cullon, Joseph. "Colonial Shipwrights and Their World: Men, Women and Markets in Early New England." PhD dissertation, University of Wisconsin, 2003.

Krivor, Michael. "Archaeological Investigation of an Eighteenth-Century British Merchant Vessel in Chub Heads Cut, Bermuda." Master's thesis, East Carolina University, 1998.

Mitchell, Amy. "A Comparison of Wood Use in Eighteenth-Century Vessels." Master's thesis, East Carolina University, 1994.

Moore, David. "Anatomy of a 17th Century Slave Ship: Historical and Archaeological Investigations of the *Henrietta Marie*." Master's thesis, East Carolina University, 1989.

Pecorelli, Harry. "Archaeological Investigation of the B & B Wreck (38B1672): An Eighteenth Century Plantation-Built Vessel, Charleston, South Carolina." Master's thesis, East Carolina University, 2003.

Reid, Phillip. "A Very Good Sailer: Merchant Ship Technology and the Development of the British Atlantic Empire, 1600–1800." PhD dissertation, Memorial University of Newfoundland, 2017.

Rose, Kelby James. "The Naval Architecture of *Vasa*, A 17th-Century Swedish Warship." PhD dissertation, Texas A&M University, 2014.

Sabick, Christopher R. "His Majesty's Hired Transport Schooner *Nancy*." Master's thesis, Texas A&M University, 2004.

Southerly, James C.W. "Cedar on the Reef: Archaeological and Historical Assessments of the Eighteenth Century Bermuda Sloop, Exemplified by the Hunter Galley." Master's thesis, East Carolina University, 2003.

VanHorn, Kellie M. "Eighteenth Century Colonial American Merchant Ship Construction." Master's thesis, Texas A&M University, 2004.

Wright, Thomas. "Ship Hydrodynamics 1710–1880." PhD thesis, Science Museum/University of Manchester Institute of Science and Technology, 1983.

Wrike, Peter. "The *Godspeed* and *Discovery* of London and Virginia, 1606–7." Unpublished bound research report, Jamestown-Yorktown Foundation, three volumes, 2002.

Index

Page numbers in italics refer to figures and tables.